Eco-hydrodynamic modelling of primary production in

coastal waters and lakes using BLOOM

Hans Los

Deltares Select Series
Volume 1
ISSN 1877-5608

This volume contains the PhD thesis and two additional papers (Chapters 4 and 6) of Hans Los, who defended his work successfully at the Wageningen University, the Netherlands, on February 4, 2009.

Deltares is a Dutch independent research institute for water, soil and subsurface issues. It was formed in 2008 from a merger of Delft Hydraulics, GeoDelft, the Subsurface and Groundwater unit of TNO and parts of Rijkswaterstaat.

Cover illustration: Water lily almost completely covered by floating masses of blue green algae.

Eco-hydrodynamic modelling of primary production in coastal waters and lakes using BLOOM

Hans Los

IOS Press

ISBN 978-1-58603-987-5
Library of Congress Control Number: 2009922275

Publisher
IOS Press BV
Nieuwe Hemweg 6B
1013 BG Amsterdam
Netherlands
fax: +31 20 687 0019
e-mail: order@iospress.nl

Distributor in the UK and Ireland
Gazelle Books Services Ltd.
White Cross Mills
Hightown
Lancaster LA1 4XS
United Kingdom
fax: +44 1524 63232
e-mail: sales@gazellebooks.co.uk

Distributor in the USA and Canada
IOS Press, Inc.
4502 Rachael Manor Drive
Fairfax, VA 22032
USA
fax: +1 703 323 3668
e-mail: iosbooks@iospress.com

LEGAL NOTICE
The publisher is not responsible for the use which might be made of the following information.

PRINTED IN THE NETHERLANDS

Contents

Abstract

In many areas nutrient loadings to aquatic ecosystems have increased considerably as a result of population growth, industrial development and urbanization. This has resulted in enhanced growth of phytoplankton, shifts in composition of the plankton community, and changes in the structure of ecosystems, which are often considered to be objectionable. To help understanding these processes and to predict future conditions, a mathematical model, BLOOM, has been developed and applied since 1977. It simulates the biomass and composition of phytoplankton and macro algae in relation to the amount of nutrients, the under water light climate and grazing. It can be applied as a relatively simple screening tool, but also as part of advanced integrated modelling systems including additional hydrodynamic, suspended matter and habitat components. The model has been extensively validated, which means that its credibility was demonstrated systematically for certain types of applications. It has been applied as a supporting management tool to a very large number of aquatic systems worldwide: lakes, channel systems, estuaries, lagoons and coastal seas, using generic coefficients (one set for fresh water, one set for marine simulations) as much as possible. The principles of the model, its validation and a number of representative applications are described in this thesis.

General Introduction

Hans Los

General introduction

The need for ecological models

Man's activities have affected the environmental conditions on earth for a long time, but never on such a global scale as witnessed during the last one or two centuries. This is partly because the human population has grown more or less exponentially, but also because industrial and agricultural production per capita grew to unprecedented levels. Direct effects include deforestation, canalization of rivers, farming, construction of artificial lakes and reservoirs etc. Often the ecological consequences are significant i.e. loss of biodiversity, accelerated extinction of certain species and in contrast a massive explosion of others, development of new habitats etc. Somewhat more indirect effects occur due to the release of various substances into the environment. These include toxic substances, green house gases, which affect the global climate, but also nutrients, which are not toxic but in contrast, promote growth of plants and algae.

Managers, politician and the general public are aware that many activities by humans affect the environment. Hence there is a request to the scientific community to forecast what is going to happen in the future based on what was learned from understanding the past. Due to the complexity of ecosystems, mathematical models might be valuable tools to organize and quantify ecological knowledge. This idea is not new. The famous logistic growth equation was already proposed in the 19th century. The first ecological predator - prey models were constructed almost a century ago (Lotka, 1925; Volterra, 1926). Even up till now these equations, are still being used widely. Many more predator - prey or parasite - prey models were also developed quite a long time ago i.e. by Nicholson and Bailey (1935); Ivlev (1955); Holling (1966). Population dynamic models such as those employed for fisheries (Beverton and Holt, 1957) were derived from demographic models by insurance companies and so are the well known Leslie matrices (Leslie, 1945; 1948). The development of modern computers paved the way to the numerical solution of more complex ecosystems models during the 1970s i.e. Cleaner by Park et al. (1979). During that same period the first primary production or more specifically eutrophication models were developed (DiToro, 1971; Vollenweider et al., 1975). Many more models were developed since then.

In conclusion: there is a long history of ecological model as tools to assist solving management questions. At the same time it should be noted that ecological models have a reputation of being complex and inaccurate perhaps even misleading. In contrast physical models are acclaimed to be much more reliable, not in the least as there is much less dispute about the mathematical equations. For instance in hydrodynamic modelling almost every existing model uses the same Navier - Stokes equations; models only differ with respect to the way these equations are numerically solved. There are many types of ecological models and they often use different mathematical equations and different techniques (see for instance Jorgensen, 2008 for a recent overview). With some exaggeration one might say there are as many ecological models as ecological modellers. Consequently ecological modellers frequently have to demonstrate that their models make sense from a scientific point of view. What makes model B better than model A?

There is a second issue. Even if a particular model is approved by the scientific community, it still is uncertain if its results will receive recognition and get approved by the managers and by the public. Regarding these two aspects of ecological models v.d. Molen (1999) makes a distinction between 'credibility' and 'acceptability' of ecological models. Credibility addresses the scientific quality of an ecological model. But a 'good' quality, regardless of how this is defined, is not sufficient to promote acceptance of the results. Even the opposite may be true: results from models with a disputable scientific background, may get accepted for instance because they are easy to understand, the model has an appealing look, the results agree with the reigning policy etc.

With respect to the credibility Los et al. (1994; 1995; 2008) in accordance with Mankin et al. (1975) point out that ecological models cannot be validated in the most restrictive sense: by providing evidence that they are correct. In stead it is proposed in these papers that ecological modellers should demonstrate that their models are good enough for a specific, well defined task. In short models should be 'fit for purpose'.

Basic modelling philosophy: the system approach

At the end of the 1970s eutrophication (high production of phytoplankton due to high discharges of nutrients) was recognized as one of the central water management issues in the Netherlands as well as elsewhere. The origin of the modelling system described here dates back to this period. Available methods such as the well known statistical Vollenweider equation (Vollenweider et al., 1975) proved to be rather inaccurate and not applicable to Dutch conditions. So in cooperation with the Ministry of transport and public works, Delft Hydraulics was commissioned to develop a set of models dealing with the eutrophication problem in 1977.

The question that had to be addressed by these early model versions was relatively simple: what will happen to the phytoplankton biomass and composition if the concentration of a particular nutrient is lowered? To that purpose the BLOOM phytoplankton module was developed. Unlike in most other existing models, different species of phytoplankton were explicitly included from the very start in BLOOM.

To deal with chemical processes, transport, feedback from the sediment etc., BLOOM was linked to other modules during the 1980s. These modelling systems could deal with more elaborate questions because the nutrient concentration in the water could be traced back to individual sources such as a sewage treatment plant or a tributary to a lake. Since the mid 1990s there was a particularly strong focus on enhancements in the physical schematizations dealt with by the models. This resulted in applications to complex networks of canals and lakes, to large lakes, to deep stratified lakes, to transitional waters and to (coastal) seas.

New (management) questions have lead to the development of new model generations and vice versa. This means that the magnitude of questions posed to the present eco-hydrodynamic models has been widened considerably. Typical modern questions include:

- What are the consequences of climate change on the ecology of the Wadden Sea?
- How much suspended matter needs to be removed from Lake Loosdrecht to enable the development of rooted macrophytes and how can this be achieved?
- Is it possible to solve the cyanobacteria problem of Lake Volkerak Zoom by flushing during the summer or can this problem be solved only by a (partial) resalination of the lake?
- How much filter feeders can be produced by aquaculture at a specific location?
- How much *Phaeocystis* will be present in the North Sea next week and how is its spatial distribution?

- How much nutrient reduction is necessary to meet the phytoplankton related targets of the Water Framework Directive or the Marine Strategy?

Using the modelling framework presented in this thesis, some of the management questions can be answered directly, but usually questions must first be translated into specific model inputs i.e. a change in some of its forcings, and the output results must be interpreted and combined with other knowledge to obtain an integrated answer to the research questions. The result of these activities is *System Knowledge*. A sufficient level of system knowledge should be one of the main results of any modelling exercise and always be the basis for any management advice.

When ecological modelling started at Delft Hydraulics more than 30 years ago, it was common practice to modify our models for each new application in order to obtain an adequate representation of reality. Sometimes equations had to be changed, often coefficients were modified. Without a strict methodology this results in time and or site specific models with questionable predictive value.

Since the mid 1980s we have therefore started developing and operating our models in a much more formalized way. The most adequate process descriptions have been gathered, coded and thoroughly tested. Next we have stored them in a well documented library of processes, which is available to the modeller (WL|Delft Hydraulics 2002; 2003). This library now includes appropriate equations for a large number of water quality and primary production related problems in various types of waters (rivers, canals, lakes, estuaries, seas etc).

The primary production model, which is now called DELFT3D-ECO and which includes BLOOM, has been subjected to several formal validation studies. A set of coefficients, which was established for 30 lakes and therefore supposed to be representative for many temporate fresh water systems, was applied to a complicated network system of ditches, canals, shallow and deep lakes with regulated water quantity management. Basically the same model equations but this time with a set of coefficients considered to be adequate for marine waters was used for a 25 year hind cast simulation of the southern North Sea.

During these validation projects we started by writing down the methodology based upon previous experience: where to look? At which variables? In what order? How much deviation is allowed? How to react when deviations exceed the limits? Should a mismatch be attributed to the forcing of the model such as the boundaries and loadings, to the physical representation i.e. the grid, or is a particular process missing or a model coefficient inappropriate (i.e. a temperature function when the model is applied to a tropical water system)?

We also divided the model coefficients into three categories:

1. System independent, fixed coefficients, which should not be changed. Most model coefficients fall into this category.
2. System dependent, fixed coefficients, which are known to be different in different types of waters but should not be varied i.e. their values are taken from a table indicating the appropriate value for a particular type of water.
3. System dependent, variable coefficients.

So rather than allowing 'fine tuning', each model application is set-up to be as generic as possible because this enhances the validity or credibility of the model. Hence in the work presented here, there is a strong focus on generality as opposed to local tuning (Chapter 5: Los and Gerritsen, 1995; Chapter 9: Los et al., 2008; Chapter 8: Blauw et al., 2009).

The human factor: the role of the modeller

There seems to be relatively little attention for the human factor of model application: who is doing the job? Given the complexity of modern ecological models, this is an important issue. To clarify this point, we could compare modern models to a music instrument such as an electric guitar. Two companies, Fender and Gibson, both produce guitars that are highly appreciated for their overall quality and not in the least their sound. But the inherit quality of these instruments only becomes manifest if they are played by a musician, who is both talented and skilled. Someone who does not know how to play will not produce an enjoyable sound just because the guitar is well made. We may take the analogy even a little further. One of the greatest guitar players ever, Jimi Hendrix, was a left hander, who played right handed guitars upside down. So his guitars were of excellent quality, but no other, normally trained guitarist would be able to play them, unless he would practice for a long time to master the peculiarities of these extraordinary guitars.

Working with complex ecological models resembles playing one of these guitars. If they are well build and have a user friendly interface, an intelligent person should be able to generate some results. But to apply them for answering serious (management) questions in a sound way also requires talent, training and experience of the modeller.

Nomenclature of models

For several reasons: their long history, different clients, applications to fresh and marine waters, licissing of the software etc. many different names and abbreviations of (sub-)models have been and sometimes are still being used. Different names appear in different documents and papers, some of which are included in this thesis and might therefore cause some confusion among the readers. To sort this out the following should be noted.

BLOOM: At the heart of all models included here is the phytoplankton model BLOOM. With respect to the computer code, the situation is quite simple as only one version exists. The model can be run in two different modes: (1) as a stand alone application and (2) as a sub model of an integrated modelling system. In its first mode, the inputs such as the amount of nutrients are directly taken from measurements. In its second mode, there is direct, internal communication between BLOOM and the other modules.

DELFT3D-ECO: The computer code of the integrated model is called DELFT3D-ECO. Again it should be stressed that only one version exists. So a primary production model application for a lake, a channel system, an estuary or a marine water system all use the same computer code. Differences exist with respect to the input only: the choice of the equations and the parameterization. The main distinction between fresh water and marine applications is a different set-up of the phytoplankton community within BLOOM. The name DELFT3D-ECO is commonly used for consultancy applications.

DBS: DBS can be regarded as a predecessor of DELFT3D-ECO. It is an integrated modelling framework consisting of DELWAQ (transport plus the aforementioned process library), SWITCH (a sediment model) and BLOOM. The name DBS was introduced for a fresh water model, but the same model code (but with a different input) was used for some marine applications as well, one of which is included in this thesis. The name DBS is no longer widely used.

GEM: At the end of the 1990s the Dutch Rijkswaterstaat commissioned Delft Hydraulics to take the lead in developing a 'Generic Ecological Model for estuaries and coastal waters' abbreviated as GEM. All major Dutch marine institutes participated in this project. Rather than developing a new code from scratch, the GEM processes were coded as an extension of the process library of DELFT3D-ECO. This means that technically GEM is a subset of DELFT3D-ECO. For historic reasons, in Dutch marine studies the name GEM is widely used, while internationally the name DELFT3D-ECO is preferred. Again: the same code is used and hence the selection of processes and the parameterization might be equal regardless of the name that is assigned to the model.

Aim of this thesis

A typical example of the relationship between management questions, the natural system and the mathematical modelling framework is shown in Fig. 1. This so called 'effect chain' scheme was developed in cooperation with many partners for the Flyland project, which investigated possibilities and consequences of a future airport in the North Sea. Effect chains for other cases will be (slightly) different, but the principle is the same. In the middle is the natural system with its relevant physical and ecological components. Management questions are generated at the level of the 'user functions' and have to be (re)formulated in such a way that the model can deal with them. The models, which are at the third level, should be validated, describing the most relevant parts of the natural system in a credible way in order to be applicable for making projections.

The work described in this thesis deals with only a part of the components of the system. Its focus is on algal biomass and algal species in relation to the abiotic components: hydrodynamics, suspended sediments and light, nutrients and bottom sediments. With respect to modelling the focus of this thesis is therefore on DELFT3D-ECO, but the hydrodynamical and sediment models are frequently discussed as well. Biological interactions, such as competition with macrophytes and grazing are an integral part of many studies and have been addressed with DELFT3D-ECO, but only one example from this type of work is included here: the competition between phytoplankton and *Chara* in Botshol (Chapter 6: Rip et al., 2007). A more extensive description of these kinds of model applications can be found elsewhere (see for instance Los, 1999; WL|Delft Hydraulics, 2007).

The aim of this thesis is the development and application of mathematical models for primary production by phytoplankton and macro algae that are 'credible' and 'fit for purpose'.

More specifically credibility is enhanced by using generic equations and fixed parameters as much as possible, by adopting well defined methodologies for validation, and by demonstrating that the results of the model fit in with other knowledge and contributes to obtaining a consistent, comprehensive view of the natural systems for which the models are applied.

Of course 'acceptability' is also an issue for the models presented here, but this is dealt with in a less explicit way in this thesis. In most cases results of studies were accepted. If this had not been the case, the line of modelling presented here would have been discontinued rather

7

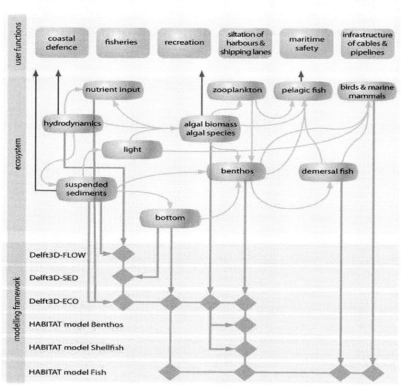

Figure 1. Effect chain for the analysis of a future airport in the North Sea.

than maintained for over 30 years. Worth mentioning within the framework of this thesis are the regular meetings and common reports by the Ospar (Oslo - Paris convention) modelling group of different member states, which are a mutual atempt by the leading eco-hydrodynamic modellers to enhance the acceptability of the results generated by each of the participants at the European level.

Outline of this thesis

In the next chapters the development and application of the eco-hydrodynamical model DELFT3D-ECO (a.k. BLOOM, DBS, GEM) is described. Although the focus is on more recent work, the historic evolution is also presented in as far as it is relevant to the aims of this thesis and to the status of the present model. As neither the model structure, nor its parameterization nor its implementation has been constant, some discrepancies may exist between the different chapters. As a general rule, the most recent papers have precedence over older papers when it comes to describing the model as it is now. At the heart of the eco-hydrodynamic model is the BLOOM module. An extended overview of this module, including the main principles and equations, is given in Chapter 1 of this thesis. A summary including some examples of early fresh water applications of the model is presented in Chapter 2. Notice that this is a discussion paper, which was written twenty years before Chapter 1. There are some inconsistencies between these two chapters in particular in relation to the objective of the BLOOM model; in line with the general remark made at the beginning of this section, Chapter 1 supersedes Chapter 2. In Chapter 3 the possibilities are described for using BLOOM as a tool to assist managers in implementing the Water Framework Directive to coastal and transitional waters.

Starting with Chapter 4, BLOOM is always used in combination with modules for transport and chemical processes in the remainder of this thesis. This chapter contains a long term hind cast simulation of DBS (DELWAQ - BLOOM - Switch) for Lake Veluwe (the Netherlands). Chapter 5 deals with the methodology for validating DBS. The general validation methodology is applied to the complex Rijnland water network in the Netherlands. Chapter 6 shows an example of a DBS application to Lake Botshol (the Netherlands); in this study phytoplankton and macrophytes are modelled simultaneously for a period of over 10 years.

Chapter 7 describes another long term hind cast simulation, but this time for a marine application. It covers a period of 20 years using a relatively refined grid of the Dutch coastal zone. In Chapter 8 the Generic Ecological Model for estuaries and seas (GEM) is introduced. It deals with the equations, the generic parameterization and validation for a broad range of water systems. Validation of complex ecological models is the main subject of Chapter 9. As an illustration the validation result of BLOOM/GEM in full 3D mode to the North Sea is presented in great detail. Chapter 10 shows the results of an inter comparison between three generations of models as applied to the North Sea. How much progress has there been over the years in terms of goodness of fit, which changes to the models were successful and which were not and on which aspects should future model developments be concentrated in order to improve their applicability for emerging management issues. The thesis is concluded with a summary, a short CV and a list of selected publications.

References

Beverton, R.J.H. and S.J. Holt, 1957. On the dynamics of exploited fish populations. Fish. Invest., London, 19: 7-533.

Blauw, Anouk N. , Hans F. J. Los, Marinus Bokhorst, and Paul L. A. Erftemeijer., 2009. GEM: a generic ecological model for estuaries and coastal waters. Hydrobiologia DOI 10.1007/s10750-008-9575-x

De Vries, I. , R.M.N. Duin, J.C.H. Peeters, F.J. Los, M. Bokhorst and R.W.P.M Laane, 1998. Patterns and trends in nutrients and phytoplankton in Dutch coastal waters: comparison of time-series analysis, ecological model simulation, and mesocosm experiments. ICES Journal of Marine Science, Vol. 55: 620-634.

Di Toro, D.M., Fitzpatrick, J.J. and Thomann, R.V., 1971. A Dynamic Model of the Phytoplankton Population in the Sacramento San Joaquim Delta. Adv. Chem. Ser., 106:131-180.

Ivlev, V.S., 1955. Experimental ecology of the feeding of fisfies (English translation by D.Scott 1961). Yale Univ. Press, New Haven.

Jørgensen, Sven Erik, 2008. Overview of the model types available for development of ecological models. Ecological Modelling, 215: 3-9.

Holling, C.S., 1966. The functional response of invertebrate predators to prey density. Mem. Ent. Soc. Can. 48: 3-86.

Leslie, P. H., 1945. On the use of matrices in certain population mathematics. Bio-metrika, 33: 183-212.

Leslie, P. H., 1948. Some further notes on the use of matrices in population mathematics. Biometrika. 35: 213-45.

Los, F.J., Ecological Model for the Lagoon of Venice. Modelling Results. WL | Delft Hydraulics report, T2162, November 1999.

Los, F.J., and M. Bokhorst, 1997. Trend analysis Dutch coastal zone. In: New Challenges for North Sea Research, Zentrum for Meeres- und Klimaforschung, University of Hamburg: 161-175.

Los, Hans and Herman Gerritsen, 1995. Validation of water quality and ecological models. IAHR Specialist Forum on Software Validation, London, September, 1995.

Los, F.J., M.T. Villars and M.R.L. Ouboter, 1994. Model Validation Study DBS in networks. WL | Delft Hydraulics, Research Report, T1210.

Los, F. J., M. T. Villars, & M. W. M. Van der Tol, 2008. A 3- dimensional primary production model (BLOOM/GEM) and its applications to the (southern) North Sea (coupled physical–chemical–ecological model). Journal of Marine Systems, 74: 259-274

Lotka, A.J., 1925. Elements of physical biology. Williams & Wilkins, Baltimore.

Mankin, J.B., O'Neill, R.V., Sugart, H.H., Rust, B.W., 1975. The importance of validation in ecosystem analysis. In: Innis, G.S. (Ed.), New Directions in the Analysis of Ecological Systems, Part 1. Simulation Councils Proc. Ser., vol. 5. Calif.: Simulations Councils, Inc., pp. 309–317. No 1, Lajolla, 132 pp.

Nicholson, A.J. and V.A. Bailey, 1935. The balance of animal populations. Part 1. Proc. Zool.Soc. London 1935, Part 3: 551-598.

Park, Richard A., Theresa W. Groden, Carol J. Desormeau, 1979. Modifications to the Model Cleaner Requiring Further Research, In: Scavia, Donald, and Andrew Robertson, (eds.), Perspectives on Lake Ecosystem Modelling. Ann Arbor Science, Ann Arbor, Michigan.

Rip, Winnie J., Maarten Ouboter, Hans J. Los, 2007. Impact of climatic fluctuations on Characeae biomass in a shallow, restored lake in The Netherlands, Hydrobiologia, Vol. 584: 415-424.

Vollenweider, R., 1975. Input-output models, with special reference to the phosphorus loading concept in limnology. Schweiz. Z. Hydrol. 37: 53-84.

Van der Molen, D. T. 1999. The role of eutrophication models in water management. Thesis. Agricultural University, Wageningen, The Netherlands.

Vollenweider, R., 1975. Input-output models, with special reference to the phosphorus loading concept in limnology. Schweiz. Z. Hydrol. 37: 53-84.

Volterra, V., 1926. Variazioni e fluttuazioni del numero d'individui in specie animali conviventi. Mem. R. Accad. Naz. dei Lincei, Ser. VI, Vol. 2 [English translation in CHAPMAN 1931].

WL | Delft Hydraulics, 2002. GEM documentation and user manual. WL | Delft Hydraulics report Z3197, Delft, The Netherlands.

WL | Delft Hydraulics, 2003. Delft3D-WAQ users manual. WL | Delft Hydraulics, Delft, The Netherlands.

WL|Delft Hydraulics, 2007. Keyzones, ecosystem scale modelling. To investigate sustainable biological carrying capacities of key European coastal zones. EU contract COOP-CT-2004-512664. Delft Hydraulics report Z3557.

Chapter 1

Description of the BLOOM model

Hans Los

Description of the BLOOM model

1. Introduction

Man's activities have affected the environmental conditions on earth for a long time, but never on such a global scale as witnessed during the last one or two centuries. Since then the amounts of various substances, which are released into the environment, have increased dramatically. This is partly because the human population has grown more or less exponentially, but also because industrial and agricultural production per capita grew to unprecedented levels.

Many of these substances are more or less toxic. But other chemicals, which are on the contrary beneficial to certain organisms, are also released in enormous quantities. Among these are several compounds of the elements nitrogen and phosphor which are required by species of phytoplankton, among others. As a result the concentrations of these plants have increased up to a level where they are considered a nuisance. This process, which is called eutrophication, is accompanied by several objectionable symptoms: it gives the water a green, turbid appearance; it can cause bad odours; it may harm other organisms because the minimum daily oxygen level can become extremely low during the night due to phytoplankton respiration; it can even cause the water to become completely deprived of oxygen (anaerobic) when a bloom declines rapidly, since the biological degradation processes consume large amounts of oxygen; it may cause clogging of filters in water transportation systems.

In the Netherlands the situation is worse than in many other countries because (1) it is densely populated, (2) still there is intensive farming, (3) it receives a major part of its water from the nutrient loaded rivers Rhine, Meuse and Scheldt and (4) most of its lakes, estuaries and coastal waters are shallow.

The total algal biomass usually consists of many species of phytoplankton belonging to different taxonomic or functional groups such as diatoms, flagellates, green algae and cyanobacteria, commonly referred to as blue-green algae. This is true for fresh-water as well as for marine systems. The species have different requirements for resources (e.g. nutrients, light) and they have different ecological properties. Some species are considered to be objectionable due to their effect on the turbidity of the water, the formation of scums or the production of toxins. For example, *Planktotrix* can achieve very high biomass levels in shallow lakes causing a very low transparency (Berger et al., 1983; Berger 1984; Zevenboom et al., 1982), and *Microcystis* is notorious for the formation of scums and has been reported to produce toxins that are harmful to animals (e.g. cattle) and men (Atkins et al., 2001; Chorus et al., 2000). In the marine environment, *Phaeocystis* is responsible for foam on beaches (Lancelot et al., 1987) and mass mortality of shellfish due to the settlement of a bloom in sheltered areas and subsequent depletion of oxygen (Rogers and Lockwood, 1990).

Already in the 1970s managers became aware that mathematical models could help to understand the complexity of eutrophication problems and to simulate the potential impacts of measures to improve conditions. However, attempts in the Netherlands to apply existing methods such as the well known Vollenweider statistical model (Vollenweider, 1975) or deterministic models such as Di Toro et al. (1971; 1977) were rather unsuccessful. In part, this was because the Dutch conditions were (and often still are) far out of the validation application range for which these models had been constructed. Therefore it seemed necessary to develop a new model. Important general requirements for such a management model were: to include (1) three macro nutrients and light, and (2) total biomass, chlorophyll, dissolved oxygen, (3) to distinguish major groups of phytoplankton, (4) to focus on predicting objectionable conditions, (5) to develop a generic set of model coefficients and discourage local fine tuning.

By the time the model development was about to start, the US Rand corporation had just finished the construction of several models to support the policy analysis of the Eastern Scheldt storm surge barrier. One of these models was called the algae bloom model, which has only been described in a report to the Dutch Ministry of Public Work (Bigelow et al., 1977). According to its purpose this model was essentially a worst case model. It computed the highest feasible bloom during a period of 10 days taking the most relevant environmental factors (light, temperature, nutrients) into account. For each period a new set of conditions was provided and simulations were performed without any recollection of the past (series of steady state conditions). To find this maximum bloom in an efficient way the mathematical set of equations was reformulated as an optimization problem which was solved using linear programming (Danzig, 1963). Being a worst case model, the total biomass was maximized in this model version. One can think of this model as a 'free allocation model': all available resources were redistributed among all phytoplankton species considered by the model at every time step of 10 days without any further restriction taken from previous circumstances.

After replacing the marine phytoplankton species by fresh water species, basically the same model was applied to a large number of lakes at the beginning of the 1980s (Los, 1980; Los, 1982a; Los, 1982b; Los et al. 1982; Los et al., 1984). This model version was successful in many aspects, but failed to reproduce two important observations. First, diatoms and green algae are often dominant in low nutrient, P limited lakes (Schreurs, 1992 and many references in his thesis). This feature could not be reproduced by early BLOOM versions since the requirement for P of cyanobacteria is among the lowest of all species of phytoplankton. Hence they would be favoured rather than hampered by nutrient reduction. Second, experimental results showed significant differences in internal concentrations of nutrients and of growth rates as a function of the availability or resources (Shuter, 1978; Zevenboom et al., 1983; Zevenboom et al., 1984). Similar variations were observed in field data for lakes with a clear dominance of a single species during a prolonged period of time (several months) (Zevenboom et al 1982; unpublished results from our own studies on many lakes). Internal stoichiometry clearly varied being highest when nutrients were high and light was the main limitation and lowest under conditions of nutrient limitation. Consequently during early model applications it was often necessary to use lake specific, species dependent characteristics such as the carbon to chlorophyll ratio in order to reproduce observed chlorophyll concentrations. Preferably these variations should be dealt with by the model itself.

So, although the Rand algae bloom model can be regarded as the origin of the present BLOOM model, BLOOM went through a long evolution and the resemblance between the present and original model is very limited. Two major lines of development can be distinguished: (1) the concept of the model itself was changed and (2) it was integrated with modules for transport, chemistry, sediment etc. In this chapter I focus largely on the model concept. In section 2 the underlying assumptions of the model are presented. The model equations are given in section 3. Section 4 deals with the applications of the model in a general way. In section 5 the BLOOM approach is discussed and compared with the approach of more traditional differential equation models. This chapter is concluded with a short summary and an appendix with details on the numerical procedure to integrate the light response curve over depth and in time. Details on BLOOM's interfacing with other modules are given in Chapter 8 (Blauw et al., 2009), Chapter 9 (Los et al., 2008), and in WL|Delft Hydraulics (1992; 2002; 2003). For reasons of simplicity this chapter is restricted to its application to fresh water systems. However, the marine version is identical with respect to the principles and equations, and differs merely in the model parameters defining the phytoplankton species. An extensive description of the marine model applications is given in Chapters 3, 9 and 10 (Los et al., 2007; Los et al., 2008; and Los et al., submitted).

2. The BLOOM model: general description

Competition principle

Competition between phytoplankton species is one of main processes in the model. In the original Rand version of the model, the species with the lowest requirement for a limiting resource would always win the competition, unless its presence was prohibited by some other factor (i.e. temperature; under water light intensity). Many phytoplankton models operate according to similar principles although the selection mechanism is different. In stead of the requirement for the limiting resource, the outcome of the competition in these models is often determined by the growth rate under conditions of a low availability of the limiting resource. In both cases competition is effectively controlled by a single, but not by the same parameter. The requirement is defined here as the amount of a resource (i.e. a nutrient) per unit of biomass of the phytoplankter necessary to remain viable.

According to the general theory on K- and r-strategies (e.g. Harris, 1986; Reynolds et al., 1983) the requirements as well as the growth or uptake rate of organisms are essential in determining the outcome of interspecific competition of phytoplankton. From observations of lakes under turbid conditions, it appears that shade adapted species i.e. cyanobacteria are often dominant whereas higher average light intensities tend to favour diatoms and green algae. Underwater light regimes are obviously complex viewed from the point of competing algae in a partially mixed environment with steep light gradients. However, from empirical case studies under a wide range of nutrient availability levels, both Schreurs (1992) and Scheffer (1998) conclude that the ratio between the euphotic and the mixing depth is an important, factor to explain which group of species will be dominant. This ratio is usually referred to as the Zeu/Zmix ratio. The euphotic depth is defined as the depth at which photosynthesis is just sufficient to compensate the losses. In practice this is often taken as the depth at which the light intensity equals 1% of the surface intensity, but in reality the surface intensity varies seasonally and the light dependence of different photo-autotrophic organism varies considerably as well. So it is important to consider the potential growth rate of individual species on a seasonal basis in stead of using a single value for Zeu; this is what is done by BLOOM. The differences in growth rates between a number of representative species can be illustrated (Fig. 1), by looking at the net growth rates estimated on the basis of lab

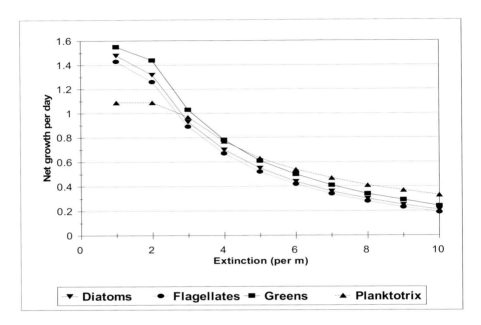

Figure 1   Net growth rate as a function of the total extinction (m^{-1}) for typical summer conditions in a 1m deep Dutch lake. Results for *Microcystis* and *Aphanizomenon* resemble those for *Planktotrix* (*Oscillatoria*) and are not shown here.

experiments (Zevenboom and Mur, 1981; Zevenboom et al., 1983; Zevenboom and Mur, 1984; Post et al., 1985) assuming a sufficient supply of nutrients as a function of the under water light intensity in a hypothetical 1m deep lake with a light intensity typical for Dutch summer conditions at a temperature of 20^0 C. The shape of these curves depends on (1) the growth versus light (P-E) curves of the species, (2) their maximum growth rate constant and (3) their respiration Notice that interspecific differences in response of phytoplankton species to the dosing of light are not accounted for in this idealised analysis.

Since according to the Lambert Beer equation the overall light intensity declines exponentially with extinction, the net growth rate constant Pn or more specifically the rate of photosynthesis, of any single species is always a declining function of the extinction coefficient. The respiration rate is often assumed to be a fraction of the maximum growth rate irrespective of the light intensity. Hence at increasing levels of the extinction, the growth rate declines and the relative importance of the respiration rate increases. Since both the rate of primary production and the respiration differs per species, their response to variations in light intensity are different as well. A comparison between individual species reveals that the curve for eukaryotic species is much steeper than for cyanobacteria. Consequently the net growth rate curves of these groups of species intersect, in this example at an extinction coefficient of about 4 m^{-1}. When the average light intensity is high (low extinction coefficient) the eukaryotic species have the highest growth rates due to their high maximum growth rate. The reverse is true at low light intensities: cyanobacteria can maintain relatively high net growth rates at decreasing light intensities. This is due to both the relatively steep initial slope of the P-E curve and to and their low respiration rate, which becomes relatively more important as the light limitation gets more severe. So according to the eco-physiological characteristics of the species, turbid conditions with a low Zeu/Zmix ratio favour cyanobacteria while clearer conditions with a high Zeu/Zmix ratio favour eukaryotic species as confirmed by other studies (Schreurs 1992; Scheffer 1998).

To deal better with the important role of light in selecting the dominant group of species, the original one parameter competition principle of BLOOM was replaced by a two parameter principle taking both the requirement and the potential growth rate into account as its selection criterion (Section 5 of this chapter; Los et al., 1988; Los et al., 2007; Loucks & Van Beek, 2005). Because a priori it is unclear what the relative importance of these factors should be, it is assumed that both of them are equally important. Defining Pn_k as the potential net growth rate constant of type k under the prevailing light conditions and n_{ik} as its requirement for resource i (i being either a nutrient or the amount of light energy at which light becomes limiting) the model considers Pn_k / n_{ik} to determine which species will become dominant. According to this principle a species A whose required amount for a limiting resource is x% higher than the requirement by species B can exactly compensate for this if A's net potential growth rate at the prevailing light climate is also x% higher. In practice this means that in model simulations opportunistic, r-selected species with high maximum growth rates dominate when the average light intensity is high for instance during the spring bloom, whereas efficient K-selected species with lower maximum growth rates and lower resource requirements dominate when the average light intensity is low and external forcings are relatively stable for instance during the summer period in eutrophic lakes.

Conditional steady states: transformation to dynamic model

In the Rand algae bloom model, a steady state was assumed for each phytoplankton species for every time step of the model, which was 10 days. In general this was a reasonable assumption since the model was applied to the Eastern Scheldt with a low turbidity and hence relatively high potential growth rates by the phytoplankton in relation to the rate of change in external forcing conditions. One of the main advantages of the steady state assumption was the possibility to apply the highly efficient Linear Programming method to quickly select the optimum combination of species and limiting factors among all possible combinations.

However, under more turbid conditions actual growth rates are small relative to maximum growth rates and they may not be sufficient to allow a complete shift in species dominance within a period of 10 days (see Fig. 1). Moreover, the time step of later versions of BLOOM was reduced because it was linked to chemical and transport modules, which consider processes at a time scale of a day or even less. Hence it is unrealistic to assume that any theoretically possible steady state combination of species and limiting factors can always be achieved during a short time interval in eutrophic waters where growth rates are low due to a high turbidity. Hence the unconditional steady state assumption of the original model had to be abandoned. Two alternatives were considered. One option was to rewrite the model into a (classical) differential equation model and solve it numerically. The second option considered was to extend the concept of the original model with a set of conditions to delimit the number of acceptable solutions to those that might be obtained given the growth and mortality rates of the species under the prevailing conditions. In other words: combinations of species, which could not reasonably be achieved within a time step of the model, would not be selected. This last alternative was adopted. If at the beginning of a time step the biomass of a species is small relative to its equilibrium, BLOOM will delimit its biomass during the next time step by its potential net growth rate. If in contrast the biomass of a species is sufficiently close to its steady state, its biomass will be determined by the availability of the resources. So rather than a 'free reallocation' model concept, BLOOM in its present form adopts a 'conditional reallocation' scheme excluding all potential combinations of species which cannot be achieved within the time step size set for solving the model.

Adaptation to varying conditions

It has been observed both under experimental as well as under field conditions that most biological species of phytoplankton adapt rapidly to variations in their external environment (i.e. Chapter 5 of Harris, 1986). The measurable characteristics of individuals of a species can therefore display a wide range of variation. Again different options are available to include interspecific variability in a model. One option would be to describe adaptation as a process by which one or several characteristics of the model species are changed. For instance the internal nutrient concentration (stoichiometry) could be modelled as a function of external and or internal conditions. In practice this approach is not trivial as many factors have to be considered and moreover interactions between different factors should also be taken into account. Relevant characteristics not only include the nutrient stoichiometry, but also the growth rate, the mortality and sedimentation rate, perhaps the susceptibility to grazing, the biomass to chlorophyll ratio etc. Hence the equations to describe adaptation may become very complicated depending on the amount of interactions that are taken into account. PCLAKE is an example of a model which adopts this kind of approach (Janse, 2005).

In BLOOM another approach was adopted taking into account that the model equations are solved by an optimization technique. Two basic units are distinguished in BLOOM. The first unit is that of functional groups which we refer to as 'species groups' or simply 'species'. A model species may be equivalent to a taxonomic species, or it can be representative for larger taxonomic units consisting of several species or genera whose ecological characteristics are sufficiently similar to be treated as a single functional group. For instance in the default configuration of the model only one 'species' of diatoms is included. The second unit is that of 'types'. Model species usually consist of several types. A type represents the physiological state of the model species under different conditions of limitation. Model species are usually divided into three different types: an N-type representing the eco-physiological condition of a species under nitrogen limitation, a P-type for phosphorus limitation and an E-type, representing the state of a species under low light conditions. This principle of modelling species and types resembles that of 'super individuals' as described by for instance Scheffer et al. (1995).

Additional types such as colonies, mixotrophic cells or nitrogen fixing cells can be explicitly included in the model. In mathematical terms types are the state variables of BLOOM. They have fixed characteristics specified in the input of the model, which however may be different from those of the other types with respect to their nutrient contents, specific extinction, growth, mortality and sedimentation rates. Since the types of a species may be selected at any ratio depending on the environmental conditions, the average characteristics of the species vary continuously in time and in space within the limits defined for the individual types. So in BLOOM dependence between controlling factors is not explicitly described by a set of equations, but in stead it is included via an appropriate selection of parameter values of the types comprising each species.

Using types provides an additional advantage with respect to the computational aspects of the model. All external processes, which act upon the phytoplankton such as transport, sedimentation and grazing, are calculated with respect to types rather than species. This means their impact on the internal stoichiometry is automatically dealt with; species wise adding up the types is sufficient to know its composition at any moment in time and at any place. It is not necessary to recalculate the internal stoichiometry and correct individual fluxes for possible changes in stoichiometry.

The formulation of the model takes into account that adaptation occurs much more rapidly than succession between species. Succession involves replacing one species by another which is controlled by the net rate of change of the species (characteristic time scale in the order of days to weeks). Adaptation is a much faster process, which according to Harris (1986) has a similar characteristic time scale as a cell division (order of hours to days). So in BLOOM (complete) adaptation that is a shift from one type of a species to another can in theory occur at any time step of the model while shifts in composition usually require a considerably longer period. Notice that usually the external factors controlling phytoplankton change rather smoothly so in most model applications transitions between types occur smoothly as well but this is due to the forcing, not the formulation of the model.

Model objective

Based on the previous description, the purpose of BLOOM may be described as:

Selecting the best adapted combination of phytoplankton types at a certain moment and at a certain location consistent with the available resources, the existing biomass levels at the beginning of a time interval and the potential rates of change of each type.

Hence although the present version of BLOOM differs significantly from its predecessor in terms of its underlying assumptions and equations, mathematically it is still formulated as an optimisation problem which can be solved by Linear Programming. The main model equations will be presented in the next sections.

3. Environmental constraints

Nutrient balance

The BLOOM model identifies the concentration of biomass, B_k, of each algae type k that can be supported in the aquatic environment characterized by light conditions and nutrient concentrations. For each algae type k, the requirements for nitrogen, phosphorus and silica (only used by diatoms) are specified by coefficients n_{ik}, the fraction of nutrient i per unit biomass concentration of algae type k.

The total readily available concentration, C_i (g.m^{-3}) of each nutrient in the water column equals the amount in the total living biomass of algae, $\Sigma_k(n_{ik}B_k)$, plus the amount incorporated in dead algae, d_i, plus that dissolved in the water, w_i. These mass balance constraints apply for each nutrient i.

$$\Sigma_k (n_{ik} B_k) + d_i + w_i = C_i \tag{1}$$

The unknown concentration variables B_k, d_i, and w_i are non-negative. All nutrient concentrations C_i are the measured or modelled total concentrations and are assumed to remain constant throughout the time period defined for the model. The system is assumed to be in equilibrium over that period. The time step is an input to the model and may be chosen to vary during the simulation to account for seasonal variations in characteristic time scales.

Nutrient recycling

A certain amount of each algae type k dies in each time step. This takes nutrients out of the live phytoplankton pool. A fraction remains in the detritus pool, and the remainder is directly available to grow new algae because the dead cells break apart (autolysis) and are dissolved in the water column. Detritus may be removed to the bottom or to the dissolved nutrient pools at rates in proportion to its concentration. Needed to model this is the mortality rate, M_k (day^{-1}), of algae type k, the fraction, f_p, of dead phytoplankton cells that is not immediately released when a cell dies, the remineralization rate constant, m_i (day^{-1}), of dead phytoplankton cells, the fraction, n_{ik}, of nutrient i per unit biomass concentration of algae type k, and the settling rate constant, s (day^{-1}), of dead phytoplankton cells.

The rate of change in the nutrient concentration of the dead phytoplankton cells, dd_i/dt, in the water column equals the increase due to mortality less that which remineralises and that which settles to the bottom.

$$dd_i/dt = \Sigma_k (f_p M_k n_{ik} B_k) - m_i d_i - s d_i \tag{2}$$

Both mortality and mineralization rate constants are temperature dependent. If the model is applied as a stand-alone (screening) tool, Equation 2 is solved under the assumption of a steady state which means its right hand side equals 0. This gives an expression relating the amount of detritus to the algal biomasses. The result of which can be substituted into Equation 1. If BLOOM is applied as a dynamic simulation model, this equation is integrated numerically.

Energy limitation

Algae absorb light for photosynthesis and growth. The response of phytoplankton to variations in light intensity is, however, complicated. In general the growth rate increases when the light intensity increases up to a certain level called Iopt: the light intensity where growth achieves its maximum value. Experimentally it is often observed that growth declines when light intensities exceed Iopt, a process called photo inhibition. Whether or not this process is important under natural conditions is a matter of debate. It may be argued that photo inhibition only occurs after a prolonged exposure to high light intensities and that this situation occurs infrequently in water bodies with turbidities as high as in many productive waters. In contrast *Microcystis* appears to get photo inhibited immediately at light intensities in access of its Iopt (Passarge, pers. comm.).

Various equations have been proposed to describe the relationship between growth and the light intensity (i.e. Steele, 1958). His equation is still frequently used in models. Some modellers use a Monod equation simply because its shape seems adequate to describe the observed growth rates as a function of the light intensities. Harris (1978) points out that different processes are responsible for different parts of the photosynthesis versus light (P-E) curves of phytoplankton. Hence preferably a mathematical equation should be adopted whose initial slope does not depend on its level of saturation or photo inhibition. Moreover, the initial part of the curve is of much more importance than the fit at high intensities since on average phytoplankton cells spend most of the time at relatively low light intensities. Therefore in BLOOM a different approach is adopted. Rather than by some functional relationship which may or may not give a good approximation of the experimental data, a table is used containing a large number of data points in order to obtain a sufficient level of accuracy. This way any set of actual measurements can be represented. Using a table also makes it possible to easily switch between light response relationships with and without photo-inhibition and to accommodate inter specific differences in response curves. In order to do this it is sufficient to use another table with input data.

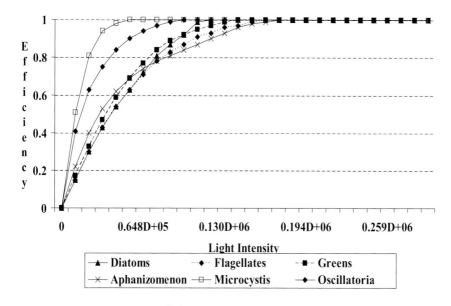

Figure 2. Growth vs. light intensity $(J.m^{-2}.h^{-1})$ curves of fresh water species in BLOOM (uninhibited).

BLOOM also takes experimental results into account indicating that the growth rate per hour is not always simply proportional to the number of hours of day light (photo period). In particular some cyanobacteria such as *Microcystis* achieve relatively high growth rates per hour when the photo period is short. This information is specified in the input of the model in the form of species specific table on an hourly basis.

In comparison to other resources (nutrients) the amount of light, to which phytoplankton is exposed under natural conditions, shows much stronger variations. First the level of irradiance varies considerably during the day. Second individual phytoplankton cells do not usually maintain a fixed vertical position in the water column, but are mixed by turbulence and other hydrodynamic processes. Third species react differently to variations in intensity per unit of time; some are well adapted to short day lengths, others seem to adjust mainly to the total daily irradiance. To account for all these variations, the growth rate is averaged over the depth and in time by a pre-processor of the model yielding an average efficiency factor E_k under all possible conditions. The averaging procedure takes the daily amplitude of the solar radiation and the length of the photo period into account. The result is put into a look up table (different per species) which is used during the actual simulation to quickly find the correct efficiency factor of each phytoplankton species for any possible light condition.

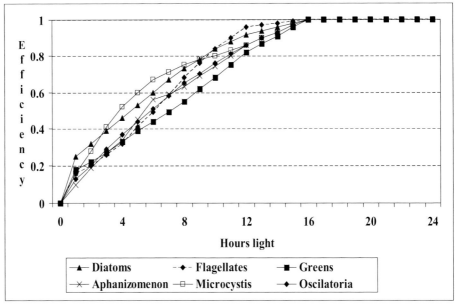

Figure 3. Day length response curves of freshwater species in BLOOM.

As Harris (1978; 1986) points out, the response of phytoplankton to variations in light intensity is a photo chemical reaction. This means that in general the same response is to be expected regardless of the value of the maximum gross growth rate Pg_k^{max} (day^{-1}), which is a function of temperature. In BLOOM Pg_k^{max} can be specified as a linear or as an exponential function. The linear function is usually adopted if the seasonal variation in observed temperatures is rather small as in tropical or in many marine systems. This type of function is also preferred for species that do not grow well at low temperatures such as *Microcystis*. Hence the following equations are implemented in the model:

$$Pg_k^{max} = P_{1k} * P_{2k}^{T}$$ (3)

or

$$Pg_k^{max} = P_{1k} * (T - P_{2k})$$ (4)

where P_{1k} and P_{2k} are type-specific model coefficients and T is the temperature. Notice that these constants have a different interpretation in the two equations.

To obtain the actual growth rate one could multiply Pg_k^{max} and E_k, but consequently Iopt would vary with temperature in the same way Pg_k^{max} varies with temperature. To correct for this problem, the light intensities in the BLOOM model are rescaled in proportion to the value of Pg_k^{max} at the actual temperature and the value of Pg_k^{max} at the temperature at which the light response curve was determined. Next the average efficiency is calculated using the rescaled light intensity (Los, 1991). This issue is rarely discussed by modellers so it is not clear whether other models adopt a similar procedure. In general this should be the case if the light response curve is first integrated over light and then multiplied with the maximum growth rate.

Other intrinsic processes affecting the energy balance of phytoplankton are respiration and natural mortality. Both are temperature dependent and an exponential relationship is assumed for both of them. Hence the respiration R_k (day^{-1}) is computed as

$$R_k = R_{1k} * R_{2k}^{T}$$ (5)

where R_1 and R_2 are type-specific model coefficients. Similarly the natural mortality M_k (day^{-1}) is computed as

$$M_k = M_{1k} * M_{2k}^{T}$$ (6)

where M_{1k} and M_{2k} are type-specific model coefficients. The overall energy budget of a phytoplankton type k can be written as:

$$\frac{dB_k}{dt} = (Pg_k^{max} * E_k - M_k - R_k) * B_k$$ (7)

where E_k is the depth and time averaged production efficiency factor (see 'Appendix: Averaging the production' for the way it is computed).

In additional to non-algal material, phytoplankton contributes to the extinction of light under water and therefore affects the efficiency E_k. As the concentration of phytoplankton increases, this absorption term increases. This means there is a negative feed back between algal biomass and the light limited growth rate: the more the biomass increases, the smaller the growth rate gets. This process is commonly denoted as 'self-shading', but one should not overlook that other species of phytoplankton contribute to the extinction as well so self shading effectively refers to the whole community, not just the individual species. Since the loss rates are usually considered to be independent of the light intensity, it follows that for each species and for each set of conditions there is a value of the time and depth averaged growth rate at which it is just balanced by mortality plus respiration plus other loss terms which may be considered in the model. This balance is obtained by putting the right hand side of Equation (7) to 0, hence

$$Pg_k^{max} * E_k = M_k + R_k \tag{8}$$

This means that for each phytoplankton type k there exists a certain value of the average light intensity and hence of the average extinction value K_k^{max} (m^{-1}) at which this is the case. When the total extinction is equal to K_k^{max}, the net growth rate of type k is exactly zero. If we take photo inhibition into account, then the light intensity can also be too high, which means the total extinction is too low for growth. This specific extinction value is K_k^{min}. The ranges between K_k^{min} and K_k^{max} differ for different algal types k because each one of them is characterized by a different set of model coefficients. Letting K_k (m^3/m/gdry) represent the specific light absorbing extinction constant for living material of algae type k, the total extinction due to all living algae is

$$KL = \Sigma_k(K_k B_k) \tag{9}$$

Added to this must be the extinction caused by detritus (dead cells), KD and the contribution of all other fractions such as inorganic suspended matter and humic substances to the extinction of the water, KW (m^{-1}). Hence

$$K_k^{min} \leq KL + KD + KW \leq K_k^{max} \tag{10}$$

The extinction per biomass unit from dead cells is usually less than that from the same amount of live cells. The amount of dead cells not yet mineralized is, from Equation 2, $\Sigma_k(fp\ M_k B_k)$. Assuming some fraction e_d (usually between 0.2 and 0.4) of the extinction rate of live cells,

$$KD = e_d \Sigma_k K_k fp\ M_k B_k \tag{11}$$

Notice that while there is a specific extinction coefficient for each individual type in the model, only a single value is adopted for dead organic algal material. If the total extinction is not within the range for a phytoplankton type k, its concentration B_k will be zero. To ensure that B_k is 0 if the total extinction is outside of its extinction range, a 0,1 binary (integer) unknown variable Z_k is needed for each phytoplankton type k. If Z_k is 1, B_k can be any non-negative value; if it is 0, B_k will be 0. This is modeled by adding three linear constraints for each phytoplankton type k.

$$KL + KD + KW \leq K_k^{max} + KM(1 - Z_k) \qquad (12)$$

$$KL + KD + KW \geq K_k^{min}(Z_k) \qquad (13)$$

$$B_k \leq BM\, Z_k \qquad (14)$$

Where KM and BM are any large numbers no less than the largest possible value of the total extinction or biomass concentration, respectively. Since the objective of maximizing the sum of all $Pn_k\, B_k$ together with Equation 8 wants to set each binary Z_k value equal to 1, only when the total extinction is outside of the extinction range K_k^{min} to K_k^{max} will the Z_k value be forced to 0. Equation 8 then forces the corresponding B_k to 0. This means that beyond its feasible range of the extinction coefficient, a species cannot maintain a positive biomass.

Growth limits

When due to transport or another physical forcing in an area with steep gradients the environmental conditions (i.e. nutrient concentrations, irradiance etc.) improve at a rate which is large relative to the potential growth rate constant of a particular phytoplankton species, it may be impossible for that species to achieve the level at which either light or some nutrient gets limiting within a single time-step of the model. To account for this situation, a constraint to delimit the maximum biomass increase within the time-interval is considered during the optimization procedure. Assuming that losses will be low during the exponential growth phase of a phytoplankton species, we make the assumption that mortality can be ignored in the computation of this growth constraint.

For all algae types k the maximum possible biomass concentration, B_k^{max} (gdry.m^{-3}), at the end of the time interval Δt (days) depends on the initial biomass concentration, B_k^{o}, (g dry. m^{-3}), the maximum gross production rate Pg_k^{max} (day^{-1}), the respiration rate constant, R_k, (day^{-1}), and the time and depth averaged production efficiency factor, E_k. Using the net production rate constant, Pn_k ($= Pg_k^{max} E_k - R_k$) (day^{-1}), for each algae type k:

$$B_k^{max} = B_k^{o} \exp[\, Pn_k\, \Delta t\,] \qquad (15)$$

The initial biomass is the concentration at time 0. It is taken as the net result of the BLOOM simulation during the previous time step and all other processes affecting the concentration of the phytoplankton type such as transport, sedimentation and grazing. Notice that it depends on the type of model application and its parameterization which of these factors are actually taken into account in a particular simulation. If the initial biomass is smaller than a certain base level, this base level is used in stead. Empirically it was found that using a base level of 1 percent of the potential maximum generally results in realistic species shifts in the model. This base level is the amount that is supposed to be always present due to germination of spores, resuspension, import from sources not considered in the model etc. If B_x^o is chosen too small, succession is too slow in the model, if the selected value of B_k^o is too large, the model shifts its composition too easily between species. Usually the model is, however, insensitive to value of this base level.

Notice that under relatively stable conditions, types are usually growth limited for a short period of time because they will become limited by a nutrient or light energy after a few days or weeks. Under fluctuating conditions, types may remain growth limited for a much longer period of time for instance because flushing prevents them from reaching a nutrient limitation.

Mortality limits

As in the case of growth, the decline of each algae species is also constrained to prevent a complete removal within a single time-step when a species is suddenly confronted with unfavourable environmental conditions. The minimum biomass value of a species is obtained assuming there is no production, but only mortality. Hence this minimum biomass, B_k^{min} (gdry.m^{-3}), of type k at the end of time interval Δt depends on the initial biomass, B_k^o (gdry. m^{-3}), of type k and the specific mortality rate constant, M_k (day^{-1}) of type k.

$$B_k^{min} = B_k^o \exp[- M_k \Delta t] \tag{16}$$

The initial biomass is computed in exactly the same way as for the growth constraint, including the result of the previous time step plus all other external processes affecting the species biomass.

These minimum values are computed for each individual algae type. However the model sums each of these minimum values over all subtypes within each species and applies it to the total biomass of the species. This way the maximum possible mortality cannot be exceeded, but transitions from one type to another remain possible.

As mortality is computed according to a negative exponential function, mathematically the minimum biomass level always remains positive, in other words a species would never disappear completely. Computationally this increases the necessary time to solve the problem because this depends on the actual number of types selected by the optimization procedure and species with a positive mortality constraint value, are automatically selected. Including a species with an insignificantly small biomass value still requires a complete matrix operation. Therefore the minimum value is replaced by zero once the value computed according to Equation (16) drops below some threshold value. So if the initial biomass is small, no minimum value is imposed on the optimization procedure. Empirically it was found that using a base level which is 10 times smaller than the base level for the growth, generally results in realistic species shifts in the model. The model is rather insensitive to this threshold value, but values close to those of the threshold level of the growth constraints should be avoided as they may lead to cyclic behaviour under certain conditions with species becoming growth and mortality limited intermittently.

Occasionally a conflict may arise between the mortality and energy constraint for instance when the light availability suddenly declines thereby demanding the new biomass to be small, while a large concentration of a type needs to be maintained according to the result of equation 16. Both conditions cannot be fulfilled at the same time. If this occurs, the mortality constraint of a species (16) has precedence over its extinction constraint (10), which is subsequently dropped from the optimization procedure. Effectively this means that types disappear at the rate of M_k (day^{-1}) under unfavourable conditions regardless of the actual light availability and will not be completely removed in a single time step even though too little light is currently available to maintain a positive growth rate.

As in the case of the growth constraints, mortality limitations occur only for short periods of time under stable condition, but may persist for a much longer period of time when conditions are variable. For instance in simulations of the Dutch coastal zone it is not uncommon to find locations at which mortality limited types are present almost permanently.

Grazing

Several possibilities exist to include the impacts of grazing in the model. The easiest, implicit method is to increase the natural mortality rate constant, which is a function of the temperature, or the sedimentation rate, which is usually constant in time. Alternatively two types of explicit formulations are available: a forcing function approach or the use of a dynamic grazer module. With the forcing function approach the biomass levels are prescribed as a function of time and space for one or several grazers. In addition their rates of filtration, digestion, food preference and their nutrient stoichiometry, are all specified in advance. Grazers may be suspended in the water (zooplankton) or attached to the sediment (filter feeders). They act upon the computed phytoplankton community taking away as much biomass as required to sustain the prescribed grazer biomass. If an insufficient amount of food is available, the grazer amount is reduced below its input value. This forcing function approach is adopted for instance to determine the carrying capacity of an ecosystem or the maximal sustainable production of shell fish in a given environment. This approach is described in more detail in Blauw et al. (2009).

Various alternative options are available in the model to simulate grazer biomasses dynamically. They resemble the forcing function approach with respect to the grazer properties, but now their biomasses are simulated as a function of time and location using a differential equation for the growth. Some grazer models were developed at Delft Hydraulics (Wijsman, 2004; Van de Wolfshaar, 2006; WL|Delft Hydraulics, 2002 & 2003), others were adopted from published models i.e. from ERSEM (Broekhuizen, et al., 1995).

Nitrogen fixation, mixotrophy and macro algae

The concept and code of BLOOM make it easy to add phytoplankton types or nutrients. To take nitrogen fixation into account, a special type can be added to the normal set for one of the cyanobacteria (i.e. *Aphanizomenon or Anabeana*). In biological terms this type may be regarded to represent heterocysts. This type requires no dissolved nitrogen, but in contrast can take up nitrogen from an additional source: atmospheric nitrogen. This source is included as a separate nutrient in the model and is assumed to be infinite. Uptake of this source is therefore never limited by its amount, but it is by the growth rate of the nitrogen fixing type, which is relatively low to account for the high energy requirement of nitrogen fixation. No other type but the N-fixer can use this nutrient. In contrast this type does not require dissolved nitrogen. As was explained previously, adaptation between the types of a species is possible. Hence the N-contents of the nitrogen fixing type can be easily passed on to the other types of the same species. So nitrogen fixation contributes to the total available amount of the nitrogen for this species. Upon its death its complete nitrogen contents, regardless of its origin, becomes detritus or is released directly by autolysis. Hence the nitrogen fixed by the N-fixing species also becomes available to other species.

Mixotrophic growth, the direct uptake of organic material by phytoplankton, is included in a similar way, but now two nutrient constraints are added: one for phosphorus and one for nitrogen. In this case the available amounts are not infinite, but taken directly from the detritus pool of the corresponding nutrient. As for nitrogen fixation, it is assumed that the net growth rate of mixotrophic types is relatively low. Unlike nitrogen fixation, mixotrophic growth does not increase the total amount of nutrients in the water system, but it increases the amounts available for those species that are capable of this type of nutrient uptake. So technically nitrogen fixation is similar to an enhanced nitrogen loading, mixotrophic growth to an enhanced mineralization of dead organic material.

Macro algae comprise another special kind of primary producers that may optionally be included in the model. Two special adjustments are made for macro algae relative to phytoplankton: (1) they are fixed at a certain location so normally they cannot be transported and (2) the available amount of light is reduced as a function of the depth and the extinction coefficient of the part of the water column above the macro alga. Up till now applications have been made for two different species of macro algae in the model: *Chara* and *Ulva*. The first may grow in fresh water lakes, the second in coastal lagoons, sheltered bays and saline lakes. For *Ulva* a special feature has been added because if the shear stress as a proxy for the wave action, exceeds a certain limit, this species is washed away and temporarily goes into suspension. While in this phase it is subjected to the same processes as phytoplankton, although its characteristics and hence its rate constants are rather different. Subsequently these floating leaves may settle again at another location if the shear stress get sufficiently low and resume being attached macro algae.

Overview

The individual balance equations ('constraints' in LP terminology) for nutrients (1) in combination with (2), for light energy (12), (13) and (14) and for growth and mortality (15) and (16) are all constructed for each time step and location. They constitute all feasible solutions to which the model applies its competition principle to select one of these. Based upon the requirements and net growth rates, BLOOM now selects one or more individual types. Using the net growth of all individual types as part of the selection criterion implies that the solution, which is selected, is the one for which the total net growth of the phytoplankton community is maximal. In other words: in mathematical terms the objective of the LP solution of BLOOM is maximizing the total net growth rate. Hence defining the net growth constant Pn_k ($= Pg_k^{max} E_k - R_k$) (day^{-1}), the objective of the model is to

$$\text{Maximize } \sum_k Pn_k B_k \tag{17}$$

In biological terms this is just the result of the competition principle which is formulated at the level of individual types, not at the community level. As a consequence of using the LP technique there is always a one to one correspondence between the number of types and limitations. Hence if n types are present, exactly n factors will be limiting and each type will be limited by a single factor. Under rather stable conditions such as in shallow lakes, n is usually small (typically between 1 and 3) during most of the season. In contrast n is often considerably larger for applications where gradients are steep such as in the Dutch coastal zone. It is not uncommon for the model to simulate the co-existence of 3 to 8 types at any given location during most of the production season.

The principle of the model was briefly described in Los et al. (1988). An extensive description covering both the equations and underlying ecological assumptions is in Los (1991). A condensed version can be found in Loucks and van Beek (2005) and in Chapter 3 (Los et al., 2007). The meaning of adopting Pn_k as coefficients of the objective function will be discussed at length in section 5 of this chapter. An overview of the model is shown in Figure 1. Applied as a screening tool only phytoplankton, dissolved nutrients and the labile form of dead algae are explicitly taken into account. Applied as part of an integrated modelling system, a number of additional compartments and fluxes are included in the model (Chapters 4-10); these are shown in grey in this figure.

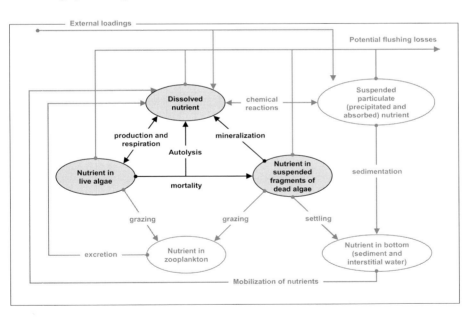

Figure 4 Overview of the BLOOM phytoplankton model. Applied as a screening tool only dissolved nutrients, nutrients in phytoplankton and in suspended detritus are explicitly considered. Compartments and fluxes in grey are additionally modelled only if the model is used as part of an integrated modelling framework.

4. Applications of the model

General validation methodology

The construction of mathematical models involves several steps as indicated in Fig. 5.

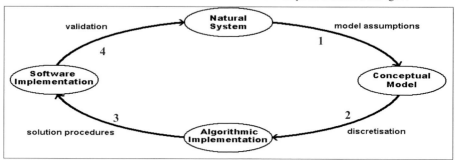

Figure 5 Overview of validation procedure mathematical simulation models.

 When describing a natural system in relationships, by default we will not describe the complete system, but only that part thereof, in which we are interested. We describe this relationship in a model (either statistically or empirical) following a cycle of steps, in which each step contains simplifications and errors due to assumptions and approximations. Types of simplifications and/or errors in the modelling cycle:

1. From the Natural System to the Conceptual Model, *assumptions and approximations* are made.
2. Discretisation of these equations, during which *truncation errors* are made, leads to an Algorithmic Implementation.
3. The software implementation is made with a *certain method and a certain machine precision.*
4. The final step in the modelling cycle is the application to a similar domain and validation of the solution by comparison with *measured data.*

If during the last step we find differences between the simulation result and the data, we have to make a decision how to proceed. Is the interpretation of the data and the model correct? Do we trust the data and are they representative? Is the model concept adequate with respect to the phenomena we want to describe? Were the appropriate equations derived? Were they translated correctly into a computational algorithm? Was this algorithm correctly implemented or do we have to conclude the code is buggy? Is the numerical procedure to solve the equations correct, is it accurate or do instabilities occur? Do we use an appropriate set of model parameters? Do we feed the model with the correct forcings for external drivers?

Of course if we find the results to be acceptable meaning that by some criterion we decide that the similarity between model and measurements is good enough, we still cannot proof that the model is correct. The agreement could be the result of compensating errors at any of the modelling steps shown here.

In this chapter of the thesis the concept of the model is the central issue. Later chapters focus on calibration and validation issues but it should be kept in mind that *all* issues mentioned in this section affect the model's results, performance and judgment. For instance in comparing the results of two models numerical aspects or the physical schemetization might actually prove to be much more import than some of the model's basic assumptions. The last Chapter of this thesis (Los and Blaas, submitted) adresses this issue in great detail.

Calibration procedure BLOOM

The number and the characteristics of the phytoplankton species are inputs to the model. The first fresh water version of BLOOM was applied in 1977 to a drinking water storage reservoir including three large enclosures in the Netherlands called Grote Rug. About ten years later the first model application for the marine environment (North Sea) was developed. With respect to the equations and assumptions, there has always been only one model version i.e. all applications use the same computer code. Applications differ with respect to the choice of coefficients in general and to the characteristics of the phytoplankton species in particular.

The initial data set for the fresh water applications was based on data from laboratory cultures particularly from the microbiological department of the Amsterdam University, where experiments were conducted for six representative species under standardised conditions (Zevenboom and Mur, 1981; Zevenboom et al., 1983; Zevenboom and Mur, 1984; Post et al., 1985). In addition to the information in these papers, many raw data from the experiments were available. Additional lab data on nutrient and growth characteristics were obtained from Rhee (1978), Rhee et al. (1980) and Rhee et al. (1981). Kirk (1975a; 1975b; 1976) provided valuable information on the specific extinction coefficients on the species. Additional information for *Planktotrix* was obtained from Zevenboom et al. (1982) and for *Microcystis* and *Aphanizomenon* from our own analysis of Grote Rug measurements (unpublished results).

Next a semi-automatic calibration procedure was developed to apply the model in a number of well known cases in which the environmental conditions showed a wide range of variation. According to this methodology (1) a range was established for each model coefficient, (2) a series of related coefficients was randomly sampled 100 times, (3) the model was run and its output was compared to the observations. The sampling procedure was not purely random, but took certain knowledge rules into account. For example the maximum growth rate of green algae should be larger than the maximum growth rate of cyanobacteria at a temperature of 20° etc. The 10 best and the 10 worst results were analysed in detail.

A recent update of this type of sensitivity work is reported by Hulsbergen (2007), who has looked particularly at the phytoplankton coefficients of the freshwater screening version of BLOOM. Results in species composition were most sensitive to the growth rate related coefficients. The sensitivity of the simulated chlorophyll-*a* and biomass levels was much smaller.

The resulting coefficient set was further tested by running the model for a number of representative cases, many of which had not been considered previously. As a result in particular the simulated species composition was improved relative to the result of the semi-auto calibration procedure. Notice that although the data sets were usually independent from those used during the calibration procedure, there is often a strong resemblance between many of these lakes. Hence 'good' or 'adequate' model results do not confirm its correctness, but only its suitability to be applied to a larger number of cases. The model validation is described in much more detail in several other chapters of this thesis.

For the marine applications basic data were gathered from Riegman (1996); Riegman et al. (1992; 1996); Riegman (unpublished results); Jahnke (1989) among others. Based on numerous model runs for the North Sea the default parameter set was established. Notice that conditions within this water system vary over a much wider range than those observed in an individual Dutch lake. Hence applying the model to only one marine system proved to be adequate to develop a generic coefficient set, which has later been used for other marine model applications. In Chapter 9: Los et al. (2007; 2008) and in Chapter 8: Blauw et al. (2009) many more detailt are given on the validation of the marine model.

Sensitivity analysis is a fundamental tool in the construction, use and understanding of models. It identifies the most important factors within a model and checks if the model resembles the system under study. Recently the North Sea GEM implementation of the model was analyzed to assess the sensitivity of a subset of the model outputs, to a subset of the input parameters (Salacinskaa et al., Submitted).

This sensitivity analysis considered 71 parameters concerning light and phytoplankton characteristics with respect to two outputs namely the maximal and the annual average chlorophyll-*a* concentration. Additionally, the change in the model response was analyzed at 49 monitoring stations representing the diversity of characteristics of the Dutch part of the North Sea.

The obtained results at different locations correspond well with local water conditions, even with a false assumption of parameter independence, which is commonly made for the sake of sensitivity analysis, and with overestimated parameter ranges. Usually chlorophyll-*a* to carbon ratios were found significant for both outputs. Moreover, in the Dutch coastal zone the model was sensitive to the phosphorus to carbon ratios of flagellates type E and P. maximum growth rates of diatoms, flagellates and *Phaeocystis* type E and of diatoms type P. Furthermore the extinction of inorganic suspended is a major source of uncertainty. However, it can be concluded that the model does not show any odd behaviour in case of variations in the input, in particular it is not highly sensitive to only one or few factors. Changes in the parameters' values do not lead to unacceptable changes in the model prediction. Hence, sensitivity analysis showed no need to revise the model structure.

The default coefficients for the stoichiometry, maximum growth, mortality and respiration rates of the fresh water and marine phytoplankton types are summarized in the next four tables. Notice that all stoichiometric coefficients are expressed here relative to the carbon contents. If the model is applied as a screening tool, coefficients in the input are expressed per unit of dry weight in stead, which is also the unit used in the previous description of the model. The conversion is done by a simple multiplication with the appropriate dry weight to carbon ratio, which is specified for each type in tables 1 and 3.

Table 1: Specific extinction coefficients and stoichiometric ratios of fresh water types defined in BLOOM.

Algal type	Specific Extinction (m²/gC)	N/C [mg/mg]	P/C [mg/mg]	Si/C [mg/mg]	Chla/C [mg/mg]	Dry/C [mg/mg]
Diatoms-E	0.27	0.210	0.0240	0.66	0.04	3.0
Diatoms-P	0.19	0.188	0.0125	0.55	0.025	2.5
Flagelat-E	0.23	0.275	0.0200	0.00	0.029	2.5
Greens-E	0.23	0.275	0.0238	0.00	0.033	2.5
Greens-N	0.19	0.175	0.0150	0.00	0.025	2.5
Greens-P	0.19	0.200	0.0125	0.00	0.025	2.5
Aphanizo-E	0.45	0.220	0.0125	0.00	0.033	2.5
Aphanizo-P	0.40	0.170	0.0088	0.00	0.025	2.5
Microcys-E	0.40	0.225	0.0300	0.00	0.025	2.5
Microcys-N	0.29	0.113	0.0275	0.00	0.017	2.5
Microcys-P	0.29	0.175	0.0225	0.00	0.017	2.5
Oscilat-E	0.40	0.225	0.0188	0.00	0.033	2.5
Oscilat-N	0.29	0.125	0.0138	0.00	0.020	2.5
Oscilat-P	0.29	0.150	0.0100	0.00	0.020	2.5

Table 2: Growth, mortality and respiration coefficients of fresh water types defined in BLOOM.

Algal type	P_1	P_2	Func. relation	M_1	M_2	R_1	R_2
Diatoms-E	0.350	1.060	Exponent	0.035	1.080	0.031	1.096
Diatoms-P	0.300	1.050	Exponent	0.045	1.085	0.031	1.096
Flagelat-E	0.350	1.050	Exponent	0.035	1.080	0.031	1.096
Greens-E	0.068	0.000	Linear	0.035	1.080	0.031	1.096
Greens-N	0.070	5.000	Linear	0.045	1.085	0.031	1.096
Greens-P	0.070	5.000	Linear	0.045	1.085	0.031	1.096
Aphanizo-E	0.190	1.083	Exponent	0.035	1.080	0.012	1.072
Aphanizo-P	0.120	1.095	Exponent	0.045	1.085	0.012	1.072
Microcys-E	0.047	3.000	Linear	0.035	1.080	0.012	1.072
Microcys-N	0.045	5.000	Linear	0.045	1.085	0.012	1.072
Microcys-P	0.045	5.000	Linear	0.045	1.085	0.012	1.072
Oscilat-E	0.045	0.000	Linear	0.035	1.080	0.012	1.072
Oscilat-N	0.034	0.000	Linear	0.045	1.085	0.012	1.072
Oscilat-P	0.034	0.000	Linear	0.045	1.085	0.012	1.072

Table 3:   Specific extinction coefficients and stoichiometric ratios of marine types defined in BLOOM.

Algal type	Specific Extinction (m²/gC)	N/C [mg/mg]	P/C [mg/mg]	Si/C [mg/mg]	Chla/C [mg/mg]	Dry/C [mg/mg]
Diatoms-E	0.24	0.255	0.0315	0.447	0.0533	3.0
Diatoms-N	0.21	0.070	0.0120	0.283	0.0100	3.0
Diatoms-P	0.21	0.105	0.0096	0.152	0.0100	3.0
Flagellate-E	0.25	0.200	0.0200	0.0	0.0228	2.5
Flagellate-N	0.225	0.078	0.0096	0.0	0.0067	2.5
Flagellate-P	0.225	0.113	0.0072	0.0	0.0067	2.5
Dinoflag-E	0.20	0.163	0.0168	0.0	0.0228	2.5
Dinoflag-N	0.175	0.064	0.0112	0.0	0.0067	2.5
Dinoflag-P	0.175	0.071	0.0096	0.0	0.0067	2.5
Phaeocyst-E	0.45	0.188	0.0225	0.0	0.0228	2.5
Phaeocyst-N	0.41	0.075	0.0136	0.0	0.0067	2.5
Phaeocyst-P	0.41	0.104	0.0106	0.0	0.0067	2.5

Table 4: Growth, mortality and respiration coefficients of marine types defined in BLOOM.

Algal type	P_1	P_2	Func. relation	M_1	M_2	R_1	R_2
Diatoms-E	0.083	-1.75	Linear	0.070	1.072	0.06	1.066
Diatoms-N	0.066	-2.0	Linear	0.080	1.085	0.06	1.066
Diatoms-P	0.066	-2.0	Linear	0.080	1.085	0.06	1.066
Flagellate-E	0.090	-1.0	Linear	0.070	1.072	0.06	1.066
Flagellate-N	0.075	-1.0	Linear	0.080	1.085	0.06	1.066
Flagellate-P	0.075	-1.0	Linear	0.080	1.085	0.06	1.066
Dinoflag-E	0.132	5.50	Linear	0.075	1.072	0.06	1.066
Dinoflag-N	0.113	4.75	Linear	0.080	1.085	0.06	1.066
Dinoflag-P	0.112	4.75	Linear	0.080	1.085	0.06	1.066
Phaeocyst-E	0.084	-3.25	Linear	0.070	1.072	0.06	1.066
Phaeocyst-N	0.078	-3.0	Linear	0.080	1.085	0.06	1.066
Phaeocyst-P	0.078	-3.0	Linear	0.080	1.085	0.06	1.066

Some results

During the 30 years of its existence BLOOM has been extensively applied to investigate conditions in aquatic systems and predict what might happen in the future. Among the fresh water systems to which the model was applied, are about 50 Dutch lakes, sometimes for as many as 10 different years. Among these are Lake IJssel, the largest lake in the country, Lake Marken, Lake Veluwe, Lake Wolderwijd, Lake Dronten, Lake Volkerak - Zoom, Lake Westeinder, Lake Braassem, Lake Kaag, Lake Reeuwijk, Lake Langeraar, Lake Loosdrecht,

Lake Breukeleveen, Lake Vinkeveen, Lake Nannewijd, Lake Sloten and Lake Zuidlaren. The model was also applied for the integrated simulation of several complex water systems consisting of channels and lakes, which are typical for the low part of the country. Among these are the so called bosom systems of Rijnland, Friesland, Groningen, Drente and the Schermer. Most of these waters are relatively shallow (1 to 5 m). The model was applied in 3D mode to the more than 20m deep Zegerplas, Nieuwe Meer and Oudekerkerplas in the Western part of the country. In the application to the Botshol wetland area *Chara* was included as an additional species (Chapter 6: Rip et al., 2007). Applications abroad include Lake Victoria (Africa), Laguna de Bay (Philippines), Lake Pyhäjärvi (Finland), upper and lower Peirce reservoir (Singapore) and several lakes and reservoirs in China and Taiwan. Because the model is so widely applied for consultancy purposes, many results are only available in the form of reports. Among the applications reported in the literature are those described by Los (1980; Los (1982a); Los et al. (1984); Chapter 2: Los et al. (1988); Brinkman et al. (1988); Los et al. (1991); Chapter 4: Van der Molen et al. (1994); Van Duin et al. (2001); Ibelings et al. (2003).

The list of marine applications is also long, particularly for the North Sea (an overview mentioning more than 30 studies is presented by Los et al., 2008). Important studies include those in support of Ospar (Ospar et al, 1996; Blaas et al., 2007), assessment of nutrient reduction programs (Chapter 7; Los et al., 1997), construction of an airport off the coast of Holland (Mare, 2001; Los et al. (2004), and extension of the Rotterdam Harbour (Nolte et al., 2005). Other Dutch applications deal with the Western Scheldt, the Eastern Scheldt, several marine lakes i.e. Lake Veere, Lake Volkerak (prediction of future marine conditions), Lake Grevelingen and the Wadden Sea. Other papers on applications of the model to the North sea include De Vries et al., (1998), Peeters et al. (1995), De Groodt et al. (1992). Applications abroad include Venice Lagoon (Los 1999; Boon et al., 2006), the coastal waters of Hong Kong, the Sea of Marmara including part of the Black Sea, a number of coastal stations in Italy, the Tagus estuary in Portugal, and the coastal waters at Merenkurkku in Finland. The physical schematizations of these applications vary from relatively simple 0D (screening mode) to complex 3D models with actual forcing of climatology. Some of these international applications are described in Chapter 3: Los et al. (2007) and in Chapter 8: Blauw et al. (2009).

Within the scope of this chapter it is impossible to show all but a fraction of its results. Typical simulated and observed levels of chlorophyll-*a* for two marine stations are shown in Fig. 6. Also included are the limiting factors. In general the model is capable of reproducing observed chlorophyll-*a* levels well, in spite of a wide variation in conditions between the systems being modelled (Chapter 3: Los et al., 2007).

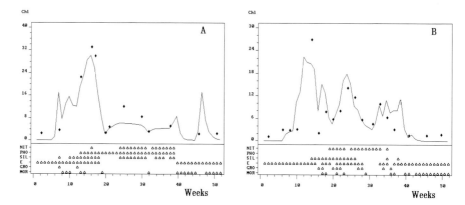

Figure 6   Observed and simulated level of chlorophyll-*a* (μg l^{-1}) at station Terschelling 4km (1998) (A) and at station Dreischor in the marine Lake Grevelingen (1998) (B). A plotted symbol in the lower area indicates a limitation by: NIT: nitrogen, PHO: phosphorus, SIL: silicate, E: light energy, GRO: growth and MOR: mortality.

A typical example of a fresh water application for Lake Veluwe is shown in Fig. 7. This case is particularly interesting since it not only illustrates BLOOM's ability to reproduce the observed downward trend in chlorophyll-*a*, but also its ability to simulate the accompanying change in phytoplankton dominance. While *Planktotrix* dominated in this lake during most of the 1970s and beginning of the 1980s, green algae and diatoms became dominant starting in 1985. Schreurs (1992) summarizes empirical results for hundreds of lakes and shows that *Planktotrix* dominates in 60 percent of the lakes in which the ratio between the euphotic depth Zeu and the mixing depth Zmix is between 0.6 and 0.8. Scheffer (1997) gives detailed information on some specific cases in which the species dominance has changed. According to his analyses a rather abrupt shift occurs when the ratio Zeu/Zmix is approximately 0.7. Below this value *Planktotrix* dominates, above it green algae and diatoms dominate.

The yearly averaged Zeu/Zmax ratio has been computed by BLOOM for 1975 through 1985. It varied from 0.20 in 1975 to 0.53 in 1983. During this entire period *Planktotrix* is completely dominant, except for a short period of diatom dominance during the spring. At a yearly averaged Zeu/Zmix ratio of 0.77 (1982) *Planktotrix* dominates 65 percent of the time, while diatoms and green algae dominate during the remainder of the simulation. So at this ratio the model is in a transitional state. Finally at a Zeu/Zmix ratio of 1.00 (1985) *Planktotrix* has completely disappeared being replaced by *Scenedesmus* and diatoms. So it may be concluded that the light regime at which the species shift occurs in BLOOM agrees very well with the general empirical data.

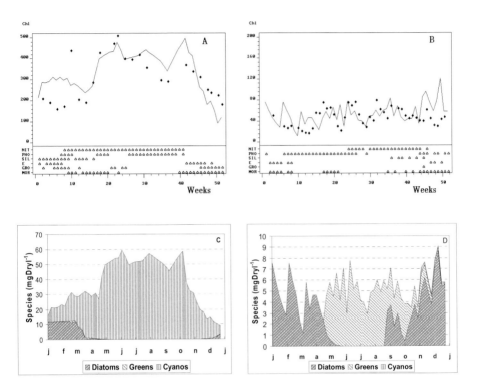

Figure 7 Observed and simulated level of chlorophyll-*a* (µg l⁻¹) in Lake Veluwe (1975 (A) and 1985 (B). A plotted symbol in the lower area indicates a limitation by: NIT: nitrogen, PHO: phosphorus, SIL: silicate, E: light energy, GRO: growth and MOR: mortality. Simulated species composition in (mgDryWeight l⁻¹) in 1975 (C) and in 1985 (D).

Many other results for individual lakes have been reported (i.e. Los, 1991). To give an impression of the overall performance of the model, Fig. 8 shows computed verses observed yearly average chlorophyll-*a* levels in more than 30 Dutch cases. The scatter is small, but there is some bias in the model results as on average the simulations are about 7 percent higher than the observations. This is mainly because during the calibration procedure for individual lakes, occasional under predictions were considered worse than over predictions.

The ability of the model to simulate annual trends is illustrated by Fig. 9, showing the observed and computed yearly average concentration of chlorophyll-*a* in two lakes. Both the observations and the model show a clear downward trend throughout the simulation period in Lake Veluwe. Annual variations appear to be random in Lake Wolderwijd during the first ten years. A downward trend is, however, also observed and simulated in this lake by the end of the 1980s.

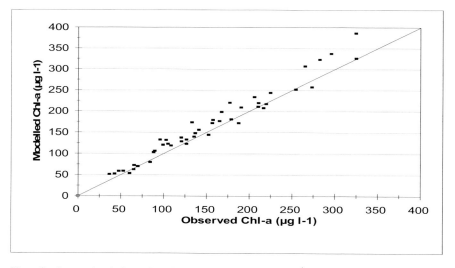

Figure 8   Computed and observed yearly averaged chlorophyll-*a* (μg l⁻¹) in more than 30 Dutch lakes. The drawn line indicates the 1:1 agreement between model and observations.

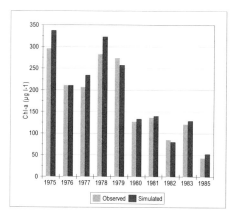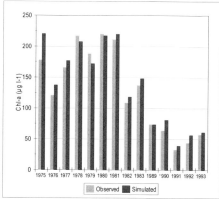

Figure 9   Computed and observed yearly average chlorophyll-*a* (μg l⁻¹) in Lakes Veluwe (left panel) and Wolderwijd (right panel) for a period of 10 to 20 years. For Lake Wolderwijd no model simulations available for years 1984 – 1988.

Fig. 10 gives another general illustration of the model behaviour and its performance. This figure shows average annual chlorophyll-*a* as a function of the concentration of one nutrient (either P or N) and the depth. The other nutrient was put at a level, where it would never be limiting. For all other forcings (silicate; meteorological conditions; back ground turbidity) representative values were adopted for a typical Dutch lake.

When nutrients are very low, they control the annual chlorophyll-*a* regardless of the depth. With increasing nutrient concentrations, light gets limiting in stead of the nutrient and simulated chlorophyll-*a* levels do not increase any further. The nutrient level at which this transition occurs, depends on the depth. In the most shallow case considered here (1m depth) the limiting nutrient remains controlling for the entire nutrient range considered here.

To validate this simulation result, the so called CUWVO lines are also included (CUWVO, 1987). These have been determined empirically by drawing an upper bound rather than a regression line through the observations collected from a large number of waters. Considering the way these lines were determined, depth should not be controlling so the lines are best compared to the simulation result for the shallow 1m deep lake. The correspondence between the lines constructed with BLOOM and the CUWVO lines is very good, indicating that BLOOM is valid over a wide range of observed nutrient conditions. Notice that both in the CUWVO relationships and in the results of BLOOM chlorophyll already becomes 0 when the amount of nitrogen is still positive. This is because total nitrogen includes some refractory organic components which are not available for uptake by phytoplankton. In addition to the information from the CUWVO lines, those derived from BLOOM also indicate at what level nutrients start getting limiting in light limited lakes.

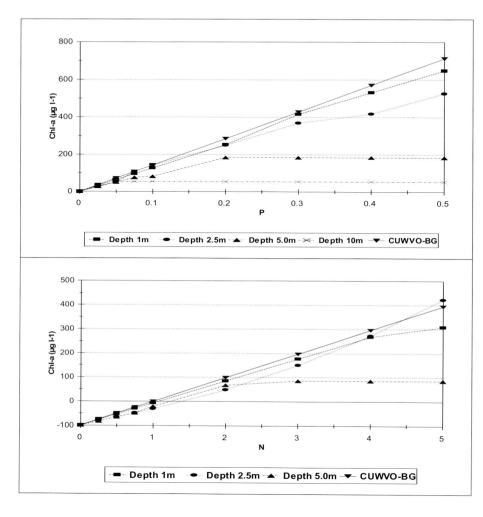

Figure 10 Response curves of simulated annual average chlorophyll-*a* (µg l⁻¹) as function of total P (upper panel) and total N (lower panel) and the water depth. CUWVO-BG are the CUWVO (1987) relationships

5. Discussion: Contrasting BLOOM to other model approaches

Nutrient depletion

It is commonly observed both under lab conditions as well as in the field that the net intrinsic rate of increase of phytoplankton species approach zero if some nutrient is depleted. Obviously, dissolved nutrients should not become negative. Most phytoplankton models use the equations developed by Monod (1942) or Droop (1973) to modify growth rates in the vicinity of nutrient limitation.

According to the Monod equation, the nutrient limited growth rate Pn equals:

$$Pn = Pn^{max} [C_i / (Ks_i + C_i)] \qquad (18)$$

where Pn^{max} is the maximum net growth rate constant, C_i is the available amount of nutrient i and Ks_i is nutrient concentration at which the net growth rate equals 50 percent of its maximum (the half saturation constant for nutrient i).

According to the Droop model, the nutrient limited growth rate constant Pn equals:

$$Pn = Pn^{max*} (1-Q_0/Q) \qquad (19)$$

in which Q is the cell quotum (= stoichiometry) of a nutrient and Q_0 is the minimum value of the stoichiometry at which the growth rate is still positive. Pn^{max*} is the maximum growth rate when Q is infinite. Notice that this Pn^{max*} is not the same as Pn^{max} as defined in the Monod model.

In contrast to the Monod equation, which only considers external nutrient levels, according to the Droop equation the growth rate depends on the internal nutrient levels. Although both equations appear to be different, they actually have similar properties. Droop (1983) already demonstrated that by defining a new variable Q' as:

$$Q' = Q - Q_0,$$

equation (19) can be rewritten as

$$Pn = Pn^{max*} [Q' / (Q' + Q_0)] \qquad (20)$$

which mathematically is similar to a Monod-type equation. The minimum stoichiometry Q_0 acts as half saturation constant and Q', the difference between the actual internal and the minimum internal concentration, is used in stead of the external nutrient concentration.

Both equations have been developed for conditions of a single species under limitation by a single nutrient. They operate on the growth rate of the species, forcing it to approach zero when the available amount of the nutrient approaches zero. Although it was not designed for conditions of light limitation, the Monod equation is often used to describe the growth rate as a function of the light intensity as well because it is a convenient equation to describe saturation processes.

In contrast to models in which differential equations are numerically integrated, BLOOM does not require an explicit equation to keep resources from getting negative. The dissolved amounts of nutrients and the maximum permissible value of the extinction which is computed from the available amount of light energy, are defined as variables in BLOOM and Danzig (1963) has already demonstrated that all variables of a Linear Programming problem are always greater or equal to zero. So in case of a resource limitation, the biomasses of the phytoplankton types are automatically set at a level at which the resource is completely exhausted. Thus in mathematical terms all three methods effectively keep resources at a non-negative level. There are differences with respect to the approach of equilibrium and the concept of limiting factors in general, which will be shortly discussed later in this section in 'The concept of limiting factors'.

Solving for equilibriums versus linear programming

In (aquatic) ecosystems usually many species compete for limited amounts of resources. Often one of the macro nutrients gets limiting and if there is an ample supply, ultimately the availability of light becomes limiting. So an important question is how the competition by different species of phytoplankton is controlled? And related: How do existing models deal with competition between multiple species for several resources?

Characteristics of individual phytoplankton species have been extensively studied in laboratories. Interpretation of the results is, however, difficult due to the differences in set-up of the experiments and because the characteristics of species vary between experiments. For instance the growth rate of a species grown under P limitation is usually not the same as under light or another limitation. In general it may be concluded that there is no simple response of phytoplankton species when offered a broad spectrum of conditions. Due to the complexity in response of individual species, the outcome of competition experiments can also be different. Still these experiments provide a lot of information on the characteristics of individual species and have as indicated in paragraph 4 of this chapter, played a major role in choosing the model parameters of BLOOM.

As might have been expected given the results of the lab studies, empirical results on field data show even more variations. Schreurs (1992) has compiled many results from field studies mostly on Dutch lakes. Different species sometimes dominate in spite of the same factor being limited at least on a seasonal basis. In contrast for instance green algae frequently dominate both under low as well as under extremely high P concentrations.

Apart from the intrinsic characteristics of the individual species, also the external forcings probably play a major role. Diurnal variations in light or grazing pressure, changes in hydrodynamics (vertical mixing; horizontal transport), in loadings, in sediment release of nutrients are all common phenomena to which real life phytoplankters are exposed, which cannot be duplicated in the lab or easily included in annually or seasonally averaged statistics. To explain the differences between various types of models, the mathematical principles will be discussed briefly.

To include the potential light limitation, usually modellers add some term for the growth rate as a function of the light intensity into equation (18):

$$Pn_k = [Pg_k{}^{max} * E_k * C_i / (Ks_{ik} + C_i)] - R_k - M_k \qquad (21)$$

where $Pg_k{}^{max}$ is the maximum gross growth rate of species k, E_k is the depth and time averaged growth efficiency factor, R_k and M_k are the respiration respectively mortality rate constant of species k. Because of the resemblance between the Monod and Droop equation, a mathematically similar result is obtained for the latter, although the parameters have a different biological interpretation.

If several species compete for a limiting resource according to equation (21) under a constant supply of resources, initially the species with the highest maximum net growth rate will dominate. Due to the uptake by phytoplankton, the available quantities of the resources are quickly depleted, which means that the Monod or Droop terms and hence the specific rate of increase of each species rapidly approach zero. It can be demonstrated that under these conditions the species with the lowest value of Ks_{ik} or Q_0 out competes all other species. So when a single nutrient is the only limitation, the result of the Monod based model depends on a single parameter; the intrinsic light dependent growth rates and the nutrient stoichiometry are unimportant. Because Monod coefficients by definition have to be estimated experimentally under conditions of low resource levels, the variation in reported values for individual species is large (see for example Van Liere et al. (1980); Schreurs (1992) and many others). So competition in this type of model is basically controlled by a single parameter with a relatively large uncertainty.

An additional complication is how to deal with several resources at the same time. One could add more Monod terms to equation (21) but since these factors are always less than 1.0, this means the actual growth rate becomes smaller even if there is a large amount of the other resources. Therefore most models adopt a minimum rule in which only the smallest Monod term is considered thereby implicitly assuming that all other terms do not affect the growth at all. In other words: it does not matter if there in an ample supply of other resources or if some of them are nearly limiting as well. Whatever assumption is adopted: it does affect the results of the model.

Depending on the choice of parameter values, interaction between resources is possible though. If for instance the species with the lowest Monod value for P requires a very high amount of N, N might become limiting too in which case there could be co-existence between two species or a shift in limitation and an associated shift in species composition.

In comparison to the Monod model, the interpretation of the parameters of the Droop model is more straightforward. Also it is easier to obtain an estimate of these coefficients from field data. The Droop model is therefore often preferred in models with several functional groups such as in PCLAKE (Janse, 2005).

In BLOOM a different approach is adopted. It is based on similar balance equations for the resources, but treats them differently and it uses a different solver (Linear Programming) and a different species selection criterion. As explained in this chapter, balance equations are constructed for all nutrients and for the permissible level of the extinction coefficient of each species to deal with light limitation. Added to these are the constraints for maximum growth and maximum decline. To visualise how BLOOM selects its types, a highly simplified example

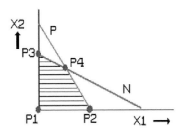

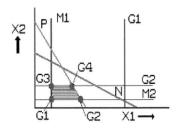

Figure 11 Illustration of Linear Programming solution of BLOOM for a simplified system with 2 species and 2 resources without (upper panel) and with (lower panel) additional growth and mortality constraints.

will be explained in detail (Fig. 11). In this example only two types are considered with biomass X1 and X2. These two types compete for two potentially limiting resources: N and P. The mass balance equations, at which each of the nutrients is completely exhausted can be depicted as straight lines. First for simplicity consider the case without growth and mortality constraints (upper panel). Since resources (in this example N and P) and biomasses should never become negative, the only range in which these conditions are met is the hatched area (P1, P2, P4, P3). This area is called the feasibility area because any point within it reflects a combination of species 1 and 2 and of the resources N and P which theoretically is possible. Fortunately, Danzig (1963) demonstrated very elegantly that the optimum solution of a Linear Program is **always** at one of the corners of the feasibility space. So in this particular example, only four points have to be considered, which makes it much easier to find a solution. P1 is a special case where biomasses of both types are 0 and nothing is limiting. Intuitively it is clear that this point does not represent the optimum condition of the plankton community. At point P2, P is limiting and only type 1 is present. At point P3 N is limiting and only type 2 is present. At point P4 N and P are both limiting and type 1 and 2 co-exit. Using simple geometric relations it is obvious that point P4 has the largest distance to the origin and therefore if the total biomass would be maximized as was the case in the early applications of BLOOM, this point represents the optimal solution selected by the model. If the potential growth rate, which was previously defined as Pn_k ($= Pg_k^{max} E_k - R_k$) is also taken into account, which is the default of the present BLOOM model version, the ratio between the distance to the origin and Pn_k is adopted as selection criterion. This means that depending on the difference in growth rate of both types, P2 or P3 might in stead of P4 be selected. If $Pn_1 \gg Pn_2$, P2 will be chosen and of course in analogy if $Pn_2 \gg Pn_1$, P3 will be selected. Notice that if the difference between Pn_1 and Pn_2 is sufficiently small, P4 will still be selected even though the biomass is not maximized. In this case growth and biomass maximization give exactly the same answer. This is because the optimum solution must be in one of the corners, but the same corner will be selected as being the optimum for a certain range of values of Pn_1 and Pn_2. Mathematically biomass maximization is a special case of growth maximization in which all Pn_x coefficients are taken as 1.0. Also notice that the equation for Pn_k does not include a nutrient dependent term so it might still be positive in spite of a nutrient limitation. This approach is a special case of the one by Tilman (1984) for finding the equilibrium of two species competing for two nutrients given the critical concentrations of the nutrients needed for positive growth. In the case of BLOOM it is assumed that the critical concentrations are zero. Finally notice that although the type (or species) composition of these possible solutions are quite distinct, the total biomass is not, which is typical for practical applications of the model.

So: there is a finite number of possible solution points to consider and growth and biomass maximization consider the same set of possible solutions, but adopt a different selection criterion for choosing one of them and therefore maight make a different choice.

As an additional complication, which can still be easily visualised, in the lower panel of Fig. 11 the growth and mortality constraints of the two types are also represented. Since these constraints operate separately for each type (more precisely for each species) in the model, they can be represented by lines perpendicular to the biomass axes. In this particular example we have assumed that both types were present during the previous time step and that the mortality and growth constraints intersect with the nutrient constraints. Taking these additional conditions into account, the feasibility space of course becomes smaller. Notice for instance that the origin is now no longer feasible. Also notice that the N mass balance line is outside of the feasible area and so P4, which was a feasible solution point in the previous example, is now no longer feasible. Because the mortality constraint of both types is positive, both of them will be present in the solution of the model. In stead of P1-P4 now G1-G4 are the possible points to consider by the selection principle. The maximum biomass is present in point G4; both types are present, type 1 is P limited and type 2 is growth limited. Using the default competition principle of the model, the model will make a choice between G2 and G4. At G2 type 2 is limited by its mortality constraint indicating that dynamically speaking the model attempts to remove it from the selected composition and type 1 is P limited. Because the growth and mortality constraints are perpendicular to the axes, point G3, although feasible, could never be optimal since in comparison to G4, type 2 has the same growth limited biomass and type 1 is now mortality rather than P limited and therefore has a lower biomass than in G4; G2 could never be selected in favour of G4. So the addition of growth and mortality constraints may affect the feasibility area, the optimum solution and the number of potentially selectable equilibrium points.

In most practical applications of the model, many more variables (phytoplankton types) and several additional resources (Si; light) are considered simultaneously. Consequently it is no longer possible to make a 2D representation of the solution space as in Fig. 11, but the same principles hold in this much more complicated case. One could think of this as an n-dimensional diamond with a lot of corners; the type selection criterion chooses between 0 and n of them, where n is the number of potentially limiting factors considered by the model.

So even though the various models obey the same mass balance equations, the selection procedure for finding a particular solution is different. This raises the fundamental questions: what is best and why?

As already stated in the 'General Introduction' of this thesis, unlike in physics where researchers often agree about principles and the equations and debate about their implementation, it is not straightforward how to define fundamental, universally applicable ecological principles. There is a vast amount of literature on these issues and a lot of debate, which is stimulating and confusing at the same time. If we limit the scope to the world of phytoplankton, we might argue that no principle applied in any of the approaches described here will probably represent the real ecosystem behaviour, neither are the equations nor the parameter values exactly correct. In the first version of BLOOM the total biomass was maximized. In traditional models the half saturation constants are decisive with respect to the selection of the dominant species. If the species with the lowest half saturation constant also has the lowest requirement, these models end up with the same solution as these early BLOOM versions. Using the potential growth rate as an alternative selection criterion means that the total biomass will be less (or equal) to the maximum as was pointed out in the discussion of Fig. 11. This is true in general; for instance for the famous logistic growth equation, the growth rate is maximal at half the equilibrium. So the question is: when will communities be close to their maximum equilibrium and when will this not be the case? Two sets of conditions may prevent communities from achieving a maximum equilibrium: (1) loss processes such as grazing, sinking etc. and (2) variable conditions. Both classes of processes lead to a reduction in biomass and hence draw the community closer to a condition at which growth rates are decisive. In fact this is but the main idea behind the general theory on r- and K-strategies: when there is room for growth, the r-selected species dominate, but if conditions remain more or less stable for a long time, K-selected species usually take over. Natural plankton communities are often exposed to (highly) variable conditions, not in the least due to the physical environment they inhabit. Eutrophic waters form a well known and widely studied exception as in these waters, often conditions are relatively stable. It is probably not a coincidence that the original competition principle of BLOOM was successful when it was mainly applied to eutrophic lakes, but proved to be less realistic compared to growth maximization when the model had to deal with the North Sea or the sanitated, much less eutrophic Lake Veluwe of the mid 1980s (Fig. 9).

In conclusion: loss processes not taken account by models and variations in environmental condition will push the plankton community away from the biomass maximum towards a condition closer to the point where the production is maximal. Therefore there is no simple answer to the question whether the growth maximum or the biomass maximum will be closer to the state approached by a system in reality; it will depend strongly on the importance of non-modelled loss processes and environmental fluctuations.

Finally the reader should keep in mind that the objective function determines the selection, not the actual constraints of the model. If the values of the objective function would be adopted from say a fuzzy logic algorithm, the model would still consider the same possible solutions and chose one of them; only its choice might be different.

Objective function and resource utilization

According to the general LP methodology, a linear function (the objective function) is either maximized or minimized. In the case of BLOOM the potential net growth rates of the phytoplankton types are maximized (equation 17). However, it has also been demonstrated (Danzig, 1963) that every LP problem, referred to as a primal problem, can be converted into a dual problem, in which the utilization of resources is minimized in stead. It has been demonstrated that mathematically both problems are in fact the same. In the case of BLOOM the dual problem is minimization of the usage of resources (nutrients; light). Because of the relationship between the primal and dual problem, as an alternative to equation 17 it is also true that in BLOOM utlization of the environmental resources is minimized. It is because of the formal relationship between the primal and dual problem that we arrive at BLOOM's competition principle being interpreted as the ratio between the net growth rates (primal problem) and the resource requirements (dual problem) of the functional types.

The concept of limiting factors

As a consequence of adopting the Linear Programming technique, limiting factors have a simple interpretation in BLOOM. The model seeks combinations of variables (types) and their associated limiting factor, which therefore have a binary (yes/no) interpretation. Biologically one could say that BLOOM strictly adopts Liebig's law. This has a large advantage if it comes to transparency of the approach. In Monod type models additional assumptions have to be made if one wants to apply Liebig's law, for instance by defining the minimum value of all Monod terms as being the limiting factor (see equation 21). But in one case it could have a value of say 0.6, in another of say 0.15. How should this difference be interpreted? In another case the lowest Monod value of say P could be 0.3 the second smallest of say N being 0.4; if the minimum rule is adopted N is not considered limiting in this example in spite of its relatively small Monod term which is even smaller than the value of 0.6 from the previous example.

Notice that for mathematical reasons there is a difference in the remaining concentration of limiting nutrients between BLOOM and a Monod based model. If a BLOOM species approaches a nutrient limitation, it will change from an energy type to the corresponding nutrient limited type, which means its growth rate is reduced in a discrete (step-wise) way. The growth rate is not reduced any further and after a while the nutrient becomes limiting as it is entirely depleted. Because of this procedure in BLOOM its concentration will be 0.0. In a Monod type model the growth rate is reduced gradually and the net growth will become 0 before the nutrient is completely exhausted so a small positive amount remains. Because the half saturation constants are usually small relative to the available amounts of nutrients, the corresponding difference in phytoplankton biomass is usually small, BLOOM's result being only a fraction higher given everything else is the same in both models.

BLOOM's very straightforward principle of limiting factors might lead to the wrong impression, that if several factors are limiting simultaneously, they are equally important. This is not the case as may be clarified by the following example. Suppose in a particular location one type is present and it is limited by P. Due to flushing a small amount say 5 percent of the total biomass of another type is imported. Further suppose that this type is not adapted to the local conditions so BLOOM will try to remove it, but due to its mortality limitation, a small amount has to be conserved and is mortality limited. So in this case both P and mortality are limiting, but the mortality limitation affects only a very small amount of the total biomass.

Potential net growth rate in objective function

On several occasions it has been pointed out in this chapter that the 'potential net growth rate' Pn_k is adopted as part of the selection criterion of BLOOM (together with the requirement). So why this choice and not for example the maximum growth rate or the net rate of increase (so including the mortality) and how does this work if a nutrient is depleted and hence the growth rates approach zero?

The basic assumption is that the model should use a value which reflects the ability of the species (more precisely the types) to take up potentially limiting resources. The maximum growth rate therefore does not seem to be a good candidate as it does not take interspecific differences in the light dependence into account, which according to Fig. 1 are important. Mortality in contrast can be regarded as something happening to the phytoplankton cell irrespective of its ability to compete for the uptake of limiting resources, for instance because it is eaten or gets attacked by a virus. In contrast respiration is more directly linked to uptake, which is why it is included in the selection criterion.

Hence in the model the following procedure is adopted. At the beginning of a time step all forcings are known. It uses the depth, total extinction, day length and irradiance to compute the depth and time averaged value of the growth versus light function (Fig. 2; see also the Appendix). This average efficiency is multiplied with the maximum growth rate constant, which is a function of the temperature, and the respiration value, which is also a temperature function, is subtracted from it. If the result is a positive number, it is put into the objective function. All algal types with a positive value directly compete with each other. If Pn_k is negative because the actual light climate is unfavourable to a particular type, a dummy value of 0.01 is put into the objective function. Types for which this is the case will either be mortality limited (so the model is trying to remove them completely, but does not yet succeed) or they will not be present at all in the optimal solution. The computed biomass of these types will never be larger than during the previous time step since this dummy value is small enough to prevent these types from winning the competition.

What happens if a nutrient is selected as one of the limiting factors? Within BLOOM this means its concentration will be 0.0 and hence according to the physiological knowledge the net growth rate of all phytoplankton types should in the limit be 0.0 as well. So is the selection criterion still valid? Two explanations could be given to say it is. First, measured dissolved nutrient concentrations may appear to be 0.0, but due to dynamical processes (transport; mineralization; sediment release; autolysis; excretion from grazers etc.) the actual fluxes remain sufficiently high to maintain a positive uptake of nutrients by the algal types keeping them close to equilibrium. Second we may assume that all types are affected in a similar way by the nutrient depletion and so the outcome of the competition just before the nutrient actually becomes 0.0 decides which type is going to win competition for this resource. In order words: there is no shift in dominance when the last free molecules are taken up.

The second explanation is actually implemented as an option in the model: a threshold can be specified for each nutrient, which is then subtracted from the available amount passed on to the optimization procedure, this might deplete all remaining nutrients completely, but next the threshold value is added to dissolved amount again, which now will never drop below this threshold. Conceptually this procedure is perhaps more elegant, numerically the difference is usually small because adopting threshold values of the same order of magnitude as the half saturation constants means they are usually small relative to the total nutrient amounts.

Species adaptation

In Section 2 of this chapter it was explained how BLOOM can deal with species plasticity in response to fluctuating conditions. Many existing models do not take these (important) processes into account or they are accounted for by adopting site-specific parameter values (local calibration). Using the BLOOM approach this is not necessary as long as the defined parameter space is sufficiently large, which it usually is (see chapters 6 and 7). So this is a strong advantage of BLOOM over many other models. Similar processes can be build into traditional models as well, but the resulting equations are complex, which may be one of the reasons this is not frequently done.

Some practical illustrations

To investigate the practical consequences of the choices made in the set-up of BLOOM, a simple competition experiment was done using BLOOM and a traditional differential equation model using Monod kinetics. Both models were run for a one year period using a constant temperature of 20 degrees centigrade, and a constant day length of 16 hours. A depth of 1 meter was adopted and a constant background extinction was used of $1.25m^{-1}$. Only one nutrient was taken into account (P) and its amount was varied between a moderately low value of 0.05 mg.l^{-1} and a high value of 0.35 mg.l^{-1}. A constant daily irradiance was adopted having either a typical summer value of 300 W.m^{-2} (total irradiance) or a relatively low value of 200 W.m^{-2}. Only two species were included: a green alga and a cyanobacterium. Typical kinetic parameters were adopted in accordance with the values given in Table 1 and 2; the same parameter values were used for both models. Hence the maximum growth rate of the green alga is higher, but its loss rates are also higher and the slope of its growth versus light curve is less steep. It requires more P per unit of biomass but has a lower specific extinction coefficient. In the case of the Monod model, the additional half saturation constant was taken as 0.0025 respectively 0.0035, but which of the two species had the lower and which had the higher value was varied as part of the experiment. Although the similarity in set-up of the models was made as large a possible, some differences remain which are visible in the results but do not affect the main conclusions. A description of the simulations is given in Table 5. The results are shown in Fig. 12; those for BLOOM are shown in the left panel, those for the Monod type model in the right panel.

Table 5:   **Differences between models in competition experiment**

Simulation	Model	P concentration	Irradiance	Species with low Monod value
A	BLOOM	low	summer average	
B	Monod	low	summer average	green
C	BLOOM	low	low	
D	Monod	low	summer average	cyano
E	BLOOM	high	summer average	
F	Monod	high	summer average	green
G	BLOOM	high	low	
H	Monod	high	summer average	cyano

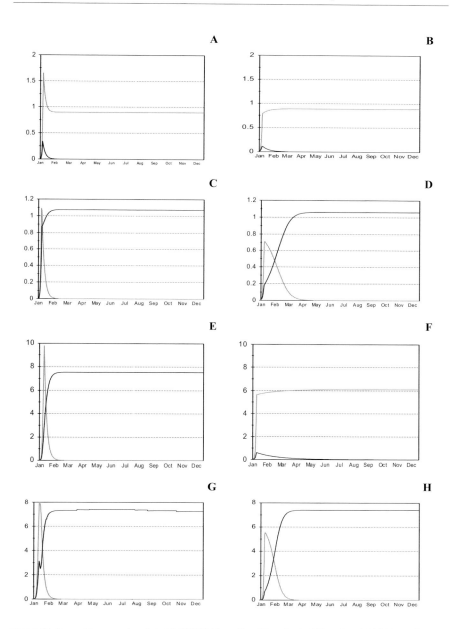

Figure 12 Competition experiments in BLOOM (left panel) and in a classical Monod model (right panel) between a green alga (green line) and a cyanobacterium.

From an analyses of the results, the following may be concluded:

- If the models predict the same species to become dominant, which is the case in seven out of eight cases, the biomass level is the essentially the same (they are not identical due to the remaining differences in set-up).

- The trajectories towards equilibrium are similar but not identical. The differences can be attributed to three factors:

  1. In the BLOOM simulation, detritus is modelled dynamically, in the (simple) Monod model adopted here a steady state with the phytoplankton biomass is assumed. This has two effects. First the light conditions are more favourable in BLOOM during the growth phase because the contribution of detritus to the extinction is smaller. Second an initial overshoot of the equilibrium value typically occurs because in BLOOM the detritus pool lags behind the phytoplankton biomass, so temporarily a larger fraction of the total amount of the nutrient is available for live phytoplankton, which is not the case in this Monod model.

  2. Growth and mortality constraints as included in BLOOM allow for more rapid transitions than the differential equations of the Monod model because they are based on an upper respectively lower bound assumption (see equations 15 and 16 in this chapter)

  3. The light integration procedure of BLOOM is much more advanced and cannot be completely emulated in the Monod model, which adopts a simple analytical procedure.

As expected, the two models differ with respect to their species selection criterion. Results of the Monod model are sensitive to the value of the half saturation constant. In BLOOM results are sensitive to the light regime, which affects the potential growth rates and hence the selection of the dominant species. When the nutrient is low and the irradiance is summer average, (case A) the green alga wins in spite of a higher nutrient requirement because the available nutrient level prohibits a high level of self shading. Hence the high potential growth rate of the green alga more than compensates for its higher P requirement. If in contrast less light is available (case C) the growth rates of both species but particularly of the green alga become lower and BLOOM switches from the green alga to the cyanobacterium. Notice that in the last simulation (case G) the light availability causes BLOOM to switch from a P limitation to a light limitation, but P is almost exhausted which explains the small difference in biomass between E and G. The species selection does not change because the cyanobacterium is more shade adapted so the model's preference for it is actually enforced by this low light level.

This last point is important in relation to the discussion on model objectives. Using the set-up of this experiment, the Monod based model cannot switch its dominant species as long as other factors are non-limiting. Depending on the choice of the half saturation value either of the two species will win over the *entire* range of nutrient levels. In contrast BLOOM adopting the summer average irradiance switches from the more r-selected green alga to the more K-selected cyanobacterium when self shading starts limiting the potential growth rates more and more. The occurrence of such shifts in reality was a major reason for exchanging BLOOM's original biomass maximization principle with the present selection principle because otherwise the species with the lowest requirement for a limiting resource would always win. One practical way to get around this problem in a Monod type model is adopting the same value for the half saturation constant of species competing for the limiting nutrient. This means that according to equation 21, now the light dependence term will decide which species wins the competition. This approach was for instance adopted by Scheffer (1998) in his algal competition experiment under lake Veluwe-like conditions. Qualitatively his model and BLOOM are similar because in both light ultimately determines the outcome of the competition. The result is dominance by cyanobacteria under turbid (= high nutrient) conditions and dominance of the green alga under less turbid (= low nutrient) conditions. If both models would adopt exactly the same parameters for the light dependence, then the point (light regime) at which the switch occurs is not exactly the same, because in BLOOM the nutrient stoichiometry is explicitly taken into account as an additional criterion. So BLOOM considers Pn_k / n_{ik} and the multiplicative Monod model by Scheffer considers Pn_k; obviously adopting the same stoichiometry coefficients n_{ik} in BLOOM would result in the two models switching at the same light climate. Another way to enhance the similarity in model results and promote a switch at the same light climate would be to change some of the coefficients of the light dependent production as part of the calibration procedure of either model.

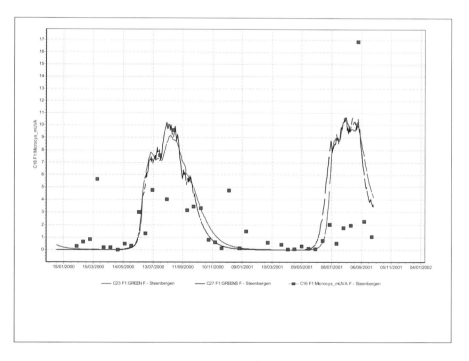

Figure 13 Comparison of total *Microcystis* biomass (mgC.l^{-1}) simulated by DBS (blue line) and the model by Verspagen (2006) (green line) for Lake Volkerak Zoom, location Steenbergen with harmonized parameters. Observations (red dots) are shown for reference purposes only since neither model application reflects its optimal calibration,

During a recent study on the future of Lake Volkerak Zoom in the Netherlands, two models were applied, one by Verspagen (2006) and DBS. Initially results by both models were quite different which raised the question whether this was due to conceptual differences or to the parameterisation. Since the model by Verspagen considers only one species (*Microcystis*), all other species were removed from BLOOM. After the elimination of all differences in parameter values, the results of the two models were very similar (Fig. 13). Notice that as in the previous experiment the way the growth and mortality constraints operate in BLOOM cause it to react more quickly during the phases of growth and decline, but the differences in this example are rather small. So while it was demonstrated in the previous example that results may be similar under nutrient limited conditions at steady state, these results demonstrate that in the case of a light limitation BLOOM and a traditional model give basically the same results during an annual simulation of a real water system if one and the same species is present.

Fig. 14 shows another comparison of a Monod type model and BLOOM. In Chapter 10 of this thesis several North Sea models are compared. Two of them: DYNAMO and GENO-NZB use the same hydrodynamics, loads and meteorological forcing (see Chapter 10 for full details). These two models mainly differ with respect to the phytoplankton kinetics. DYNAMO employs the classic Monod approach; BLOOM adopts its own optimization algorithm with the potential net growth rates as objective function. Of course on a temporal scales the results are not identical, which is the reason why BLOOM has superseded DYNAMO, but on an annually averaged bases the difference between the total phytoplankton biomass (mg C.l^{-1}) simulated by each of the two models is very small.

In conclusion: Both in the simplified competition experiments as well as in the complex 'real life' applications, there are no significant differences in total biomass between the Monod based models and BLOOM. Differences mainly occur with respect to species composition due to the difference in selection criterion. Related to this: the seasonality is likely to be different and often also the simulated chlorophyll-*a* concentration since the switches between types in BLOOM results in a relatively large variations in simulated carbon to biomass ratios in this model.

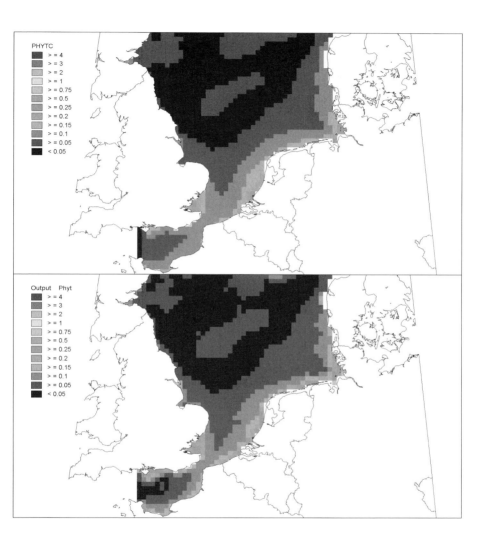

Figure 14 Comparison of annual average total phytoplankton biomass (mgC.l^{-1}) simulated by DYNAMO (Monod model; upper panel)) and BLOOM (lower panel) for the year 2003. For details on model set-up see Chapter 10 of this thesis.

6. Summary and conclusion

BLOOM is a model for predicting the dynamics of phytoplankton communities. Focusing on management applications, early versions were already applied for practical purposes by the end of the 1970s. The model has since then evolved with respect to its fundamental principles as well as the coefficients used. It has been applied to a large number of fresh water and marine systems covering a range of external conditions of more than two orders of magnitude for some factors.

The model aims to capture as much as possible of the functional diversity of phytoplankton communities. However, rather than modeling all known species, BLOOM specifies several 'ecotypes' within each major phytoplankton group to mimic functional diversity resulting in nature from species diversity and from adaptation through physiological flexibility.

BLOOM differs from many existing models in that it uses a rather strict Liebig's Law approach: each ecotype is limited only by the resource with the lowest availability. Also, instead of using a Monod or Droop equation to growth rates as a function of nutrient levels, BLOOM does not let nutrient shortage reduce growth rates, other than setting growth to zero when the concentration of a dissolved nutrient has become zero. With respect to shading effects, all species for which light is below a fixed specific tolerance limit are set to zero. The higher the total biomass permitted by the availability of nutrients, the higher the potential turbidity and hence the smaller the number of species that can still maintain a positive energy balance. Light is limiting to at most one of the species in the model.

Another difference to many other ecological models is the use of linear programming to predict the state of the system at the next time-step. The algorithm first defines the different possible states at which one of the nutrients or light halts growth of one of the ecotypes. Subsequently in accordance to the general LP methodology, from those states, the one is selected at which the the potential growth rate of all ecotypes is maximal and the requirement for the resources is minimal. It can be shown analytically that effectively a high potential growth capacity as well as a low requirements for nutrients and light are equally weighted in determining the algal composition of the predicted steady state of the system.

To prevent unrealistically fast jumps towards such steady state solutions, BLOOM also computes the biomass increment each species can potentially realize in the given time-step from the temperature and light dependent 'potential net growth'. If the result is smaller than the steady state solution, the growth-limited new state is assigned. Similarly, the model imposes a limit on mortality, to prevent unrealistically rapid declines.

All computations are made at each time-step for all phytoplankton types, and the resulting assemblage is typically a mix of various types, some of which are resource limited, and some of which are controlled by growth and mortality rates. Since a trade-off between resource requirement and potential growth rate is assumed, r-selected species are predicted to dominate under dynamic conditions where potential growth rates are high, while K-selected species are predicted to dominate under more stable conditions. Overall, the relative weight given to growth rates produces some bias towards r-select species compared to traditional equilibrium solutions. On the other hand, such equilibrium solutions may be argued to be biased towards

K-select species given that a model will typically not capture all loss processes and environmental fluctuations well. The fact that predictions by BLOOM match observed plankton communities so well for a wide range of systems, suggests that indeed the weighting of growth rates in the algorithm for selecting amongst the set of potential solutions is useful for capturing some aspects of complex reality that tend to be missed in other ways.

Although the approach in BLOOM may seem radically different from that taken in most other competition models, comparisons between BLOOM predictions and predictions from traditionally solved sets of equations in simple as well as complex systems illustrate that the particular approach taken in BLOOM involving the linear programming procedure and the strict Liebig's approach do not result in markedly different outcomes with respect to the total biomass. The main merits of the approach are its computational speed and stability and its straightforward way in dealing with limiting factors and variable characteristics of members of the plankton community, shown to produce realistic predictions of total biomass as well as species composition over a wide range of conditions and systems.

References

Atkins, R., Rose, T., Brown, R. S. and Robb, M., 2001. The Microcystis cyanobacterial bloom in the Swan River - Feb. 2000. Water Science and Technology 43: 107-114.

Berger, C., 1984. Consistent blooming of Oscillatoria agardhii gom. in shallow hypertrophic lakes, Verh. Internat. Verein. Limnol., 22: 910-916.

Berger, C. and Bij de Vaate, A.,1983. Limnological studies on the eutrophication of Lake Wolderwijd A shallow hypertrophic Oscillatoria dominated lake in the Netherlands, Schweiz. Z. Hydrol., 45: 459-479.

Bigelow, J.H., J.G. Bolten, and J.C. De Haven, 1977. Protecting an Estuary from Floods--A Policy Analysis of the Oosterschelde: Vol. IV, Assessment of Algae Blooms, A potential Ecological Disturbance, The Rand Corporation, R-2121/4-NETH, April 1977.

Blaas, Meinte, Xavier Desmit and Hans Los, 2007. OSPAR Eutrophication Modelling Workshop 2007. WL | Delft Hydraulics report Z4351, Delft, The Netherlands.

Blauw, Anouk N. , Hans F. J. Los, Marinus Bokhorst, and Paul L. A. Erftemeijer., 2009. GEM: a generic ecological model for estuaries and coastal waters. Hydrobiologia DOI 10.1007/s10750-008-9575-x

Boon, J.G., X. Desmit and M. Blaas, 2006. Ecological model of the Lagoon of Venice. Part II: Part II: Setup, calibration and validation. WL | Delft Hydraulics report Z3733, Delft, The Netherlands.

Brinkman, J.J. , S. Groot, F.J. Los and P. Griffioen. 1988. An integrated water quantity and quality model as a tool for water management. application to the province of Friesland, the Nehterlands. Verh. Internat. Verein. Limnol. 23: 1488-1494.

Broekhuizen, N., Heath, M.R., Hay, S.J., Gurney, W.S.C., 1995. Modelling the Dynamic of the North Sea's Mesozooplankton, in: Netherlands Journal of Sea Research 33 (3/4): 381-406.

Chorus, I., Falconer, I. R., Salas, H. J. and Bartram, J. 2000. Health risks caused by freshwater cyanobacteria in recreational waters. Journal of toxicology and Environmental Health, Part B, 3: 323-347.

CUWVO, 1987. Vergelijkend onderzoek naar de eutrofiering in nederlandse meren en plassen. Resultaten van de derde eutrofleringsenquete. Coordinatiecommissie Uitvoerinig Wet Verontreiniging Oppervlaktewateren, Werkgroep VI (in Dutch).

Danzig, G.B., 1963. Linear programming and extensions. Princeton University Press, Princeton N.J.

De Groodt, E.G. , Los, F.J., Nauta, T.A., Markus A.A. and Vries,I. de , 1992. Modelling cause-effect relationships in eutrophication of marine systems. In Colombo et al. (eds), Olsen & Olsen.

De Vries, I., Duin, R.N.M., Peeters, J.C.H., Los, F.J., Bokhorst, M. and Laane. R.W.P.M., 1998. 'Patterns and trends in nutrients and phytoplankton in Dutch coastal waters: comparison of time-series analysis, ecological model simulation and mesocosm experiments.' ICES Journal of Marine Science, 55: 620-634.

Di Toro, D.M., Fitzpatrick, J.J. and Thomann, R.V., 1971. A Dynamic Model of the Phytoplankton Population in the Sacramento San Joaquim Delta. Adv. Chem. Ser., 106:131-180.

Di Toro, D.M., Thomann, R.V., O'Connor, D.J., and Mancini, J.L., 1977. Estuarine phytoplankton models-verification analysis and preliminary applications. In: Goldberg, E.D. (Ed.), The sea. Marine modeling, New York.

Droop, M.R., 1973. Some thoughts on nutrient limitation in algae. Journal Phycology, 9: 264-272.

Droop, M.R., 1983. 25 years of algal growth kinetics. A personal view. Botanica Marina, Vol. XXVI: 99-112.

Harris, G.P., 1978. Photosynthesis, Productivity and Growth: The Physiological Ecology of Phytoplankton. Arch. Hydrobiol. Ergebn. Limnol., Vol. 10, No. I-IV: 1-171.

Harris, G.P., 1986. Phytoplankton ecology. Chapman and Hall, London.

Hulsbergen, R.P., 2007. Onzekerheidsanalyse algenmodellering WL|Delft Hydraulics, Report Q4282 (in Dutch).

Ibelings, Bas W., Marijke Vonk, Hans F.J. Los and Diederik T. v.d. Molen and Wolf M. Mooij, 2003. Fuzzy modelling of Cyanobacterial waterblooms, validation with NOAA-AVHRR satellite images, Ecological Applications, 13(5): 1456-1472.

Jahnke, J., 1989, The light and temperature dependence of growth rate and elemental composition of *Phaeocystis globosa Scherffel* and *P. Pouchetti (Har.) Lagerh.* in batch cultures. Neth. J. of Sea Research, 23(1): 15-21.

Janse, J.H., 2005. Model studies on the eutrophication of shallow lakes and ditches. PhD-thesis Wageningen University, 376 p.

Kirk, J.T.O., 1975a. A theoretical analysis of the contribution of algal cells to the attenuation of light within natural waters, I. General treatment of suspensions of pigmented cells, New Phytologist, Vol. 75, No. 1: 11-20.

Kirk, J.T.O., 1975b. A theoretical analysis of the contribution of algal cells to the attenuation of light within natural waters, II. Spherical cells, New Phytologist, Vol. 75, No. 1: 21-36.

Kirk, J.T.O., 1976. A theoretical analysis of the contribution of algal cells to the attenuation of light within natural waters, III. Cylindrical and speroidal cells, New Phytologist, Vol. 77, No. 1: 341-358.

Lancelot, C., Billen, G., Sournia, A., Weisse, T., Colijn, F., Veldhuis, M.J.W., Davies, A. and Wassman, P., 1987. Phaeocystis blooms and nutrient enrichment in the continental coastal waters of the North Sea. Ambio, 16:38-47.

Los, F.J., 1980. Application of an Algal Bloom Model (BLOOM II) to combat eutrophication, Hydro-biological Bulletin Vol. 14, 1/2: 116-224.

Los, F.J, 1982a. An Algal Bloom model as a tool to simulate management measures. In: Barica, J. and Mur, L.R. (eds.), 'Hypertrophic Ecosystems', Junk, Dr. W. BV Publishers, The Hague-Boston-London.

Los, F.J., 1982b. Mathematical Simulation of Algae Blooms by the Model BLOOM II, WL | Delft Hydraulics, R1310-7.

Los, F.J., 1992. Procesformuleringen DBS. T542, WL|Delft Hydraulics Delft, The Netherlands (in Dutch).

Los, F.J., 1999. Ecological Model for the Lagoon of Venice. Modelling Results. WL | Delft Hydraulics report, T2162, November 1999.

Los, F.J., Bakema, A.H. and Brinkman, J.J.,1991. An integrated freshwater management system for the Netherlands, Verh. Internat. Verein. Limnol. 24: 2107-2111.

Los F.J. and M. Blaas, (submitted). Complexity, accuracy and practical applicability of different model versions. Journal of Marine Systems, Ammer 2008 special issue.

Los, F.J., and M. Bokhorst, 1997. Trend analysis Dutch coastal zone. In: New Challenges for North Sea Research, Zentrum for Meeres- und Klimaforschung, University of Hamburg: 161-175.

Los, F.J. and J.J. Brinkman., 1988. Phytoplankton modelling by means of optimization: A 10-year experience with BLOOM II, Verh. Internat. Verein. Limnol., 23, pp. 790-795.

Los, Hans and Herman Gerritsen, 1995. Validation of water quality and ecological models. IAHR Specialist Forum on Software Validation, London, September, 1995.

Los, F.J., N.M. de Rooij, J.G.C. Smits, and J.H. Bigelow, 1982. Policy Analysis of Water Management for the Netherlands: VOL. VI, Design of Eutrophication Control Strategies, Rand Corp. & WL | Delft Hydraulics, N-1500/6-NETH.

Los, F.J., J.G.C., Smits and N.M. de Rooij, 1984. Modelling eutrophication in shallow Dutch lakes, Verh. Internat. Verein. Limnol. 22: 917-923.

Los, F. J., S. Tatman & A. W. Minns, 2004, FLYLAND - A Future Airport in the North Sea? An Integrated Modelling Approach for Marine Ecology. 6th International conference on hydroinformatics. World Scientific Publishing Company ISBN 981-238-787-0.

Los, F. J., M. T. Villars, & M. W. M. Van der Tol, 2008. A 3- dimensional primary production model (BLOOM/GEM) and its applications to the (southern) North Sea (coupled physical–chemical–ecological model). Journal of Marine Systems, 74: 259-294

Los F.J. and J.W.M. Wijsman, 2007. Application of a validated primary production model (BLOOM) as a screening tool for marine, coastal and transitional waters. Journal of Marine Systems, 64: 201-215.

Loucks, D.P. & E. Van Beek (eds), 2005. Water Resources Systems Planning and Management - an introduction to methods, models and applications. Chapter 12: 'Water quality modelling and prediction'. Studies and Reports in Hydrology, UNESCO publishing, ISBN 92-3-103998-9.

Monod, J. Recherches sur la Croissance des Cultures Bacteriennes. Herman, Paris.

Nolte, A., P. Boderie, & J. van Beek, 2005. Impacts of Maasvlakte 2 on the Wadden Sea and North Sea coastal zone. Track 1: Detailed modelling research. Part III: Nutrients and Primary Production. WL Delft Hydraulics, report Z3945, November 2005.

OSPAR, Villars, M., de Vries, I., Bokhorst, M., Ferreira, J., Gellers-Barkman, S., Kelly-Gerreyn, B., Lancelot, C., Menesguen, A., Moll, A., Patsch, J., Radach, G., Skogen, M., Soiland, H., Svendsen, E. and Vested, H.J., 1998. Report of the ASMO modelling workshop on eutrophication issues, 5-8 November 1996, The Hague, The Netherlands, OSPAR Commission Report, Netherlands Institute for Coastal and Marine Management, RIKZ, The Hague, The Netherlands.

Peeters, J.C.H., Los, F.J., Jansen, R., Haas, H.A., Peperzak, L. and De Vries, I., 1995. The oxygen dynamics of the Oyster Ground, North Sea. Impact of eutrophication and environmental conditions. Ophelia, 42: 257-288.

Post, A.F., De Wit, R. and Mur, L.R., 1985. Interactions between temperature and light intensity on growth and photosynthesis of the cyanobacterium Oscillatoria agardhii. J. Plankton Res., 7:487-495.

Reynolds, C.S., S.W. Wiseman, B.M. Godfrey & C. Butterwick, 1983. Some effects of artificial mixing on the dynamics of phytoplankton populations in large limnetic enclosures. Journal of Plankton Research 5: 203-234.

Rhee, G-Yull, 1978. Effects of N : P atomic ratios and nitrate limitation on algal growth, cell composition, and nitrate uptake, Limnol. Oceanogr., Vol. 23, No. 1: 10-25.

Rhee, G-Yull and Ivan J. Gotham, 1980. Optimum N:P ratios and coexeistence of planktonic algae. J. Phycol., Vol. 16: 486-489.

Rhee, G-Yull, and Ivan J. Gotham, 1981. The effect of environmental factors on phytoplankton growth: Temperature and the interactions of temperature with nutrient limitations, Limnol. and Oceanogr., Vol. 26, No. 4: 635-648.

Riegman, R., Noordeloos, A.A.M., and Cadee, G., 1992. *Phaeocystis* blooms and eutrophication of the continental coastal zones of the North Sea. Marine Biology, 112: 479-484.

Riegman, R., De Boer, M. and De Senerpont Domis, M., 1996. Growth of harmful marine algae in multispecies cultures. Journal of Plankton Research, 18 (10): 1851-1866.

Riegman, R., 1996. Species Composition of Harmful Algal Blooms in Relation to Macronutrient Dynamics, In: Allan D. Cembella and Gustaaf M. Hallegraeff, Physiological Ecology of Harmful Algal Blooms, Donald M. Anderson, NATO ASI Series, Vol. 41., Springer Verlag, 1996.

Rip, Winnie J., Maarten Ouboter, Hans J. Los, 2007. Impact of climatic fluctuations on Characeae biomass in a shallow, restored lake in The Netherlands, Hydrobiologia, Vol. 584: 415-424.

Rogers, S.I. and Lockwood, S.J., 1990. Observations of coastal fish fauna during a spring bloom of *Phaeocystis pouchetii* in the Eastern Irish Sea. J. Mar. Biol. Assoc. UK., 70:249-253.

Salacinskaa, K., G. El Serafy, H. Los, A. Blauw. Sensitivity analysis of the 2-dimensional application of the Generic Ecological Model (GEM) to the North Sea with respect to algal bloom prediction (Submitted).

Scheffer Marten, Sergio Rinaldi, Alessandra Gragnani, Luuc R. Mur, and Egbert H. van Nes, 1997. On the dominance of filamentous Cyanobacteria in shallow, turbid lakes. Ecology, 78(1): 272-282.

Scheffer, Marten, 1998. Ecology of Shallow Lakes. Chapman & Hall.

Schreurs, H., 1992. Cyanobacterial dominance. Relations to eutrophication and lake morphology. PhD Thesis, University of Amsterdam, The Netherlands.

Shuter, B.J., 1978. Size dependence of phosphorus and nitrogen subsistence quotas in unicellular microorganisms, Limnol. Oceanogr., Vol. 23, No. 6: 1248-1255.

Steele, J.H., 1958. Plant production in the Northern North Sea. Rapp. Cons. Explor. Mer., Vol. 144: 79-84.

Tilman, G. David,1984. Plant dominance along an experimental nutrient gradient, Ecology, Vol. 65, No. 5: 1445-1453.

Van der Molen, D.T., F.J. Los & M. Van der Tol, 1994. Mathematical modelling as a tool for management in eutrophication control of shallow lakes. Hydrobiologia 275/276: 479-492.

Van de Wolfshaar, K.E., 2006. A biomass based size-structured mode for higher trophic levels; a grazer model for Mytilus edulis. WL | Delft Hydraulics report Z3515, Delft, The Netherlands.

Van Duin, Elisabeth H.S., Gerard Blom, F. Johannes Los, Robert Maffione, Richard Zimmerman, Carl F. Cerco, Mark Dorth, and Elly P.H. Best, 2001. Modeling underwater light climate in relation to sedimentation, resuspension, water quality and autotrophic growth, Hydrobiologia, 444: 25-42.

Van Liere, Louis and Luuc Mur, 1980. Occurrence of *Oscillatoria agardhii* and some related species, a survey. In: Barica, J. and Mur, L.R. (eds.), 'Hypertrophic Ecosystems', Junk, Dr. W. BV Publishers, The Hague-Boston-London.

Vollenweider, R., 1975. Input-output models, with special reference to the phosphorus loading concept in limnology. Schweiz. Z. Hydrol. 37: 53-84.

Verspagen, Jolanda, 2006. Bethic-pelagic couplin in the population dynamics of the cyanobacterium *Microcystis*. PhD thesis, Utrecht University, the Netherlands.

Vries, I. de , R.M.N. Duin, J.C.H. Peeters, F.J. Los, M. Bokhorst and R.W.P.M Laane, 1998. Patterns and trends in nutrients and phytoplankton in Dutch coastal waters: comparison of time-series analysis, ecological model simulation, and mesocosm experiments. ICES Journal of Marine Science, Vol. 55: 620-634.

Wijsman, J.W.M., 2004. Een generiek model voor hogere trofie niveau's. WL | Delft Hydraulics report Z3515, Delft, The Netherlands.

WL|Delft Hydraulics/MARE, 2001. Description and model representation of an artificial island and effects on transport and ecology. Delft Hydraulics Report WL2001103 Z3030.10.

WL | Delft Hydraulics, 2002. GEM documentation and user manual. WL | Delft Hydraulics report Z3197, Delft, The Netherlands.

WL | Delft Hydraulics, 2003. Delft3D-WAQ users manual. WL | Delft Hydraulics, Delft, The Netherlands.

Zevenboom, W., Bij De Vaate, A. and Mur, L. R. 1982. Assesment of factors limiting growth rate of *Oscillatoria agarhii* in hypertrophic Lake Wolderwijd, 1978, by use of physiological indicators, Limnol. and Oceanogr., 27: 39-52.

Zevenboom, W. and Mur, L.R., 1981. Amonium-Limited Growth and Uptake by *Oscillatoria agardhii* in chemostat cultures. Arch. Microbiol., 129: 61-66.

Zevenboom, W. and Mur, L.R., 1984. Growth and phtosynthetic response of the cyanobacterium *Microcystis aeruginosa* in relation to photoperiodicity and irradiance. Arch. Microbiol., 139: 232-239.

Zevenboom, W., Post, A.F., Van Hes, U. and Mur, L.R., 1983. A new incubator for measuring photosynthetic activity of phototrophic microorganisms, using the amperometric oxygen method. Limnol. Oceanogr., 28: 787-791.

Appendix: Averaging the production

The light intensity encountered by, and hence the efficiency of a phytoplankton cell is not constant but varies with the water depth, turbidity and time. To account for these variations, the model must compute the average efficiency E_k in a certain period. The method to compute E_k in the present model is similar, but not identical, to the method described in Section 5.4 of the Algae Bloom report (Bigelow et al., 1977).

Assuming that the water is well mixed from the surface to the mixing depth $Zmax$ we can compute the light intensity, $I(z)$, at depth z according to the Lambert-Beer equation:

$$I(z) = I_s \, EXP \, (-Kz) \tag{1}$$

where

- I_s is the light intensity in Joules/m^2/h just below the water surface,
- K is the extinction coefficient per m.

The average efficiency of an algal cell at depth z is $E[I(z)]$.

For the moment leaving out the subscript k for simplicity, to compute the depth averaged efficiency EDEP Bigelow et al. (1977) made the assumption that each cell spends an equal amount of time at every depth from the surface to the mixing depth. We shall arrive at the same set of equations making a more lenient assumption namely that the concentration of phytoplankton cells is the same at every depth. This can be shown to be a sufficient condition.

Assuming (1) a constant surface light intensity, (2) a particular day length (number of hours of day light) DL, and (3) the same light adaptation for all cells, the average efficiency EDEP can be computed as:

$$EDEP = \frac{1}{Zmax} \int_0^{Zmax} E[I_s * EXP(-K.z)] \, dz \tag{2}$$

Introducing a new variable s:

$$s = K.z - \log I_s, \text{ hence: } z = \frac{s + \log I_s}{K}$$

Also: $dz = ds/K$
and: $s = K*Zmax - \log I_s$ $(z=Zmax)$
          $s = -\log I_s$ $(z=0)$

we can transform (2) into:

$$EDEP = \frac{1}{K*Zmax} \int_{-\log I_s}^{K*Zmax-\log I_s} E\,[\,EXP(-s)\,]\,ds \qquad (3)$$

Next define:

$$F(v) = \int_0^v E\,[\,EXP(-s)\,]\,ds \qquad (4)$$

Then obviously:

$$EDEP = \frac{F(K*Zmax - \log I_s) - F(-\log I_s)}{K*Zmax} \qquad (5)$$

This equation only holds for a constant light intensity I_s and for a day length DL. Thus the functions E and F (and the next function G) might have been indexed by DL, but we have not done this for simplicity. We shall later deal with the day length and first consider variations in light intensity with time. Writing $Is(t)$ as a function of time, define:

$$G(v) = \frac{1}{24} \int_0^{24} F[\,v - \log I_s(t)\,]\,dt \qquad (6)$$

Then the time and depth averaged efficiency EAVG is:

$$EAVG = \frac{G(K*Zmax) - G(0)}{K*Zmax} \qquad (7)$$

Eq. (7) only holds for one specific function $I_s(t)$ in other words: only for the intensity pattern of one particular day. So theoretically, we must recalculate Eq. (7) for each day. However, Bigelow et al. (1977) have shown that two important changes of $I_s(t)$ are possible without recalculation of Eq. (7):

1. A change in the number of day light hours, but with the same intensity pattern $I_s(t)$.
2. A change in intensity at each instant by a constant fraction w.

Any combination of the two can also be accomodated. Strictly speaking the average intensity pattern is a function of season, and moreover the actual intensity at any instant depends on stochastic events such as clouding. There is no practical way to deal with all of these factors, however, and there is no reason to expect a major impact on the average efficiency, because phytoplankton cells spend most of the time at light intensities below Iopt, where the response of E(I) to changes in irradiance is fairly linear.

As shown in Fig. 3, different species of phytoplankton react differently to a change in day length. Denote this relation by Pn(Iopt,DL). We shall now make the important assumption that Pn(Iopt,DL) holds at every other light intensity I as well. We can then compute the growth rate constant Pn at any arbitrary day length DL and light intensity I as

$$Pn(I,DL) = \frac{Pn(Iopt,DL)}{Pn(Iopt,DLopt)} Pn(I,DLopt) \tag{8}$$

Notice that by definition the growth rate constant at both optimal light intenstity and day lengths equal the maximum net growth rate: $Pn(Iopt,DLopt) = Pn^{max}(T)$.

Of course the following holds equally well:

$$Pn(I,DL) = \frac{Pn(I,DLopt)}{Pn(Iopt,DLopt)} Pn(Iopt,DL) \tag{9}$$

Let

$$L(DL) = \frac{Pn(Iopt,DL)}{Pn(Iopt,DLopt)} \tag{10}$$

and define:

$$H(v) = \frac{1}{24} \int_0^{24} F[\, v - \log I_s(t)\,]\, dt \tag{11}$$

for a day with length 24 hours. We can now express G(v) as

$$G(v) = \frac{L(DL)}{L(24)}\, H(v) \tag{12}$$

Substituting Eq. (12) into Eq. (7) we find

$$EAVG = \frac{L(DL)}{L(24)}\, \frac{H(K*Zmax) - H(0)}{K*Zmax} \tag{13}$$

In analogy to Bigelow et al. (1977) on page 35 it is obvious that Eq. (13) is replaced by

$$EAVG = \frac{L(DL)}{L(24)}\, \frac{H(K*Zmax - \log w) - H(-\log w)}{K*Zmax} \tag{14}$$

for a day whose intensity pattern is $w*Is(t)$ rather than $I_s(t)$. EAVG is the type specifiic average growth efficiency factor E_k. As may be notes from Eq. (14) it is proportional to the type-specific day length function (Fig. 3) and inversely proportional to the product of the total extincion K and the mixing depth $Zmax$. See also Fig.1 for a graphical illustration on how this relation affects the shape of the growth vs. light function.

There is a complication when different vertical segments are considered by the model (i.e. in 3D mode). The extinction K and the light intensity at the surface might be accurately computed for each vertical segment, but what value should be adopted for the mixing depth $Zmax$? Since the computational procedure is set up in such way that the model is applied to each individual segment seperately, it would normally take the depth of this segment as being $Zmax$ and feed that into Eq. (14). This basically means that during the computation of the depth averaged production, it is assumed there is no transport across the vertical boundaries of the model segment. In other words: the light intensity encountered by the phytoplankton is assumed to vary between the value at the top and at the bottom of each segment. If according to the hydrodynamic conditions mixing across these boundaries is significant during a time step, this should be accounted for in the primary production calculation as well. To correct for this problem the depth averaging procedure is extended in the following way. Eq. (14) is solved for each vertical segment in a column of water using $Zmax$, but its result is stored rather than assumed to yield the correct value for E_k. Next a conservative tracer is released in each vertical segment to compute the vertical excursion of the phytoplankton across the segment boundaries during the time step. The proportion of time this tracer spends in each vertical segment is then applied to determine an additional weight function, which can be multiplied by the segment wise computed results of Eq. (14) giving the correct value for E_k in each vertical segment (Leo Postma, Deltares, unpublished).

Chapter 2

Phytoplankton modelling by means of optimization:

A 10-year experience with BLOOM II

F.J. Los & J.J. Brinkman, 1988

Verh. Interim. Verein. Limnol. 23: 790-795

| Verh. Interim. Verein. Limnol. | 23 | 790-795 | Stuttgart, August 1988 |

Phytoplankton modelling by means of optimization: A 10-year experience with BLOOM II

F. J. Los and J. J. BRINKMAN

With 4 figures in the text

Introduction

The Netherlands have suffered heavily from eutrophication problems at least since the beginning of the 20th century because
1. About two thirds of the annual freshwater supply is provided by the River Rhine.
2. Sediments contribute considerably to the annual nutrients loadings of most Dutch lakes.

The Dutch reacted rather slowly to these problems, perhaps because eutrophication has always been so common. Still it is obvious that conditions in many lakes deteriorated during the 1960s and 1970s. As a result eutrophication started to receive the attention it deserved: (1) national and regional monitoring networks were established covering practically every major water system, (2) engineering solutions such as chemical precipitation of phosphorus were tested at some sewage treatment plants and at several drinking water reservoirs, (3) universities and other institutes started laboratory experiments (chemostats; micro-cosms), (4) modelling began.

Modelling approach

As in many others countries it soon became obvious that eutrophication is a complicated problem, involving physical, biochemical, and biological aspects. Moreover, there was a strong need for methodologies to support the decision-making process. The first models to be applied were the OECD models (Vollenweider et al. 1980). The computed chlorophyll levels for the present situations, however, were far below the observations. This, of course, is not too surprising as very few of the shallow Dutch lakes meet any of the presumptions of the model.

The next step was to copy deterministic eutrophication models, most of which were originally developed in the USA for the Great Lakes (DiToro et al. 1975). Application of these models was not successful either. Again the computed chlorophyll levels were too low. Moreover, the predicted annual patterns (the classical two peaks) were much too regular.

Both types of model could not provide answers to the questions, which troubled the water managers. They were and of course still are, particularly interested in those periods, when phytoplankton biomasses reach their maxima, because this is exactly the period with the most serious management problems.

Ten years ago we therefore started a completely different modelling approach. We revised and extended a salt water model of the US RAND corporation (Bigelow et al. 1977) and developed a model called BLOOM II. This model uses an optimization technique called linear programming to compute the maximum total biomass concentration of several phytoplankton species at equilibrium in a certain time-period consistent with the environmental conditions.

To compute values for the environmental constraints, the model needs information on the concentrations of total available nutrients, temperature, the influx of solar radiation and certain lake-specific characteristics (depth and turbidity). These conditions can all be determined directly or indirectly from measurements and are sufficient for the model's calibration and validation.

Early model versions (Los 1982) strongly focussed on predicting objectionable peak values, which is exactly the point where the other models had failed. To this purpose a steady state was assumed with a one week time-step, which is justified by the high potential growth rate of phytolankton species under natural conditions during major parts of the year. A representative example of the performance of this model version is shown in Fig. 1.

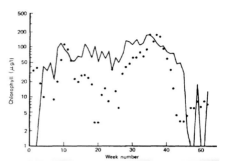

Fig. 1. Predicted (drawn line) and observed (dots) chlorophyll in the Grote Rug storage water reservoir No. 2 using an early version of BLOOM II.

Conclusion: peak values were indeed computed correctly, but in periods between blooms the computed biomass levels were usually much higher than those observed. From the very start the model often computed dominant phytoplankton groups correctly and even the dominant species often agreed with the observations in peak situations.

Obviously these periodic over-predictions are in complete accordance with the model's purpose. However, we had to extend the scope of the model in later years, to predict the impacts of sanitation measures. To this purpose BLOOM II was integrated with a sophisticated chemical model CHARON (Los et al. 1984), computing the surface water chemistry (nutrients among others). The first attempts to integrate both models completely failed, as the common deviations between computed and measured biomass levels made it impossible to obtain a feasible solution with the chemical model.

We therefore wondered whether it would be possible to improve BLOOM's overall performance, at the same time maintaining its ability to predict peaks and dominant species. To this purpose the model was completely revised since 1982.

Many specific coefficients (growth rates; day length dependence; growth vs light curves; nutrients requirements) were updated using measurements in chemostats and an incubator at the Microbiological Department of the Amsterdam University.

The model was extended with two important options to constrain the biomass of each phytoplankton species by its level at the previous time-step and by the current growth and mortality rates. In other words: complete switches in dominance can only occur when they are permitted by the dynamics of each species.

We added variable stochiometry to the model. To this purpose each phytoplankton species is split into several (usually 3) types. Depending on the external conditions the model can select any combination of types and therefore any allowable stochiometry, including the dry weight to chlorophyll ratio, for each species.

To extent the scope of BLOOM II we coupled it with the chemical model CHARON and to the general water quality model DELWAQ. The latter makes it possible to apply the model to any kind of hydraulic system (even 3-dimensional) provided that water movements are known.

The impact of these modifications is shown in Fig. 2, which compares the performance of the 1979 and 1985 versions of the model for one particular case. An extended documentation will appear shortly (Los 1991).

It may be concluded that the current model version tends to predict actual chlorophyll levels throughout the year. In a few cases the early model versions were definitely better. This is due in particular to the use of smaller and generally (so not always) more realistic growth rate coefficients by the current model.

Results

After its initial calibration BLOOM II was applied to about 30 different lakes for between 1 and 9 consecutive years, making a total of over 100 cases. Usually the model is first validated with

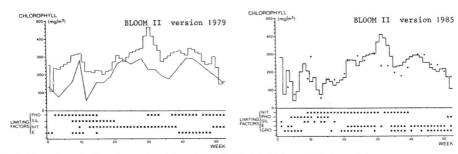

Fig. 2. Comparison of early and present model version applied to Lake Veluwe, 1976. Computed chlorophyll levels (µg/l) shown as histograms.

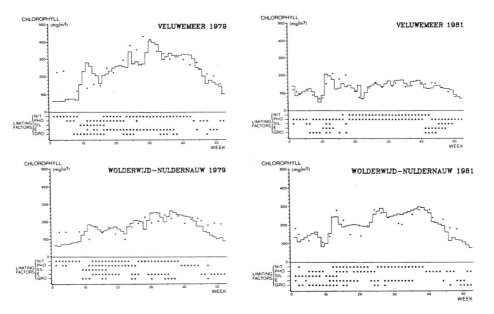

Fig. 3. Computed (histogram) and measured (diamonds) chlorophyll concentrations (µg/l) and limiting factors in Lakes Valuwe and Wolderwijd, 1979 and 1981.

measured nutrient data. Next phytoplankton and nutrients are validated using the integrated BLOOM II-CHARON model. This model combination is also applied to predict the impacts of sanitation measures.

The performance of the stand-alone version of BLOOM II is illustrated by the results for two twin lakes: Lake Veluwe and Lake Wolderwijd. These shallow lakes (average depth about 1.5 m) were created in the 1960s, and rapidly became eutrophic due to high nutrient loadings. Management of Lake Veluwe changed drastically starting in 1980. First the external loadings of phosphorus were reduced. Second the lake is now flushed several times each winter with nutrient-poor water.

BLOOM II was applied to both lakes for the period of 1975 through 1983. The results for two representative years (1979 and 1981) are shown in Fig. 3.

Conclusion: the computed and measured chlorophyll levels agree very well in both lakes. Notice that the observed and computed levels in the two lakes are similar in 1979, but in 1981 the levels in Lake Veluwe are considerably below those in Lake Wolderwijd. The model predicts dominance of the blue-green alga *Oscillatoria agardhii* in all cases, which agrees with the observations.

It is interesting to note that the actual improvement of Lake Veluwe is less spectacular than suggested by the chlorophyll data. The computed and observed reduction in bio-mass (dry weight in the model, wet weight in the observations) is much smaller. This is due to a change in limiting factors and a corresponding change in stochiometry. Before 1980 light and nitrogen are the main limiting factors, after 1980 phytoplankton is limited by phosphate and sometimes nitrogen. These severely nutrient limited phytoplankters have less chlorophyll per unit of biomass than before 1980.

Discussion

Most biological models are based on simple first order differential equations of the type:

$$\frac{dx}{dt} = (p + q + r + ...) x$$

Some of the coefficients p, q, r, ... are simple rate constants or functions of external conditions such as the temperature. Other terms, however, are also included in similar equations for other groups of organisms. Thus in general there are several coupled equations, for example for phytoplankton and zooplankton. Analytical solutions cannot usually be obtained for this kind of coupled equations, but that by itself is no longer a problem in the computer age. However, these models tend to be over-structured.

BLOOM II is essentially based upon similar equations, but they are not solved in this model. All its equations are transformed into an optimization problem, which is solved by the linear programming method. Early versions of the model were rather unpretentious: they simply gave the kind of answers in which water managers are interested. These versions were based upon a steady state assumption for phytoplankton growth, using a small number of rather loose environmental constraints.

As we have explained earlier: the constraints of the current model version are much more restrictive. Moreover, we have added constraints that take the dynamics of each species into account. As a result the current model version computes actual rather than (theoretical) maximum biomass levels, although it is still based upon the same maximization principle. The results of this model version, of which only a very small fraction could be shown here, suggest that this optimization principle is valid under a broad range of conditions. So we may indeed wonder if there is a biological background for the success of the model. Is there a tendency for species to maximize their biomass (offspring) on the expense of other species? Is this principle perhaps just a formal expression of the Darwinian principle of "Survival of the fittest"?

Some opponents of the model have argued, that the model should maximize growth rather than total biomass. The rationale behind this is that rapidly growing species have a higher affinity for limiting resources and hence have better opportunities to claim what little there is for themselves and their offspring. The potential impacts are shown in Fig. 4, which shows the nett growth rates of three phytoplankton species as a function of the extinction under fixed

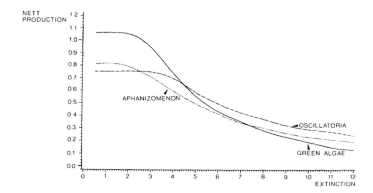

Fig. 4. Net growth rate of several representative phytoplankton species as a function of extinction (hence light intensity) under standardized conditions.

conditions of temperature, day length and solar radiation for a 1m deep lake. The curve for other eukaryotic species is almost similar to the one for the green alga, the curve for *Microcystis* resembles the one for *Oscillatoria*. The general shape of the curves shown in Fig. 4 hold under a broad range of conditions. Only temperature has a significant impact on this picture.

The growth rate of the eukaryotic species is relatively large when the extinction is small due to their high maximum growth rate. As the extinction increases, their net growth rates level off considerably faster than those for blue-green algae, because (1) their maintenance energy is larger and (2) their growth rates are relatively small at low light intensities. Maximizing growth instead of total biomass is therefore advantageous to eukaryotic species at low extinction levels (high average light intensities), but favors blue-greens when the extinction is high. We have made a large number of computations to compare both maximization principles. In general the results of maximizing biomass agree better with the observed biomass levels and compositions. However, maximizing growth has one strong advantage: it predicts dominance of eukaryotic species under oligotrophic conditions, as is usually observed. Maximizing total biomass predicts dominance of blue-greens.

In spite of considerable effort and model improvements during the past ten years, BLOOM II is still being updated and extended. Some current developments are (1) the first serious applications of a 2-dimensional model version to water networks (combined with DELWAQ and CHARON), (2) a complete revision of the simplistic formulation of mortality, which is currently a function of temperature, but in the future will be computed as a function of the age distribution as well and (3) re-evaluation of model concepts (growth versus biomass maximization).

Acknowledgement

The authors thank various departments within the Dutch ministry of Public Works, who sponsored the development of the models (WABASIM project), cooperated during various other projects and provided the necessary data.

References

BIGELOW, J. H., BOLTEN, J. G. & DEHAVEN, J. C., 1977: Protecting an estuary from floods — A policy analysis of the Oosterschelde. IV: Assessment of algae blooms, a potential ecological disturbance. — Rand Corp.

DITORO, D. M., O'CONNOR, D. J., THOMANN, R. V. & MANCINI, J. L., 1975: Phytoplankton-zoo-plankton-nutrient interaction model for Western Lake Erie. — Systems analysis and simulation in ecology, III: Academic Press, New York.

LOS, F. J., 1982: Mathematical simulation of algae blooms by the model BLOOM II. — Report R1310-7, Delft Hydraulics.

LOS, F. J., 1991: Mathematical simulation of algae blooms by the model BLOOM II. Part 2. Modifications and extensions of the model. — Report T68, Delft Hydraulics.

LOS, F. J., DE ROOIJ, N. M. & SMITS, J. G. C, 1984: Modelling eutrophication in shallow Dutch lakes. — *Verb. Internat. Verein. Limnol.* 22: 917—923.

VOLLENWEIDER, R. A. & KEREKES, J. J., 1980: Synthesis Report. Cooperative programme on monitoring inland waters (Eutrophication Control). — Organization of Economic Cooperation and development (OECD), Paris.

Authors' address:

Delft Hydraulics, Rotterdamseweg 185, P.O. Box 177,2600 MH Delft, The Netherlands

Chapter 3

Application of a validated primary production
model (BLOOM) as a screening tool for
marine, coastal and transitional waters

F.J. Los & J.M.W. Wijsman, 2007
Journal of Marine Systems Vol. 64: 201-215.

Application of a validated primary production model (BLOOM) as a screening tool for marine, coastal and transitional waters

F.J. Los *, J.W.M. Wijsman [1]

WL | delft hydraulics, P.O. Box 177, 2600 MH Delft, The Netherlands

Received 31 August 2005; received in revised form 8 February 2006; accepted 2 March 2006
Available online 3 July 2006

Abstract

In order to manage aquatic systems, it is necessary to apply methods relating the environmental variables and system-state parameters with external factors that affect the system. External factors can be natural (i.e. the movement of water) or partly-anthropogenic (i.e. nutrient loads). In addition to the national authorities, who have been implementing environmental policies for several decades, the EU is presently implementing the Water Framework Directive (WFD) aimed at establishing a new set of standards for the ecological and water quality of water systems. Among these are the phytoplankton biomass and composition. Phytoplankton affects turbidity, oxygen depletion, total productivity of the system and the occurrence of (harmful) algal blooms. A range of methods is available to relate phytoplankton to the controlling environmental conditions. Among these are statistical relations for instance of the Vollenweider type as well as deterministic simulation models. At the end of the 1970s, a generic deterministic phytoplankton module called BLOOM was developed, which has since been applied to a wide range of fresh water and marine systems. Here we test the applicability of this model as a screening tool for coastal waters. We conclude that the model is able to reproduce observed chlorophyll levels adequately under a wide range of conditions. Subsequently the model is applied to demonstrate the potential impacts of reductions in nitrogen, phosphorus or both nutrients simultaneously. Depending on which factors are initially controlling, the impacts of these reductions vary considerably both between locations and during the season. While this type of application lacks explicit relations between nutrient concentrations and external loadings, it does consider a number of relevant conditions in a consistent way and requires remarkably little data and effort. It is therefore a valuable screening tool.

Keywords: Phytoplankton modelling; Nutrient reduction; Coastal waters; Water Framework Directive; Chlorophyll-a

1. Introduction

In order to manage aquatic systems, it is necessary to apply methods relating the environmental variables and system-state parameters with external factors that affect the system. External factors can be natural (e.g. the movement of water) or partly-anthropogenic (e.g. nutrient loads). In addition to the national authorities, who have been implementing environmental policies for several decades, the EU is presently implementing the Water Framework Directive (WFD) (European Commission, 2000).

The WFD provides a framework for water management, with the aim of achieving a 'good chemical and

* Corresponding author. Tel.: +31 15 285 8549; fax: +31 15 285 8582.
E-mail address: hans.los@wldelft.nl (F.J. Los).
[1] Present address: Netherlands Institute for Fisheries Research, P.O. Box 77, 4400 AB Yerseke, The Netherlands.

ecological status' in fresh and near shore coastal waters throughout Europe. The implementation of the WFD involves a step-wise approach determining for each water body: (1) its status under undisturbed 'reference conditions', (2) its present status, (3) what could be considered as a 'good status' and finally (4) what should be done to achieve this status. With respect to the ecological status of transitional waters, the 'quality elements' of importance are phytoplankton, macrofauna and macrophytes. Fish are considered for fresh but not for the marine systems. To obtain a uniform implementation throughout Europe, all water bodies are classified according to the same principles, but this still leaves room for variations between water bodies in different countries or even within a country. It is also necessary to determine which part of the pressures are anthropogenic and which part can be considered as natural. By the year 2015, all aquatic systems should comply to these standards. While the WFD is restricted to a small part of the coastal zone, OSPAR (www.ospar.org) and the new Marine Strategy also consider offshore areas.

Hence the phytoplankton biomass and primary production are important indicators for managers of water systems. Phytoplankton affects turbidity, oxygen depletion, total productivity of the system and the occurrence of (harmful) algal blooms. When biomass levels are considered to be too high, nutrient reduction is a proven method to reduce levels in both marine and freshwater systems. However, it is not always obvious which nutrient should be reduced and sometimes results are disappointing. The effectiveness of nutrient reductions can be addressed by statistical data analysis; i.e. empirical relations between total P loadings and algae concentrations has been successfully done using the well known Vollenweider type relationships (Vollenweider, 1975). Because many more data are available on measured concentrations as compared to loadings, often empirical relationships directly relate algal biomass to P or N concentrations.

Alternatively, these questions can be addressed by deterministic modelling, where algae concentration and primary production can be calculated from measured or modelled nutrient concentrations. During the last decades, many simulation models for primary production have been developed with different characteristics, purposes and degree of validation (Di Toro et al., 1971, 1977; Baretta et al., 1995; Ebenhöh et al., 1997; Lancelot et al., 2000; Fulton et al., 2004; Janse, 2005) to mention just a few of them. At the end of the 1970s, a generic phytoplankton module called BLOOM was developed (Los et al., 1984; Los, 1991, 2005). Since then the model has been applied extensively both in hind

cast mode to explain what has happened in the past as well as in forecast mode to simulate the possible impacts of future conditions including management scenarios (Los and Brinkman, 1988; De Groodt et al., 1992; Van der Molen et al., 1994; Peeters et al., 1995; Los and Bokhorst, 1997; De Vries et al., 1998; Villars and DeVries, 1998; WL | Delft Hydraulics/MARE, 2001; Van Duin et al., 2001).

Different modes of complexity are possible with BLOOM ranging from a straightforward 0-D screening tool to a more detailed 3-D eco-hydrodynamic model. Applying models with different levels of complexity are valuable and each has various pros and cons. For a manager of a coastal water system it is sometimes necessary to obtain a rapid impression of the status of a particular water body and its sensitivity to certain pressures. How much phytoplankton biomass can be sustained? What factors seem to be controlling? In this paper, we present the results of the application of the BLOOM module for phytoplankton concentration and composition as a quick-scan tool. This is an example of a least complicated, 0-D application. As such BLOOM computes the total biomass, its division into major functional groups and limiting factors based on measured concentrations of nutrients, irradiance and temperature. Transport is not explicitly included, but it does affect the measured forcing by nutrients and background turbidity of the model. With this type of model, a first assessment of management measures can be made by varying the nutrient concentration levels, but there is no direct linkage to internal or external nutrient loadings or to anthropogenic or natural sources. An experienced user with a properly organised database can set up and apply this tool within a few hours. Many examples of this type of analysis have been reported by Los (1991). In this paper the reliability of the model results is tested by comparison against field observations for a variety of marine, coastal and transitional waters. Furthermore, model applications are shown to evaluate the effectiveness of nutrient reduction as a tool to decrease chlorophyll-*a* concentrations.

2. The model

Algal blooms usually consist of various species of phytoplankton belonging to different taxonomic or functional groups such as diatoms, flagellates, green algae and cyanobacteria, commonly referred to as blue-green algae. They have different requirements for resources (e.g. nutrients, light) and they have different ecological properties. Some species are considered to be harmful due to their effect on the turbidity of the water, the formation of scums or the production of toxins. For example, *Oscillatoria* can achieve very high biomass levels in shallow lakes causing a very low transparency, *Microcystis* is notorious for the formation of scums and has been reported to produce toxins that are harmful to animals (e.g. cattle) and men. In the marine environment, *Phaeocystis* is probably responsible for foam on beaches (Lancelot et al., 1987), and mass mortality of shellfish due to the settlement of a bloom in sheltered areas and subsequent depletion of oxygen (Rogers and Lockwood, 1990). To deal with these phenomena, it is important to distinguish between different types of phytoplankton in a model. The phytoplankton module BLOOM is based upon the principle of resource competition between different species. Note that the use of the term species in this paper is a flexible term. Sometimes a model species is equivalent to a biological species, but the term species could also refer to a number of biological species, grouped in larger ecological units, which are supposed to have similar characteristics. For example the group of diatoms, which consist of various biological species is regarded as one model species or plankton functional type. Most biological species of phytoplankton adapt rapidly to changes in their external environment. Individuals of a single species can therefore display a significant range of variation. To incorporate this phenomenon, each species of BLOOM is represented by several (pheno-)types. A type represents the eco-physiological state of the species under various possible conditions of limitation. Typically different types are considered for nitrogen, phosphorus and light limitation respectively. Hence for example an N-type has the characteristics of a species grown under prolonged conditions of nitrogen limitation. In a similar way P and E-types are defined. Occasionally additional types such as colonies or nitrogen fixing cells are explicitly included in the model. Types are the basic variables of this module. The number of types and their characteristics are inputs, so they can be easily adjusted for different kinds of water systems. Since types differ with respect to all characteristics included in the model, a shift between types not only implies a shift in nutrient stochiometry, but also in other characteristics such as the growth, mortality, sedimentation and respiration rates and in the carbon to chlorophyll ratio.

Once the model has computed the biomasses of the types, they are summed up to compute the biomass of each species. Often more than one type of a species is present at a particular moment in time, in theory all of the types of a species can be present simultaneously. The formulation of the model takes into account that adaptation occurs much more rapidly than succession between species.

The model considers the growth rate and the requirements for all potentially limiting environmental factors to determine the optimum combination of types using the linear programming method (Danzig, 1963). The nutrient and algae biomass concentrations at the beginning of the simulation period and the temperature and light intensity during the period are assumed known. The model must be solved for successive time periods in which the nutrient levels and initial biomass concentrations can be changed in accord with the solution of the previous time step. The optimization procedure distributes the available resources among all chosen algae types yielding a new composition of algae type biomass concentrations.

Typically, BLOOM considers between 3 and 10 representative algae species. For example, consider the following four (groups) of species: diatoms, micro-flagellates, dinoflagellates and *Phaeocystis*. These algae groups can be divided into three types based on their limiting nutrient or energy. Hence a total of 12 different algae types could be defined in this example.

Each distinct species subtype (from now on called type) is denoted by the index k. The BLOOM model identifies the concentration of biomass, B_k, of each algae type k that can be supported in the aquatic environment characterized by light conditions and nutrient concentrations. It can be demonstrated that finding the best adapted types at any moment in time is equivalent to maximizing the total rate of primary production given a number of environmental conditions (constraints). Defining the gross growth constant Pg_k (day^{-1}), the objective of the model thus is to

$$\text{Maximize } \sum_k Pg_k\, B_k \qquad (1)$$

For each algae type, the requirements for nitrogen, phosphorus and silica (only used by diatoms) are specified by coefficients n_{ik}, the fraction of nutrient i per unit biomass concentration of algae type k.

The total readily available concentration, C_i (g m^{-3}) of each nutrient in the water column equals the amount in the total living biomass of algae, $\sum_k(n_{ik}B_k)$, plus the amount incorporated in dead algae, d_i, plus that dissolved in the water, w_i. These mass balance constraints apply for each nutrient i.

$$\sum_k (n_{ik} B_k) + d_i + w_i = C_i \qquad (2)$$

The unknown concentration variables B_k, d_i, and w_i are non-negative. All nutrient concentrations C_i are the measured or modeled total concentrations and are assumed to remain constant throughout the time period defined for the optimization model. The system is assumed to be in equilibrium over that period. The time step is an input to the model and may be chosen to vary during the simulation period to account for seasonal variations in characteristic time scales.

2.1. Nutrient recycling

A certain amount of each algae type k dies in each time step. This takes nutrients out of the live phytoplankton pool. A fraction remains in the detritus pool, and the remainder is directly available to grow new algae because the dead cells break apart (autolysis) and are dissolved in the water column. Detritus may be removed to the bottom or to the dissolved nutrient pools at rates in proportion to its concentration. Needed to model this is the mortality rate, M_k (day^{-1}), of algae type k, the fraction, f_p, of dead phytoplankton cells that is not immediately released when a cell dies, the remineralization rate constant, m_i (day^{-1}), of dead phytoplankton cells, the fraction, n_{ik}, of nutrient i per unit biomass concentration of algae type k, and the settling rate constant, s (day^{-1}), of dead phytoplankton cells.

The rate of change in the nutrient concentration of the dead phytoplankton cells, dd_i/dt, in the water column equals the increase due to mortality less that which remineralizes and that which settles to the bottom.

$$dd_i/dt = \sum_k (f_p\, M_k\, n_{ik}\, B_k) \; - m_i\, d_i - s\, d_i \qquad (3)$$

Both mortality and mineralization rate constants are temperature dependent. If the model is applied as a screening tool, Eq. (3) is solved under the assumption of a steady state which means its right-hand side equals 0. This gives an expression relating the amount of detritus to the algal biomasses. If BLOOM is applied as a dynamic simulation model, this equation is integrated numerically.

2.2. Energy limitation

Algae absorbs light for photosynthesis and growth. Energy becomes limiting through self-shading when the total light absorption consisting of a non-algal part and an algal part, exceeds the maximum at which growth is just balanced by respiration and mortality. For each algae type k there exists a specific extinction value K_k^{max} (m^{-1}) at which this is the case. The light intensity can also be too high, which means the total extinction is too low (photo-inhibition) for growth. This specific extinction value is K_k^{min}. The ranges between K_k^{min} and K_k^{max} differ

for different algal types k because each one of them is characterized by a different set of model coefficients. Among others a different light response curve for growth is used for each species in the model in the form of a table, through which a curve is fitted which is integrated numerically to account for diurnal variations in light intensities over depth due to mixing and in time. Letting K_k (m^3/m/g dry) represent the specific light absorbing extinction constant for living material of algae type k, the total extinction due to all living algae is

$$KL = \sum_k (K_k\, B_k) \qquad (4)$$

Added to this must be the extinction caused by dead cells, KD and the contribution of all other fractions such as inorganic suspended matter and humic substances to the extinction of the water, KW (m^{-1}). Hence

$$K_k^{min} \leq KL + KD + KW \leq K_k^{max} \qquad (5)$$

The extinction from dead cells is usually less than half of that from live cells. The amount of dead cells not yet mineralized is, from Equation 3, $\sum_k (f_p\, M_k\, B_k)$. Assuming some fraction e_d (usually between 0.2 and 0.4) of the extinction rate of live cells,

$$KD = e_d \sum_k K_k\, f_p\, M_k\, B_k \qquad (6)$$

If the total extinction is not within the range for an algae type k, its concentration B_k will be zero. To ensure that B_k is 0 if the total extinction is outside of its extinction range, a 0,1 binary (integer) unknown variable Z_k is needed for each algae type k. If Z_k is 1, B_k can be any non-negative value; if it is 0, B_k will be 0. This is modeled by adding three linear constraints for each algae type k.

$$KL + KD + KW \leq K_k^{max} + KM\,(1 - Z_k) \qquad (7)$$
$$KL + KD + KW \geq K_k^{min}\,(Z_k) \qquad (8)$$
$$B_k \leq BM\, Z_k \qquad (9)$$

Where KM and BM are any large numbers no less than the largest possible value of the total extinction or biomass concentration, respectively. Since the objective of maximizing the sum of all $Pg_k\, B_k$ together with Equation 9 wants to set each binary Z_k value equal to 1, only when the total extinction is outside of the extinction range K_k^{min} to K_k^{max} will the Z_k value be forced to 0. Equation 9 then forces the corresponding B_k to 0. This means that beyond its feasible range of the extinction coefficient, a species cannot maintain a positive biomass.

2.3. Growth limits

When the environmental conditions improve at a rate which is large relative to the potential biomass increase of a particular phytoplankton species, it may be impossible to achieve the level at which either light or some nutrient gets limiting within a single time-step of the model. To account for this situation, a constraint to delimit the maximum biomass increase within the time-interval is considered during the optimization procedure. Assuming that losses will be low during the exponential growth phase of a phytoplankton species, mortality is ignored in the computation of this growth constraint.

For all algae types k the maximum possible biomass concentration, B_k^{max} (g dry m^{-3}), at the end of the time interval Δt (days) depends on the initial biomass concentration, B_k^{0}, (g dry m^{-3}), the maximum gross production rate Pg_k^{max} (day^{-1}), the respiration rate constant, R_k (day^{-1}), and the time and depth averaged production efficiency factor, E_k. Using the net production rate constant, Pn_k ($=Pg_k^{max}E_k-R_k$) (day^{-1}), for each algae type k:

$$B_k^{max} = B_k^{o} \exp\{ Pn_k \Delta t \} \qquad (10)$$

If the initial biomass is smaller than a certain base level, this base level is used in stead. Empirically it was found that using a base level of 1% of the potential maximum generally results in realistic species shifts in the model.

2.4. Mortality limits

As in the case of growth, the mortality of each algae species is also constrained to prevent a complete removal within a single time-step when conditions get worse. The minimum biomass value of a species is obtained when there is no production, but only mortality. The minimum biomass, B_k^{min} (g dry m^{-3}), of type k at the end of time interval Δt depends on the initial biomass, B_k^{0} (g dry m^{-3}), of type k and the specific mortality rate constant, M_k (day^{-1}) of type k.

$$B_k^{min} = B_k^{o} \exp\{-M_k \Delta t \} \qquad (11)$$

These minimum values are computed for each individual algae type. However the model sums each of these minimum values over all subtypes within each species and applies it to the total biomass of the species. This way the maximum possible mortality cannot be exceeded, but transitions between limit types remain possible.

As mortality is computed according to a negative exponential function, the minimum biomass level is always positive, in other words a species can never disappear completely. To prevent that insignificantly small biomass values are maintained in the model, the minimum value is replaced by zero once the value computed according to Eq. (11) drops below some base level. Empirically it was found that using a base level, which is 10 times smaller than the base level for the growth, generally results in realistic species shifts in the model.

The mortality constraint of a species (11) has precedence over its extinction constraint (9). Hence in case of a conflict when the mortality constraint demands a certain biomass level to be maintained which exceeds the maximum permitted by the available amount of light, the extinction constraint is dropped from the optimization procedure. Effectively this means that types disappear at the rate of M_k (day^{-1}) under unfavorable conditions and will not be completely removed in a single time step even though too little light is currently available to maintain a positive biomass level.

2.5. Competition between species

In biological terms the competition in the BLOOM model is governed according to the following principle. The algal types defined in the input compete with each other for all potentially limiting resources taking the existing biomass into account. The outcome of the competition for a potentially limiting resource is determined by the ratio between the gross growth rate constant and the requirement for that resource. Hence species with very high growth rates may outcompete more efficient, but slowly growing species, or very efficient species may outcompete species with a higher potential growth rate but a much higher requirement for that particular resource. In practice this means that opportunistic species with high growth rate usually dominate when total available nutrients are low and the average light intensity is high, whereas efficient species with lower potential growth rates and lower resource requirements dominate when total available nutrient levels are high and the average light intensity is low (high level of self-shading).

The principle of the model was briefly described in Los et al. (1984). An extensive description covering both the equations and underlying ecological assumptions is in Los (1991). A condensed version can be found in Los (2005). An overview of the model is shown in Fig. 1. Applied as a screening tool only phytoplankton, dissolved nutrients and the labile form of dead algae are explicitly taken into account. A number of additional compartments and fluxes are part of the model, but these were not considered here and are shown in grey.

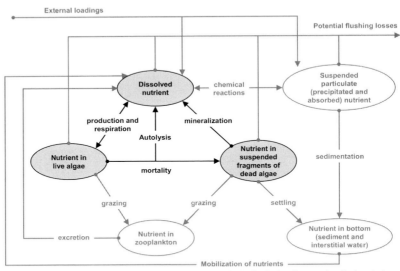

Fig. 1. Overview of the BLOOM phytoplankton model. Applied as a screening tool only dissolved nutrients, nutrients in phytoplankton and in suspended detritus are explicitly considered. Compartments and fluxes in grey can optionally be included but were not in the cases included in this paper.

The number and the characteristics of the phytoplankton species are inputs to the model. Data for about 20 different marine and fresh water species have been collected over the years based on literature, laboratory experiments (Zevenboom and Mur, 1981; Zevenboom et al., 1983; Zevenboom and Mur, 1984; Post et al., 1985; Riegman, 1996; Riegman et al., 1992; 1996; Riegman (unpublished results); Jahnke, 1989) and previous model applications. Depending on the problem, sometimes only major groups are included such as diatoms, greens and blue-greens, and sometimes individual species are modelled such as *Aphanizomenon* or *Phaeocystis*. In by far the majority of model applications, these characteristics are kept at their present default values. They are not tuned to improve the model fit because in so doing they could lose their generality.

3. Implementation of the model

3.1. BLOOM application for screening

In order to test the applicability of the BLOOM model as a 0-D screening tool, it has been applied to six different locations in the Netherlands (Fig. 2, Table

1). The selected stations are located in different marine, coastal and transitional waters in the Netherlands ranging from station Dreischor in the salt water lake Grevelingen to Terschelling 235 (Doggerbank) in the central part of the North Sea. The model is run with a time step of 1 week. To apply the model the following forcing needs to be specified on a weekly basis: water depth, water temperature, surface irradiance, total available nutrient levels (N, P, Si), background extinction coefficient plus the extinction due to suspended matter and humic substances. Data on surface irradiance levels were obtained from the Royal Dutch Meteorological Institute. Each system was assumed to have a specific depth. The other data for all locations, have been extracted from the DONAR database that is available on the internet (www.waterbase.nl). The data have been sampled in the framework of the MWTL programme (Monitoring Programme of the National Water Systems) of the National Institute for Coastal and Marine Management (RIKZ). The year 1998 has been selected since a good data coverage was available for this year at all locations. Data were retrieved on the following parameters: NH_4, NO_2, NO_3, PO_4, SiO_2, Chl-a, suspended solids and salinity.

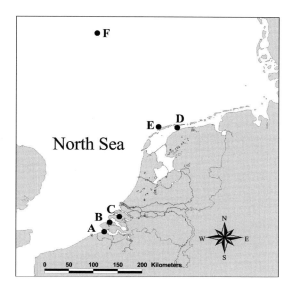

Fig. 2. Overview of the modelled locations. (A) Vlissingen; (B) Wissekerke; (C) Dreischor; (D) Dantziggat; (E) Terschelling 4; (F) Terschelling 235 (Doggerbank).

Estimation of the particulate organic nitrogen (PON) and phosphorus (POP) presents a problem since insufficient data are available for total nutrients. Moreover, measured levels of total nutrients include refractory components that are not readily available for phytoplankton growth and hence should not be included in the model's input. Since the purpose of the model application was to test its applicability as an easy to set-up screening tool, a uniform, simple assumption was used to estimate the amount in nutrients in phytoplankton and detritus based on the measured amount of chlorophyll. It is assumed that that 1 g chlorophyll-a corresponds to 7.5 g N and 0.75 g P in phytoplankton. This corresponds to a g C/ Chl-a

Table 1
Locations of the modelled monitoring stations in this paper

	Bassin	Latitude	Longitude
Vlissingen	Westerschelde	51°24'43.2"N	3°33'56.2"E
Wissekerke	Oosterschelde	51°36'05.7"N	3°43'14.0"E
Dreischor	Lake Grevelingen	51°42'52.6"N	3°59'57.6"E
Dantziggat	Wadden Sea	53°24'04.1"N	5°43'37.1"E
Terschelling 4	Coastal North Sea	53°24'50.9"N	5°09'00.6"E
Terschelling 235	Central North Sea	55°10'15.2"N	3°09'26.7"E

ratio of 50 and N/C and P/C ratios of 0.15 and 0.015, respectively. In spite of a large range of variation, these ratios can be considered as typical for the species included in the model, which have been derived from the laboratory experiments described at the end of the previous chapter. Furthermore, it is assumed that for each g PON and POP in algae, also 1 g is present in the form of labile detritus. This 1-to-1 ratio is approximately the annual average computed by the model. No doubt an improved estimation is possible for individual locations, but for this model application the validity of this simple approach was tested and considered accurately enough. As a result, total for phytoplankton available N can be estimated using the following function:

$$TotN = NH_4 + NO_2 + NO_3 + 2 \cdot (7.5 \cdot Chl\text{-}a) \qquad (12)$$

and total P is estimated as:

$$TotP = PO_4 + 2 \cdot (0.75 \cdot Chl\text{-}a) \qquad (13)$$

For silicate, measured dissolved silicate levels have been used with a minimum concentration of 0.05 g l^{-1} throughout the year.

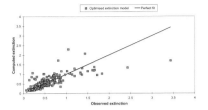

Fig. 3. Computed and measured extinction values (optimized extinction model). From van Gils and Tatman (2003).

Because primary production is strongly influenced by light availability and can even become limited if there is too little light, the calculation of the light conditions in the water is an important process. The availability of light is a function of the solar irradiation on the water surface of light within a certain wave length range (photosynthetically available radiation: PAR) and of the extinction due to absorption and scattering of the light inside the water column. The extinction of light in water is described by the Lambert–Beer model where the extinction coefficient can be related to the absorption and scattering properties of the water constituents. In this BLOOM application, the extinction coefficient is calculated according to an empirical model (Van Gils and Tatman, 2003), which compared to other previously reported relationships provided the best possible fit to the extinction measurements from all locations that are sampled by the RIKZ. The best model in terms of explaining the variability in the extinction coefficient

(Kd), is a four parameter model based on POC, salinity and two fractions of TSM ash weight (below and above 15 mg l^{-1}):

$$Kd = 0.067 + 0.081\left(19.4 - \frac{S}{1.8}\right) + 0.30 \, POC \qquad (14)$$
$$+ \, 0.036 \, SS_1 + 0.005 \, SS_2$$

where S is the salinity of the water (−), POC is the measured particulate organic carbon concentration (mg C l^{-1}) and SS$_1$ and SS$_2$ are the concentration of small and large suspended sediment particles, respectively. It is assumed that the concentration of small sediment particles (SS$_1$) is 15 mg l^{-1} with a relative extinction coefficient of 0.036 m^2 g^{-1} while the remainder of the total suspended sediment is considered as the coarse fraction with a specific extinction coefficient of 0.005 m^2 g^{-1}. The background extinction according to Eq. (14) equals 0.067 m^{-1}, the second term represents the extinction due to the dissolved humic substances. It is taken from the fraction of fresh water. This term vanishes when the salinity is 34.92. The fit of this extinction model to the data is shown in Fig. 3, which is taken from the original report by van Gils and Tatman (2003). To obtain the non-algal related part of the extinction coefficient, POC is put equal to 0 in Eq. (14). During the model simulation the contribution of live phytoplankton and labile detritus is computed by BLOOM. Weekly radiation data for 1998 were derived for location de Kooy, near Den Helder. Photosynthetic active radiation (PAR) was calculated from the total radiation by multiplication with a factor of 0.45. For all locations, the same radiation data were used. Missing values in time for a particular forcing function have been obtained by linear interpolation.

Table 2
Specific extinction coefficients and stochiometric ratios of types defined in BLOOM

Algal type	Specific extinction (m^2/g C)	N/C (mg/mg)	P/C (mg/mg)	Si/C (mg/mg)	Chla/C (mg/mg)	Dry/C (mg/mg)
Diatoms-E	0.24	0.255	0.0315	0.447	0.0533	3.0
Diatoms-N	0.21	0.070	0.0120	0.283	0.0100	3.0
Diatoms-P	0.21	0.105	0.0096	0.152	0.0100	3.0
Flagellate-E	0.25	0.200	0.0200	0.0	0.0228	2.5
Flagellate-N	0.225	0.078	0.0096	0.0	0.0067	2.5
Flagellate-P	0.225	0.113	0.0072	0.0	0.0067	2.5
Dinoflag-E	0.20	0.163	0.0168	0.0	0.0228	2.5
Dinoflag-N	0.175	0.064	0.0112	0.0	0.0067	2.5
Dinoflag-P	0.175	0.071	0.0096	0.0	0.0067	2.5
Phaeocyst-E	0.45	0.188	0.0225	0.0	0.0228	2.5
Phaeocyst-N	0.41	0.075	0.0136	0.0	0.0067	2.5
Phaeocyst-P	0.41	0.104	0.0106	0.0	0.0067	2.5

Original data based on laboratory experiments (references in main text). These were adjusted during previous validations of 2D and 3D North Sea model applications.

The model was applied to the selected locations in a 0-D mode. Species groups included in the model and its coefficients were adopted from previous 2- and 3-dimensional applications to the North Sea (Los and Bokhorst, 1997; Blauw et al., 2009). They were kept the same here for each individual station, only the forcing functions of the nutrient concentrations, water depth and non-phytoplankton related contribution to the extinction were varied. The selected species groups are diatoms, microflagellates, dinoflagellates and Phaeocystis. The main stochiometric coefficients used for this application are shown in Table 2. The time series of the calculated chlorophyll-*a* concentrations are compared with the observed chlorophyll-*a* concentrations.

3.2. Sensitivity analysis

The previously described BLOOM phytoplankton module was applied to establish the relations between phytoplankton biomass (e.g. chlorophyll-*a*) and physico-chemical quality elements and pressures (e.g. nutrients and light conditions) for the Dutch coastal waters. These relations are illustrated in the form of 'response curves', which depict the chlorophyll-*a* concentration as a function of different nutrient and/or light conditions (Fig. 4). By defining criteria for the desired phytoplankton biomass indicator level, to represent 'Good' (G) or 'Moderate' (M) status, the corresponding required pressure reduction in nutrients (e.g. R1 and R2) can be estimated from the response curve. Values for 'G' and 'M' have not yet been set for the Dutch coastal and transitional waters.

In general, the 'pressures' of relevance for phytoplankton are nitrogen, phosphorus, silicate and light availability (as proxies of nutrient loading/status and turbidity). Thus, separate response curves can be made for each of these factors. Although in reality the

response of water bodies to changes in loadings will be complicated, as a first assessment response curves for nutrient concentrations are shown as percentage reductions with respect to a baseline concentration, ranging from 0% to 90% for N, for P and for N and P together. As such the response curves show how the phytoplankton will respond to a decrease in nutrients, which may correspond to a particular management strategy. All other model settings remained unchanged. For each simulation, the summer-averaged chlorophyll-*a* concentration was calculated where the summer was defined as 1 April–1 October.

4. Results

4.1. Model validation

The chlorophyll-*a* concentration is an important indicator for the state of eutrophication of marine systems. In the temperate North Sea system however, chlorophyll-*a* concentrations show large seasonal fluctuations throughout the year, with higher concentrations during spring and summer and lower concentrations during the winter. This is mainly because the light availability in winter is too low to support primary production in most areas. The onset of the spring bloom is determined by an increase in available light and varies considerably depending on the solar radiation, the depth and the non-algal part of the extinction such as the TSM concentration. The spring bloom is often limited by the available amount of phosphorus and/or silicate while the summer bloom in many stations is still nitrogen limited. Near shore, however, nitrogen limitation has become a rare phenomenon in Dutch waters since the end of the 1990s, due to the extremely high N/P ratio of the river loads. In autumn light becomes the main limitation again and biomasses decline to small values.

In this validation section, we compare the calculated chlorophyll-*a* concentrations with the observed values. The confidence in the model increases if there is a good agreement between model and observations in both the seasonal patterns and average concentrations.

In general, the seasonal variation and the absolute concentrations of chlorophyll-*a* are well described by the model at these locations (Fig. 5). Highest concentrations are observed at location Dantziggat, located in the Wadden Sea, with average concentrations during the summer half-year of 24 µg l^{-1} (Table 3) and a peak concentration of 57.6 µg l^{-1}, measured on April 21. Summer biomass levels are mainly controlled by

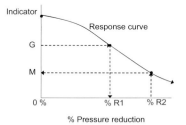

Fig. 4. Generic response curve illustrating the relation between the selected indicator (e.g. chlorophyll concentration) and % pressure reduction (e.g. nutrient concentration).

nitrogen and phosphorus at this location. Relatively high concentrations in winter are probably due to benthic rather than to pelagic primary productivity. Lowest concentrations are observed at Terschelling 235, at the Doggerbank in the central North Sea, with average chlorophyll-*a* concentrations of 1.5 µg l^{-1}. Nitrogen is the main limitation at this station, but summer phosphorus levels are also very low. The underestimation of the average chlorophyll-*a* concentrations at location Wissekerke is mainly due to the high peak concentration (35.2 µg l^{-1}) that is observed at the end of April, but not simulated by the model. Considering the size of this peak and the average light intensity at this station, it is unlikely that it could be produced locally, hence this peak is probably imported from adjacent parts

of the North Sea where the spring bloom starts earlier. At all locations, the winter concentrations of chlorophyll-*a* are underpredicted by the model. In most cases, the timing of the Spring phytoplankton peak is well modelled. Only at location Vlissingen, in the Westerschelde estuary, it appears that the increase in chlorophyll-*a* starts too late.

An example of the simulation results for individual species at the location Dreischor is given in Fig. 6. Diatoms dominate in spring and autumn, flagellates dominate in early summer, dinoflagellates during late summer. Simulated *Phaeocystis* levels are consistently low at this location. This pattern is controlled by the seasonal variation in limiting factors and the characteristics of the model species with respect to these factors.

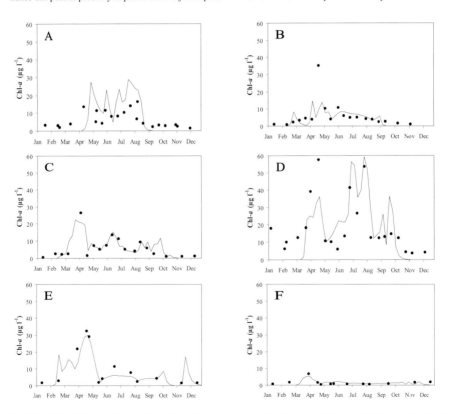

Fig. 5. Modelled (solid lines) and observed (dots) chlorophyll-*a* concentrations (µg l^{-1}) at the locations Vlissingen (A), Wissekerke (B), Dreischor (C), Dantziggat (D), Terschelling 4 (E) and Terschelling 235 (F).

Measurements for a direct comparison are lacking, but the general succession of species seems reasonable.

4.2. Sensitivity analysis: effect of nutrient reduction

Response curves for the nutrient reductions are shown in Fig. 7. In the coastal location Terschelling 4 the response curve shows a fairly linear relation between decreased chlorophyll concentration and phosphorus reduction (Fig. 7E). With low levels of N reduction (0–30% reduction), there is a limited response in chlorophyll concentration. This indicates that phosphorus rather than nitrogen is the main limiting factor to the phytoplankton growth at this location. The same is true for many other Dutch coastal stations (not shown here).

Of the selected stations representing estuaries, marine lakes and Wadden Sea, location Dreischor in Lake Grevelingen shows a limited response for P reduction in the range of 0–50%. This indicates that Phosphorus is not the limiting factor in the phytoplankton growth at these locations (Fig. 7C). A 50% reduction of the N concentration is much more effective in this system, resulting in a decrease in summer averaged chlorophyll-a of more than 30%. At location Vlissingen on the other hand, a reduction of P concentrations seems to be more effective than reduction of N, indicating that phytoplankton growth at this location during the summer is more limited by P than by N. At the locations Wissekerke, Dantziggat and Terschelling 235, the response curves for N and P are comparable, implying that both nutrients limit the average algal biomass to the same extent. It should be pointed out that even if the response in chlorophyll appears to be linear, the reduction percentage is usually considerably smaller than the corresponding reduction in the affected nutrient. This demonstrates that the model is adjusting

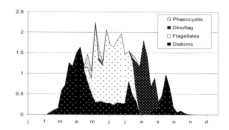

Fig. 6. Simulated species composition (mg C l^{-1}) at location Dreischor (Lake Grevelingen).

its nutrient to chlorophyll ratio by shifting among the simulated phytoplankton types. This kind of response is maintained until the most efficient phytoplankton type has been selected and only one nutrient is constantly limiting. This usually occurs with a reduction in the order of 70%. From there on the simulated reduction in chlorophyll is similar to the reduction of this nutrient.

5. Discussion

In Introduction it was pointed out that assessing the present status and its response to changes in external pressures are important elements of the implementation of the WFD. The purpose of this model application is to demonstrate that it is possible to obtain an acceptable first impression of these aspects for coastal water systems by applying the BLOOM model as a screening tool. To that purpose the simulated chlorophyll-a concentrations by the standard version of the model were compared to the measurements at a number of stations where conditions in terms of nutrients and light vary considerably. Most forcing conditions could be directly obtained from measurements, other required some assumptions. To keep these assumptions as generic as possible we used the same method to estimate the forcings everywhere thus preferring robustness of the approach over tuning to local conditions.

The results presented here demonstrate that it is indeed possible to obtain an acceptable overall agreement between simulated and observed chlorophyll-a levels particularly during the summer half year. Some deviations occur in the Wadden Sea (station Danziggat), for which winter levels are obviously underpredicted. This is most probably due to the lack of microphytobenthos in the model, which according to De Jonge

Table 3

Overview of the modelled and observed chlorophyll-a concentrations (µg l^{-1})

	Yearly averaged		Summer averaged	
	chlorophyll-a concentrations (µg l^{-1})		chlorophyll-a concentrations (µg l^{-1})	
	Modelled	Observed	Modelled	Observed
Vlissingen	6.3	6.4	12.6	8.4
Wissekerke	3.4	5.7	6.1	7.7
Dreischor	5.7	5.9	8.7	8.3
Dantziggat	15.1	18.3	27.8	24.0
Terschelling 4	7.0	9.5	9.0	12.8
Terschelling 235	1.2	1.5	1.6	1.4

The summer is defined from 1st April to 1st October.

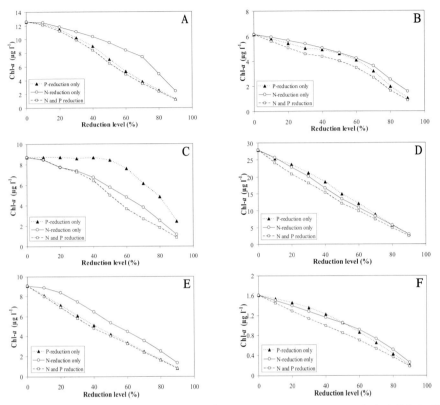

Fig. 7. Predicted summer averaged chlorophyll-*a* concentrations (µg l⁻¹) resulting from P-reduction (dotted lines), N-reduction (solid lines) and both N- and P-reduction (broken lines) for the locations Vlissingen (A), Wissekerke (B), Dreischor (C), Dantziggat (D), Terschelling 4 (E) and Terschelling 235 (F). Note the different scaling on the y-axis.

(1992) is a major source of chlorophyll-*a* in the Wadden Sea.

Other factors responsible for some deviations between simulations and observations are (1) the absence of advective transport processes in the 0-D model, (2) the assumption of complete vertical mixing under all conditions, (3) the estimation method for available nutrients and background turbidity and (4) the absence of grazers in the model. It is important to note that the model simulations presented here already cover a wide range of conditions in which the highest and lowest average chlorophyll-*a* values vary by a factor of 20 (Table 3). This increases the credibility of the simulation results for nutrient reductions.

Considering that BLOOM was applied in 0-D mode, the resemblance between simulated and measured chlorophyll levels may look surprising because most of these stations are affected by the tide and by other horizontal transport processes. The reason why the model results do not suffer heavily from a lack of advective transport is that its main forcings are periodically updated based on in situ measurements. Hence if the nutrient, salinity or TSM levels have changed due to some event which is not explicitly included in the 0-D box model, these new levels are used to generate the input conditions for the next model time step. As long as the potential net growth rates of the phytoplankton types in the model are in the same order

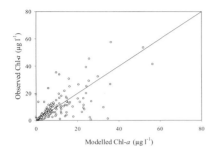

Fig. 8. Observed and modelled summer chlorophyll-*a* concentrations (μg l⁻¹) at for all stations. The line indicates the 1:1 relation between observed and modelled chlorophyll levels. The R² of this relation is 0.40.

of magnitude as the observed rate of change in forcing, the model will be able to adjust to the new conditions rapidly enough to track the observed changes in chlorophyll levels.

Inhomogeneous vertical mixing is sometimes observed, but this is only a temporary, not a dominant phenomenon at any of the stations reported here. Hence it seems that this source of error is of minor importance to the locations considered here.

In Eqs. (12) and (13) the term which estimates the organic part of the available nutrients is obviously a simplification of reality, ignoring temporal and spatial variations in stochiometric coefficients and seasonal variations in the ratio between labile detritus and live phytoplankton. No attempt was made to adjust these estimations to local conditions or to vary them seasonally because applying the same equation everywhere is considered to be more attractive from a management point of view. For the same reason Eq. (14), which was used to estimate the non-algal part of the extinction coefficient, was uniformly applied. Considering the overall results of the model, errors in the estimation of nutrients and turbidity in general seem to be acceptable.

Lack of grazing is another potential source of error. Unfortunately little quantitative information is available on the grazing rates of filter feeders in Dutch marine waters. Obviously at some locations, notably in the Wadden Sea filtration by mussels is important, but the importance of this source of error cannot be quantified.

It should be noted that most deviations between model and observations occur in the winter season. The present application is, however, focused on the summer half year so the performance during that part of the year is most important. As a further illustration the summer half year simulation results and measurements are plotted against each other in Fig. 8. In spite of some deviations particularly in the range between 20 and 40 μg l⁻¹ of chlorophyll, there are no systematic errors in the model results, which confirms that the overall fit is acceptable for its application as a screening tool.

The definition of the summer half year period taken as a basis for the assessment of the model output is to some extent arbitrary. It may be argued that this period starts too late considering that the spring bloom occurs earlier at some but not at all locations. To test the sensitivity of the conclusions we have redrawn Fig. 8 and recomputed the response curves for the period March 1 till September 1. Although the results are not identical, the differences are insignificant.

As an additional form of validation BLOOM has been applied in a similar way to a number of sites from Italian and Portuguese coastal waters as part of the Rebecca project. The same methodology was applied to estimate the model forcings (Eqs. (12)–(14)), but day length and irradiance data were adapted to comply with local conditions. The result for the oligotrophic station Miramare in the Northern part of the Adriatic Sea is shown in Fig. 9 as an example. Observed and simulated chlorophyll-*a* levels at this location are in the order of 1 μg l⁻¹, which is similar to the values simulated with extreme nutrient reductions for most of the Dutch stations. In particular the summer half year values are well reproduced, hence the model presented here still holds when nutrient levels are far below the present values in Dutch coastal and transitional waters. So the model is valid even at these low nutrient concentrations.

In comparison to statistical methods, the application of BLOOM as a screening tool has several advantages.

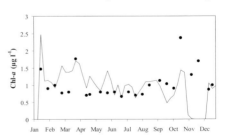

Fig. 9. Modelled (line) and observed (dots) chlorophyll-*a* concentrations (μg l⁻¹) at Miramare in the Northern part of the Adriatic sea.

The model concept and its coefficient are generic and do not have to be adapted for local conditions of European waters. The model contains explicit formulations on the relationship between the effect variables (e.g. chlorophyll-*a*, composition of the phytoplankton community) and many relevant environmental conditions. These include not only the pressure proxies (i.e. nutrient levels) but also other factors, which may not be controlled, such as the light intensity or water temperature. Also, because simulations are performed on a weekly basis, seasonal variations in controlling factors are explicitly taken into account. In most of the computations, at least two different factors become limiting in different parts of the year. For instance at

location Dreischor, the spring bloom in March and at the beginning of April is limited by P (Fig. 10). As a result, N reduction by 50% has no effect on the size of the spring bloom. During the summer however, the primary production becomes limited by N, resulting in lower chlorophyll-*a* concentrations for the 50% N-reduction scenarios. A 50% P-reduction even results in slightly higher chlorophyll-*a* concentrations in June, which is due to a change in simulated phytoplankton composition.

For the Dutch coastal zone, a 2-D/3-D primary production model is operational: the GEM model, which is a rather detailed eco-hydrodynamic model which includes the BLOOM module (Blauw et al., in press). The validation result of the screening version of BLOOM has been compared to GEM (Fig. 11). The resemblance in results of the two models is apparent. This is reassuring as it indicates that the same conclusions on the status of a water system can be drawn regardless whether the screening or a detailed model version is applied. A choice for either model

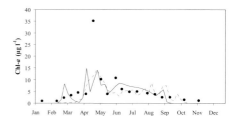

Fig. 11. Comparison of the calculated chlorophyll-*a* concentrations resulting from the 0-D BLOOM model (solid line) and the 2-D GEM model (broken line) for the location Wissekerke in the Oosterschelde. Dots indicate the observations.

version can thus be made based upon the purpose and the availability of data.

In general it is recommended to apply a screening model as described here during an initial phase to assess the present conditions and the sensitivity to different pressures. Occasionally this analysis may even be sufficient for instance in water bodies where light remains controlling even if nutrient concentrations are reduced down to the maximum attainable level. Whenever additional information on individual sources or an enhanced level of detail on physical or biological processes is considered to be of importance, it is recommended to set-up a 2- or 3-D eco-hydrodynamic model.

Acknowledgements

The authors wish to thank Tony Minns and Jos van Gils for critically reviewing the first version of the manuscript. We also thank Anouk Blauw for contributing to the work for the Dutch locations and Alessandro Carletti of the JRC in Ispra for providing the data for the Miramare case. We thank the two anonymous reviewers who made a number of valuable comments to improve the papers. This study is part of the EU Rebecca project, which is jointly funded by the EU 6th Framework Programme (SSPI-CT-2003-502158) and the research programmes of the 19 collaborating organisations.

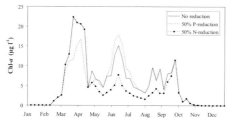

Fig. 10. Predicted chlorophyll-*a* concentrations (μg l⁻¹) at the location Dreischor. The solid line represents actual situation for 1998 without nutrient reduction. The broken line represent the predictions with 50% P-reduction and the broken line with dots are for 50% N-reduction compared to the actual situation.

References

Baretta, J.W., Ebenhöh, W., Ruardij, P., 1995. The European regional seas ecosystem model, a complex marine ecosystem model. Neth. J. Sea Res. 33, 233–246.

Blauw, Anouk N. , Hans F. J. Los, Marinus Bokhorst, and Paul L. A. Erftemeijer., 2009. GEM: a generic ecological model for estuaries and coastal waters. Hydrobiologia DOI 10.1007/s10750-008-9575-x

Danzig, G.B., 1963. Linear Programming and Extensions. Princeton University Press, Princeton N.J.

De Groodt, E.G., Los, F.J., Nauta, T.A., Markus, A.A., DeVries, I., 1992. Modelling cause–effect relationships in eutrophication of marine systems: an integral approach. Marine Eutrophication and Population Dynamics. In: Colombo, et al. (Ed.), Proceedings of the 25th EMBS. Olsen and Olsen, Fredensborg, Denmark. ISBN: 87-85215-19-8: 55–58.

De Jonge, V., 1992. Physical processes and dynamics of micro-phytobenthos in the Ems estuary (The Netherlands). PhD thesis University of Groningen, The Netherlands.

De Vries, I., Duin, R.N.M., Peeters, J.C.H., Los, F.J., Bokhorst, M., Laane, R.W.P.M., 1998. Patterns and trends in nutrients and phytoplankton in Dutch coastal waters: comparison of time-series analysis, ecological model simulation and mesocosm experiments. ICES J. Mar. Sci. 55, 620–634.

Di Toro, D.M., Fitzpatrick, J.J., Thomann, R.V., 1971. A dynamic model of the phytoplankton population in the Sacramento San Joaquim Delta. Adv. Chem. Ser. 106, 131–180.

Di Toro, D.M., Thomann, R.V., O'Connor, D.J., Mancini, J.L., 1977. Estuarine phytoplankton models—verification analysis and pre-liminary applications. In: Goldberg, E.D. (Ed.), The Sea. Marine modeling, New York.

Ebenhöh, W., Baretta-Bekker, J.G., Baretta, J.W., 1997. The primary production module in the marine ecosystem model ERSEM II, with emphasis on the light forcing. J. Sea Res. 38, 173–193.

European Commission, 2000. Directive 2000/60/EC of the European Parliament and of the Council of 23 October 2000 establishing a framework for Community action in the field of water policy. Off. J., L 327, 1–73.

Fulton, E.A., Smith, A.D.M., Johnson, C.R., 2004. Biogeochemical marine ecosystem models: I. IGBEM—a model of marine bay ecosystems. Ecol. Model. 174, 267–307.

Jahnke, J., 1989. The light and temperature dependence of growth rate and elemental composition of Phaeocystis globosa Scherffel and P. pouchetti (Har.) Lagerh. in batch cultures. Neth. J. Sea Res. 23 (1), 15–21.

Janse, J.H., 2005. Model studies on the eutrophication of shallow lakes and ditches. PhD-thesis Wageningen University. 376 p.

Lancelot, C., Billen, G., Sournia, A., Weisse, T., Colijn, F., Veldhuis, M.J.W., Davies, A., Wassman, P., 1987. Phaeocystis blooms and nutrient enrichment in the continental coastal waters of the North Sea. Ambio 16, 38–47.

Lancelot, C., Hannon, E., Becqevort, S., Veth, C., De Baar, H.J.W., 2000. Modeling phytoplankton blooms and carbon export production in the Southern Ocean: dominant controls by light and iron in the Atlantic sector in Austral spring 1992. Deep-Sea Res. I 47, 1621–1662.

Los, F.J., 1991. Mathematical Simulation of Algae Blooms by the Model BLOOM II Version 2. WL | Delft Hydraulics, Delft, p. T68.

Los, F.J., 2005. An algal biomass prediction model. In: Louks, D.P., Van Beek, E. (Eds.), Water Resources Systems Planning and Management—An Introduction to Methods, Models and Applications. UNESCO.

Los, F.J., Bokhorst, M., 1997. Trend analysis Dutch coastal zone. New Challenges for North Sea Research. Zentrum for Meeres- und Klimaforschung, University of Hamburg, pp. 161–175.

Los, F.J., Brinkman, J.J., 1988. Phytoplankton modelling by means of optimization: a 10-year experience with BLOOM II. Verh. - Int. Ver. Theor. Angew. Limnol. 23, 790–795.

Los, F.J., Smits, J.G.C., De Rooij, N.M., 1984. Application of an Algal Bloom Model (BLOOM II) to combat eutrophication. Verh. - Int. Ver. Theor. Angew. Limnol. 22, 917-923.

Peeters, J.C.H., Los, F.J., Jansen, R., Haas, H.A., Peperzak, L., De Vries, I., 1995. The oxygen dynamics of the Oyster Ground, North Sea. impact of eutrophication and environmental conditions. OPHELIA 42, 257-288.

Post, A.F., De Wit, R., Mur, L.R., 1985. Interactions between temperature and light intensity on growth and photosynthesis of the cyanobacterium Oscillatoria agardhii. J. Plankton Res. 7, 487-495.

Riegman, R., 1996. Species composition of harmful algal blooms in relation to macronutrient dynamics, In: Allan D. Cembella and Gustaaf M. Hallegraeff, Physiological Ecology of Harmful Algal Blooms, Donald M. Anderson, NATO ASI Series, vol. 41, Springer Verlag, 1996.

Rogers, S.I., Lockwood, S.J., 1990. Observations of coastal fish fauna during a spring bloom of Phaeocystis pouchetii in the Eastern Irish Sea. J. Mar. Biol. Assoc. U.K. 70, 249-253.

Riegman, R., Noordeloos, A.A.M., Cadee, G, 1992. Phaeocystis blooms and eutrophication of the continental coastal zones of the North Sea. Mar. Biol. 112, 479-484.

Riegman, R., De Boer, M., De Senerpont Domis, M., 1996. Growth of harmful marine algae in multispecies cultures. J. Plankton Res. 18 (10), 1851-1866.

Van der Molen, D.T., Los, F.J., Van Ballegooijen, L., Van Der Vat, M.P., 1994. Mathematical modelling as a tool for management in eutrophication control of shallow lakes. Hydrobiology 275/276, 479-492.

Van Duin, E.H.S., Blom, G, Los, F.J., Maffione, R., Zimmerman, R., Carl, F.C., Dorth, M., Best, E.P.H., 2001. Modeling underwater light climate in relation to sedimentation, resuspension, water quality and autotrophic growth. Hydrobiology 444, 25-42.

Van Gils, J., Tatman, S., 2003. Light penetration in the water column. MARE WL2003001, Z3379.

Villars, M.T., DeVries, I. (Eds.), 1998. Report of the ASMO modelling workshop on eutrophication issues. 5-8 November 1996. The Hague, The Netherlands. OSPAR Commission. 86 pp.

Vollenweider, R., 1975. Input-output models, with special reference to the phosphorus loading concept in limnology. Schweiz. Z. Hydrol. 37, 53-84.

WL | Delft Hydraulics/MARE, 2001. Description and model representation of an artificial island and effects on transport and ecology. Delft Hydraulics Report WL2001103 Z3030.10. Available at www.Flyland.nl, search for Author "MARE Combinatie".

Zevenboom, W., Mur, L.R., 1981. Ammonium-limited growth and uptake by Oscillatoria agardhii in chemostat cultures. Arch. Microbiol. 129, 61-66.

Zevenboom, W., Mur, L.R., 1984. Growth and photo synthetic response of the cyanobacterium Microcystis aeruginosa in relation to photoperiodicity and irradiance. Arch. Microbiol. 139, 232-239.

Zevenboom, W., Post, A.F., Van Hes, U., Mur, L.R., 1983. A new incubator for measuring photosynthetic activity of phototrophic microorganisms, using the amperometric oxygen method. Limnol. Oceanogr. 28, 787-791.

Chapter 4

Mathematical modelling as a tool for management in eutrophication control of shallow lakes

D. T. van der Molen, F. J. Los, L. van Ballegooijen &
M. P. van der Vat, 1994

Hydrobiologia 275/276: 479-492

Hydrobiologia **275/276**: 479-492, 1994.
E. Mortensen et al. (eds), Nutrient Dynamics and Biological Structure in Shallow Freshwater and Brackish Lakes.
©1994   *Kluwer Academic Publishers. Printed in Belgium.*

Mathematical modelling as a tool for management in eutrophication control of shallow lakes

D. T. van der Molen[1], F. J. Los[2], L. van Ballegooijen[1] & M. P. van der Vat[2]
[1] *Institute for Inland Water Management and Waste Water Treatment, Lelystad, The Netherlands;*
[2] *Delft Hydraulics, Delft, The Netherlands*

Key words: modelling, eutrophication, Lake Veluwe, water management

Abstract

The eutrophication model DELWAQ-BLOOM-SWITCH is developed to be a functional tool for water management. Therefore it includes nutrients, algal biomass and composition as well as water transparency. A module describing the interaction between water and bottom gives the model the flexibility to deal with measures, such as a decrease of the external phosphorus loading and flushing with water differing in composition from the lake water. This paper focuses on the functional aspects of the model, the results of an application on Lake Veluwe, The Netherlands, and the implications for water management.

With one set of coefficients DBS reproduces the most important characteristics of Lake Veluwe for a period of two years before measures (reduction of the external loading and flushing during the winter months) and eight years after the measures. The phosphorus concentration decreased and became growth limiting for algae instead of nitrogen and light. Both in measurements and modelling results, the algal composition changed from blue-green algae dominance to green algae and diatom dominance. Lake Veluwe had a relatively short transient phase after reduction of external loading, because high nitrate concentrations in the flushing water inhibited a long period with high phosphorus releases from the bottom.

Model calculations were carried out to investigate the effects of fish stock management and optimization of flushing. Both measures are promising.

Introduction

Eutrophication is one of the main topics in water quality management of freshwater lakes since the last few decades. An excessive amount of nutrients causes abundant algal growth, which in turn reduces the natural and economical value of lakes. Several measures are taken to combat eutrophication (e.g. Cullen & Forsberg, 1988; Boers & Van der Molen, 1993), but only in a few cases the desired quality targets are achieved. After a reduction of the external loading, the internal loading from the bottom is able to supply algae with phosphorus until it reaches equilibrium with the lowered external loading (Lijklema, 1986; Sas, 1989; Van der Molen & Boers, 1994); according to a survey of 27 Danish lakes this may take 4-16 years (Jeppesen *et al.,* 1992). Therefore, additional measures are introduced. Examples are dredging or chemical treatment of sediments (e.g.

97

Bjork, 1985; *Cooke et al.*, 1986; Boers *et al.*, 1992) and fish stock management (e.g. Benndorf, 1990).

Eutrophication is a complicated problem with several interrelated aspects. Important factors are nutrient levels, biomass and composition of algae and water transparency. This makes it difficult for a water quality manager to choose among alternative management scenarios to improve the water quality. A sound decision requires a great deal of knowledge concerning the system. In addition to measures, application of models provides a possible way of dealing with these complicated systems.

Simple, empirical models (e.g. Vollenweider, 1975) and expert opinions can give an indication of the results that may be expected from measures, but have serious shortcomings for understanding lake ecosystems (Van der Molen & Boers, 1994). Hence, the earliest operational deterministic eutrophication models were already developed in the early seventies (DiToro *et al.*, 1971). The transient phase of a lake after reduction of the external loading can only be described by dynamical models, with inclusion of the nutrient dynamics in the bottom. Besides nutrients, algal biomass and species composition as well as water transparency are constraints for a management eutrophication model. Considering the significance of all these pieces of information, the relations between them and the necessity to create an operational tool for water managers, we have developed the DELWAQ-BLOOM-SWITCH (DBS) model system.

In this paper we present the model. Specifying all model equations and process coefficients would require too much space and is beyond the scope of this paper. Therefore we shall only mention the main processes and coefficients and focus on the results of application and the possibilities offered by DBS. We will illustrate this using the application of the model to Lake Veluwe, The Netherlands, as an example. Lake Veluwe is a shallow, eutrophic lake in which from 1979 onwards several measures have been taken. Optimization of the current measures and application of additional measures are points of interest for water managers. Relevant data are available from

before and after the measures, making the construction of reliable input possible as well as validation of the results.

Modelling approach

General set-up

DBS includes modules for physical-, (biochemical- and biological processes (Los, 1992; Los, 1993). DBS is one of several modelling systems using the general DELWAQ framework (Postma, 1988). DELWAQ serves four major purposes:

- it contains the physical schematization,
- it calculates transport of substances as a function of advective and dispersive transport, processes and loads,
- it accumulates fluxes and computes resulting concentrations for each time-step; a large number of numerical solution schemes are included in DELWAQ and can be selected by the modeller constructing a DELWAQ based model,
- it produces output in a standardized way, which can easily be processed further, creating graphs and statistics.

The actual water system is represented within DBS by means of one or several segments. Segments are computational units, which can be considered homogenous with respect to the processes included in the model. Segments can be arranged 0-, 1-, 2- or 3 dimensionally. Water transports between segments must, however, be known in advance. For complicated systems with many segments these transports might be derived from dedicated models. For simple (lake) systems the user might simply specify the water flows to and from the system. Internally DBS multiplies fluxes with concentrations to obtain masses across internal and external boundaries.

Several modules have been developed and tested previously. Computations concerning phytoplankton are performed by the BLOOM II model, which has been shown to be accurate and reliable in both fresh and salt water systems (Los *et al.*, 1984; Los *et al.*, 1988; Los, 1991).

Included within DBS is also a light module which computes the components of the total extinction caused by non-living suspended matter as well as the Secchi depth (transparency) of the water (Buiteveld, 1990). The contribution of phytoplankton to the total extinction is computed by BLOOM II. The module that describes the interactions between water and bottom (SWITCH) was developed specifically for DBS, although it was also tested in a stand alone version (Smits & Van der Molen, 1993). To operate the model system we have simultaneously developed a 'user-friendly' interface.

DBS creates a detailed report for each individual segment containing all concentrations plus additional information (i.e. on limiting factors, light regime, depth, etc.) and the size of each flux through the system. Output can be inspected in the form of graphs and tables.

Kinetic modules

DBS considers the cycles of carbon, oxygen and three nutrients: nitrogen, phosphorus and silicon. Each of these elements can appear in several different compartments:

- inorganic nutrients (nitrogen and phosphorus are subdivided into several fractions of inorganic nutrients),
- dead organic material (two groups of detritus),
- phytoplankton (several groups),
- grazers (zooplankton and mussels),
- bottom algae (diatoms),
- detritus at the bottom-water interphase (we call this 'complex' detritus as many organic, inorganic, biochemical and biological processes occur in a complicated way within this thin boundary layer),
- bottom detritus,
- inorganic bottom nutrients.

With the exception of grazers the amount in each compartment is computed by DBS. Often this is done by solving a series of differential equations numerically. Several equations, however, are solved analytically, some processes are assumed to be at steady state and the phytoplankton pro-

duction is calculated using an optimization technique. Chemical, physical and biological processes result in fluxes from one compartment to another. The basic cycles for the three nutrients are identical. The most important fluxes within DBS and some additional information are presented in (Fig. 1).

DBS considers three main phytoplankton groups: diatoms, green algae and blue-green algae. Each of these groups is further differentiated into types, according to their growth-limitation. To compute the amount of each type, an optimization technique is used to maximize the total net production of all types in a certain period of time, consistent with the environmental conditions and the existing biomass levels. The environmental conditions include water temperature, solar radiation, depth, background extinction and the concentrations of available nutrients. Each type in the model is characterized by a different set of physiological parameters such as growth and mortality rates and nutrient requirements. The model selects the optimal composition of phytoplankton species, based upon their physiological characteristics and the available resources.

Although formulations are available to model zooplankton (e.g. Jorgensen, 1983), results are rarely validated with observations. Therefore, the biomass of grazers is an input to DBS. It can be provided in the form of observed (or any other series of) numbers or in the form of a function. The model does check whether the computed biomass of phytoplankton and various forms of detritus can sustain the amount of grazers specified in the input. If not, it will reduce the amount of grazers to the highest possible level that can be supported by the system.

Two types of dead organic matter are distinguished, detritus and a pool named 'other organic nutrients'. The mineralization of both types of detritus results in the regeneration of nutrients. Mineralization is described as a first order process and the mineralization rate is dependent on the nutrient to carbon ratio of the material. Higher ratio's result in increased mineralization rates. 'Other organic nutrients' mineralize very slowly and sedimentation and burial of this fraction are

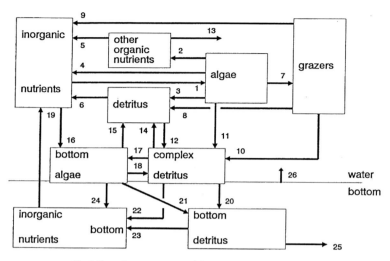

Fig. 1. The main compartments and fluxes in DBS; numbers refer to:

1. Uptake of inorganic nutrients by phytoplankton. DBS assumes that ammonia is (completely) depleted first before the phytoplankton switch to nitrate.
2. Phytoplankton mortality flux to other organic substances.
3. Phytoplankton mortality flux to detritus.
4. Phytoplankton mortality flux to inorganic nutrients (autolysis). The sum of fluxes 2 + 3 + 4 equals the natural phytoplankton mortality.
5. Mineralization of other organic substances. It is assumed that this process is very slow and that it only takes place in the water column.
6. Mineralization of detritus. The actual rate depends not only on temperature, but also on the composition of detritus. It is maximal when there are sufficient nutrients relative to carbon.
7. Mortality of phytoplankton due to grazing.
8. Production of detritus due to zooplankton grazing (faeces).
9. Release of inorganic nutrients by grazers.
10. Production of complex detritus due to grazing by mussels (faeces).
11. Sedimentation of phytoplankton.
12. Sedimentation of detritus in the water phase.
13. Sedimentation of other organic nutrients, which become refractory.
14. Resuspension of complex detritus.
15. Resuspension of bottom phytoplankton.
16. Uptake of inorganic nutrients by bottom algae. This process is activated if process 17 cannot supply enough nutrients.
17. Uptake by bottom algae of dissolved nutrients released through mineralization of complex detritus in the water-bottom interphase.
18. Mortality flux of bottom algae to complex detritus.
19. Total flux of inorganic bottom nutrients to the water. This flux is influenced by all processes included within the bottom module of DBS.
20. Burial of complex detritus.
21. Burial of bottom algae.
22. Mineralization of complex detritus. For conceptual reasons this flux is handled by the bottom module; nutrients are not directly released into the surface water.
23. Mineralization of bottom detritus.
24. Part of bottom algae mortality flux to inorganic nutrients (autolysis).
25. Burial of bottom detritus. This process actually removes nutrients from the system.
26. Sediment growth. This results in dilution of the contents in sediment; usually it is set zero.

considered as removal of nutrients from the system.

For phosphorus in the overlying water we have included adsorption and desorption of phosphate, as well as degradation and sedimentation of adsorbed phosphorus. Nitrification and denitrification are assumed to take place in the bottom only.

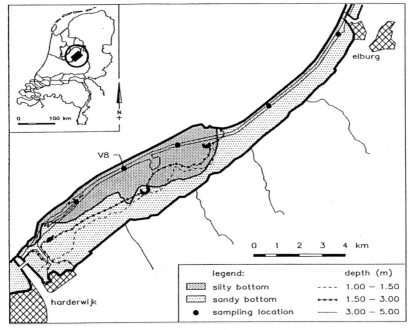

Fig. 2. Lake Veluwe.

The bottom is divided into an 'active' and an 'inactive' layer. The active layer is usually assumed to be 0.1 m and is divided into four layers: the aerobic layer, the denitrifying layer, the upper reduced layer and the lower reduced layer. Layers can change in thickness according to the circumstances; the thickness of the aerobic layer and the denitrifying layer are derived from the differential equations of oxygen and nitrate in the bottom, respectively. The mineralization rate of bottom detritus is lower for the reduced layers than for the oxidized layer. Nitrate is formed in the oxidized layer by first order nitrification of ammonium. The removal of nitrogen by denitrification of nitrate in the denitrifying layer is also described as a first order process. For phosphorus this bottom module includes adsorption and desorption processes (according to the Langmuir isotherm) and precipitation and dissolution of phosphorus minerals (first order kinetics with the difference between the actual concentration and the saturation concentration as a driving force). These processes also vary over the four bottom layers. Hence the phosphorus flux to the surface water varies significantly as a function of the redox situation in the bottom.

Model application

Case study Lake Veluwe

Lake Veluwe (The Netherlands) is an artificial lake which was formed when the Flevoland polder was created in 1956 (Hosper & Meijer, 1986). The lake has a total surface area of $32.8 \, 10^6 \, m^2$; approximately half of the lake is very shallow (depth less than 1 m) with a sandy sediment; the other half is deeper (average depth 2 m) with a more silty sediment (Fig. 2). The water is assumed to be well mixed, so the system is modelled as one segment.

In the second half of the sixties the lake quality deteriorated, because of increased nutrient loading. This resulted in high chlorophyll *a* concentrations and an almost permanent bloom of blue-green algae (predominantly *Oscillatoria agardhii*) from 1970 onwards. In 1979 the following restoration measures were taken: from February onwards external phosphorus loading was reduced from 3 to $1.0\text{-}1.5 \, gP \, m^{-2} \, y^{-1}$ by phosphorus elimination at the sewage treatment plant discharging its effluent to the lake. Furthermore, flushing the lake with water from the Flevoland polder (poor in algae and phosphorus, but rich in calcium and nitrate) during the winter, decreased the retention time of dissolved compounds from 0.35 to 0.15 years. In the second half of the eighties, the lake was flushed in summer as well, decreasing the retention time of dissolved compounds from 0.50 to 0.25 years (Jagtman *et al.*, 1992).

The performance of DBS was checked for the Lake Veluwe case for the years 1978 through 1987. As no data on loadings were available for 1984 and from 1988 onwards, we ran the model separately for two periods: 1978 through 1983 and 1985 through 1987. The initial conditions for 1978 were obtained from measurements; the initial conditions for 1985 were taken from the computations at the end of 1983. We used a single set of model coefficients for the entire period of time. So, effectively DBS was applied to the entire ten year period.

Almost all coefficients for the water phase had been obtained during previous studies for many different lakes, including Lake Veluwe. These studies include laboratory and field work (Zevenboom *et al.*, 1982; Brinkman & Van Raaphorst, 1986) as well as model studies in which similar formulations were used (Los *et al.*, 1984; Los *et al.*, 1988; Los, 1991). For the recently developed bottom module minor modifications of the standalone calibration-/-validation, as reported by Smits & Van der Molen (1993), were necessary. As this paper focuses on the results of the model application and the implications for management, only the values of the most important coefficients are given here (Table 1).

To illustrate the possibilities this model offers for management purposes, two scenario's were studied in addition to the reference case: fish stock

Table 1. Values of some important coefficients of the model DBS (rates at 20 °C; DM = dry matter, PW = pore water).

Coefficient	Value/range	Unit
Growth rate algae	0.68-1.36	d^{-1}
Mortality rate algae	0.16-0.23	d^{-1}
Respiration rate algae	0.05-0.12	d^{-1}
C:P ratio algae	42.1-88.9	gg^{-1}
N:P ratio algae	3.64-8.00	gg^{-1}
C:hlorophyll ratio algae	25.0-50.0	gg^{-1}
Mineralization rate detritus C, N and P	0.12-0.18	d^{-1}
Sedimentation rate detr.	0.05	md^{-1}
Maximum nitration rate zooplankton	1.5	$m^{-3}g^{-1}Cd^{-1}$
Diffusion coefficients O_2, PO_4, NO_3, NH_4	$4.20\text{-}9.33 \, 10^{-5}$	m^2d^{-1}
Adsorption capacity bottom	0.4-0.8	$gPkg^{-1}DM$
Ortho P saturation concentration bottom	0.05	$gPm^{-1}PW$
Mineralization rate oxidized bottom	0.03	d^{-1}
Nitrification rate bottom	50.0	d^{-1}
Denitrification rate bottom	50.0	d^{-1}

management and optimization of flushing with polder water. Fish stock management might be promising, because the phosphorus loading and concentration is sufficiently low to expect success of this intervention (Benndorf, 1990; Jeppesen *et al.*, 1990). Reconsideration of flushing is necessary, because from 1985 onwards the phosphorus concentration in the lake approached the concentration in the flushing water. Optimization was established by flushing only when the phosphorus concentration in the other inlet water was significantly higher than the concentration in the flushing water. The water level was maintained.

Results

In the two years (1978 and 1979) prior to the onset of the sanitation measures both computed and observed levels of chlorophyll range between 100 mg m^{-3} in winter and about 400 mg m^{-3} in summer (Fig. 3). There is a significant drop in 1980 and a more gradual decline during the remainder of the 1980s. During the last year (1987) computed and observed chlorophyll levels are well below 100 mg m^{-3} all year long. The agreement between computed and observed concentrations varies to some extent. For example computed levels for 1983 fall below those for 1982, but the opposite is true in the measurements. Still

it may be concluded that the model does reproduce the trend over this ten year period of time sufficiently well.

In addition to biomass both the observed and computed (Fig. 4) phytoplankton species composition changed dramatically during this ten year period. Before 1980 blue-green algae, mainly *Oscillatoria agardhii*, dominated all year long, except for a short period of diatom dominance in spring. In the early summer of 1982 green algae replaced the blue-greens for the first time, but blue-green algae returned later and remained dominant in 1983 and 1984. The computed result for 1983, suggesting a summer dominance of green algae, is incorrect. From 1985 to 1987, however, diatoms and green algae were dominant and blue-green algae (re)appeared infrequently. For two representative years (1985 and 1986) the computed species are presented with measurements (Fig. 5). The model does not reproduce the reappearance of diatoms in autumn 1985 correctly. However, in general the computed species composition agrees well with the observed trend.

As a consequence of the improved light conditions near the bottom, bottom algae increased dramatically from 1985 onwards. The model computes that from 1985 to 1987 almost 20 percent of the algal carbon can be found on the bottom. Measurements based on chlorophyll indicate that in the shallow part this can be more than 50 percent (Van der Molen & Helmerhorst,

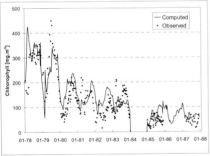

Fig. 3. Computed and measured chlorophyll in Lake Veluwe.

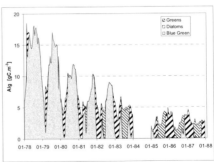

Fig. 4. Computed algal composition in Lake Veluwe.

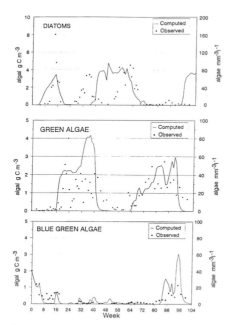

Fig. 5. Computed (left axis) and observed (right axis) algal species in Lake Veluwe for 1985-1986.

1991), meaning that model result is probably an underestimation.

The decrease in algal biomass is mainly due to a dramatic decrease in the amount of phosphorus in the water phase (Fig. 6). In 1978 and 1979 the available amount of phosphorus exceeded the requirements of phytoplankton. Algal growth was usually limited by lack of nitrogen (summer) and light energy. Starting in 1980, the levels of phosphorus became progressively more restraining to the phytoplankton community, although nitrogen limitations still occurred in the summer. During the last three years when phosphorus levels decreased to about 0.1 g m^{-3}, phosphorus was in fact the only important limiting factor. Observed and computed levels of ortho phosphorus (not shown here) were mostly equal to zero. As for

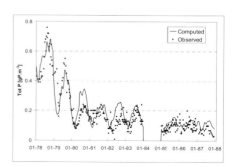

Fig. 6. Computed and measured total phosphorus in Lake Veluwe.

chlorophyll, the model results agree sufficiently well with the observations. The largest deviation occurs in 1980: the year immediately following the onset of the sanitation measures. The real system reacted even more dramatically than the model in this transient year.

The phosphorus content of the bottom changed only slowly, but varied within the limits of the measurements (not shown). However, sediment data are scarce and difficult to interpret (Smits & Van der Molen, 1993).

Computed levels of nitrate (Fig. 7) and ammonia (Fig. 8) agreed well with observations. Notice that the period of nitrate depletion during summer became shorter during this ten year period, indicating that nitrogen became progressively less im-

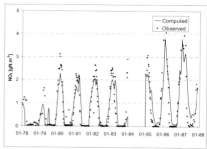

Fig. 7. Computed and measured nitrate in Lake Veluwe.

portant as a limiting factor. As was the case for phosphorus there was also a decreasing trend in the computed and observed values for Kjeldahl nitrogen (total nitrogen minus nitrate) during this ten year period, though less pronounced (Fig. 9). This can be explained from the fact that some, but not all fractions of Kjeldahl nitrogen declined significantly. In later years less nitrogen was present as a component of phytoplankton and detritus, but the amounts in 'other organic nitrogen' and ammonia generally remained constant. Before the measures, denitrification was limited by nitrate production from nitrification in the sediment. As a result of the flushing of Lake Veluwe with nitrate-rich water, nitrate became an important fraction of total nitrogen. From 1980 onwards the model computed increased denitrification as a result of the enhanced nitrate diffusion into the sediment.

In the results of the model application we have seen that DBS is able to reproduce some important characteristics of the system. Similar or even better agreement with measurements exists for other substances such as silicon and oxygen. An accurate reproduction of concentrations can, however, also be the result of compensating errors in individual fluxes. This is a general problem with models: many concentrations can be validated, but few fluxes can. Moreover, most of these processes cannot be measured very accurately. For 1983 and the period from 1987 on-

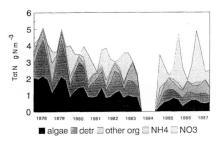

Fig. 9. Computed nitrogen fractions in Lake Veluwe (three monthly averaged).

wards, infrequent measurements of the phosphorus release from the bottom were available (Fig. 10). The phosphorus release was measured under laboratory conditions using the continuous flow system with the temperature ranging from 10-21 °C (Brinkman & Van Raaphorst, 1986; Boers & Van Hese, 1988). The measurements are in the same range as the model results, although the maximum value in 1983 is not reproduced. Both measurements and simulation show a decrease in release rates. The calculated sediment oxygen demand is also comparable with measurements in 1983 and 1987 (not shown).

The effect of fish stock management was simulated by increasing the zooplankton biomass.

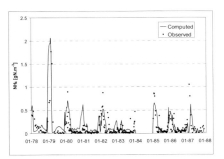

Fig. 8. Computed and measured ammonium in Lake Veluwe.

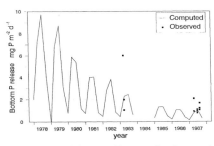

Fig. 10. Computed (three monthly averaged) and measured phosphorus release from the bottom in Lake Veluwe.

Although the zooplankton biomass was increased by a factor of 4, the effective increase of zooplankton grazing was only a factor of 2.5, due to food limitation. No preference for certain algal species was included. Zooplankton biomass compares well with data from Boers *et al.* (1991), although after removal of 75% of the fish stock in Lake Wolderwijd, The Netherlands, no increase of the grazing capacity was observed (Meijer *et al,* 1994). The optimized flushing resulted in a significant decrease of the total yearly water load, although there was an increase in the spring.

The results of the simulation offish stock management showed that both nitrogen and phosphorus concentrations were lower when compared to the reference calculations and, consequently, chlorophyll was decreased and transparency increased. Biomass of blue-green algae diminished as a consequence of the increased grazing. About 75% of algal bound carbon was present in bottom algae. This compares well with measurements in a comparable lake after fish stock management (Van der Molen & Helmerhorst, 1991).

The results from the optimized flushing regime calculations indicate the importance of the phosphorus concentration in the inlet water. Dilution with the inlet water in periods when the phosphorus concentration was high in the lake decreased the total algal biomass, but the percentage of blue-green algae increased slightly. The percentage of bottom algae did not change.

Discussion

An important phenomenon is the change in algal species dominance from blue-greens to greens and diatoms. According to the model, this is not a direct but an indirect result of declining phosphorus levels. Blue-green algae require less phosphorus per unit of biomass than green algae, so one might expect them to remain dominant as phosphorus becomes increasingly limiting. The decrease in biomass, however, also results in improved light conditions and hence in higher potential growth rates of all species and of green

algae and diatoms in particular (due to the form of their growth versus light relationship). Between 1982 and 1985 the growth rates of green algae and diatoms become sufficiently high to displace the blue-green algae, but competition was intense. The additional reduction in available phosphorus and hence biomass in 1985-1987 was sufficient to provide greens and diatoms a clear advantage over blue-greens. Still, the balance between blue-green algae on one hand and green algae and diatoms on the other is unstable. Initial and climatic conditions, e.g. ice coverage, were also found to be important (Jagtman *et al.,* 1992).

The biomass of bottom algae is possibly underestimated, although their increasing importance is reproduced in the model results. Probably the improved light conditions near the bottom and relatively high availability of nutrients are favourable for this group of algae.

The decrease of algal biomass in the waterphase results in better light conditions. On the other hand, green algae and diatoms have a higher turn-over rate and produce detritus which degrades more slowly, hence an increase of detritus can be expected (Gunnison & Alexander, 1975). The net effect of an increased detritus concentration and a decreased chlorophyll concentration was a minor increase in transparency (Fig. 11).

As already mentioned, the reduction of the concentration of available phosphorus is the main cause for the decline of phytoplankton biomass. In Table 2 we distinguish the main processes within the waterphase and fluxes across the bottom-water interphase. In 1978 the external loading was high, just as in the previous years. Losses of phosphorus to the bottom exceeded the release from the bottom: the bottom became a sink for phosphorus. The retention of phosphorus (quotient of the external loading minus outflow and the external loading) was approximately 55%. The system began to change in 1979. The external loadings to the lake declined and the amount of phosphorus in the water decreased significantly. Interesting enough, the net exchange with the bottom was small: yearly averaged retention was less than 10%, while in summer the retention became negative due to a relatively high

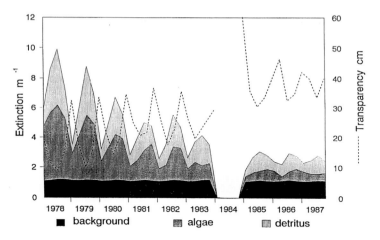

Fig. 11. Computed extinction (cumulative) and transparency in Lake Veluwe (three monthly averaged).

internal loading. The organic flux to the bottom was still high, but inorganic sedimentation was much lower than in 1978.

In the period 1980 through 1983 the retention increased to 40%. Notice, that the inorganic sedimentation of phosphorus was less than 10% of the organic sedimentation. The trends for the 1980-1983 period continued during the 1985-1987 period. The phosphorus release from the bottom decreased by a factor of 7 compared to

Table 2. Main processes (yearly averaged) of the phosphorus cycle (direct losses from the bottom such as seepage and burial to deeper layers are omitted).

Process rate [mg P m^{-2} d^{-1}]	Time period			
	78	79	'80-'83	'85-'87
External loading	8.7	4.5	4.2	4.2
Outflow	-3.9	-4.1	-3.0	-2.5
To bottom (inorg)	-4.1	-0.8	-0.2	-0.2
To bottom (org)	-7.1	-5.0	-3.4	-2.2
From bottom	5.8	4.6	2.3	0.8
Total water	-0.6	-0.8	-0.1	+ 0.1
Total bottom	-5.4	-1.2	-1.3	-1.6

1978 and was also small compared to the external loading. Since the inorganic phosphorus sedimentation during these periods was insignificant, the bottom processes were dominated by the mineralization of settled and buried detritus. The computed retention was about 40%. This is less than the value derived from the computation with mass balance equations, because the phosphorus concentration and consequently the outflow were slightly overestimated.

The phosphorus release from the bottom decreased more than was proportional to the external loading. The model computes an increase of the thickness of the oxidized layer as a result of the decreased sedimentation of organic matter and the bottom uptake of nitrate from the overlying water. So, as a consequence of the flushing, the recovery of Lake Veluwe took place in a few years, which is short when compared to many other lakes (e.g. Cooke *et al.,* 1986; Sas, 1989; Jeppesen *et al.,* 1992). Therefore, additional measures to reduce the release of phosphorus from the bottom, such as dredging and chemical fixation of phosphorus, do not seem useful in the case of Lake Veluwe.

More likely, effects may be expected from a further decrease of the external phosphorus loading, since phosphorus is the growth limiting factor. Two additional measures are examined in this paper: fish stock management and optimization of the flushing rate. Increased grazing, assumed to be the result of the fish stock management, resulted in a lowered algal biomass and consequently in a decreased detritus concentration. This caused a remarkable increase of transparency. Transparency was almost doubled and was limited by particular, inorganic matter. As this is not included in the model, we assumed that this was not influenced by the measures. The increased algal turn-over resulted in increased organic sedimentation, which was favourable for the bottom algae. However, these results are partly hypothetical. Meijer *et al.* (1994) showed that a 75% removal of the fish stock in the large and shallow neighbouring Lake Wolderwijd resulted in a significant decrease in primary production, while grazing only increased slightly. Meijer *et al.* (1994) suggest that a high zooplankton biomass could not sustain due to predation by the mysid shrimp Neomysis integer and that the lowered primary production was due to reduced internal loading. This clearly illustrates the fact that food-web interactions are not yet fully understood and that including such interactions in mathematical models must be done with caution.

Flushing lost its diluting effect to a large extent, but concentration of flushing in periods with relatively high phosphorus concentrations in the other inlet water had positive effects. Both chlorophyll and total phosphorus decreased. Due to the lowered total flushing rate, the percentage of bluegreens increased, because a high water retention time is advantageous to blue-green algae with their low growth and mortality rates. Total nitrogen decreased, as a result of reduced nitrate loading. Low nitrate concentrations resulted in a less oxidized state of the sediment, but due to the lowered organic sedimentation, no increase of the phosphorus release from the bottom was computed. Organic sedimentation was low in the reference calculations when compared to the sedimentation calculated in the simulation of fish

stock management, resulting in nutrient limitations of bottom algae. Only 20 percent of total algal biomass consisted of bottom algae.

Conclusions

The eutrophication model DBS is capable of reproducing the most important water quality indicators. With one set of coefficients the model was able to simulate a 10 year period, in which the external phosphorus loading of Lake Veluwe decreased significantly and flushing was applied. The model calculates concentrations in the bottom and fluxes through the system which are comparable to the available measured data. The coefficients were derived from previous laboratory, field and model studies, except for some bottom coefficients.

The measures taken resulted in decreased algal biomass as well as in a change in the algal composition from blue-green algae dominance towards green algae and diatom dominance. Flushing lowered the phosphorus concentration, but increased the nitrate concentration in the lake. According to the model results, phosphorus became the growth limiting factor instead of nitrogen and light energy.

After the measures the phosphorus release from the bottom strongly decreased, more so than the external loading. Apart from the lowered input of organic material, high nitrate concentrations in the flushing water also caused the oxidized layer of the bottom to become thicker and reduced internal loading. This implies that dredging and phosphorus fixation in the bottom cannot be expected to have much effect.

Therefore calculations were made to study the effects of fish stock management and optimization of the flushing. If an increase of the grazing capacity by a factor of 2-3 can be realized, dissolved nutrient levels, as well as chlorophyll and detritus concentrations lowered significantly. The transparency will increase by nearly a factor of 2. However, reduction of the fish stock in a large lake will probably not always result in an increase of grazing capacity and the chain of events caused by fish stock management measures is still not fully understood. The effect of flushing only when the phosphorus concentration in the inlet water is high, is also clear, though less spectacular. Due to low organic sedimentation it is expected that the lowered total flushing will not impead further improvements of Lake Veluwe, through increased internal loading. These examples point out that DBS can be helpful in understanding the processes in lake ecosystems and that the model is a powerful tool for reviewing various management strategies.

References

Benndorf, J., 1990. Conditions for effective biomanipulation; conclusions derived from whole-lake experiments in Europe. Hydrobiologia 200-201/Dev. Hydrobiol. 61: 187-203. Bjdrk, S., 1985. Lake restoration techniques. In Proceedings
    International Congress Lake Pollution and Recovery. Rome: 281-292. Boers, P. C. M., L. Van Ballegooijen & E. J. B. Uunk, 1991.
    Changes in phosphorus cycling in a shallow lake due to food web manipulations. Freshwat. Biol. 25: 9-20. Boers, P. C. M. & D. T. Van der Molen, 1993. Control of entrophication in lakes: the state of the art in Europe. Eur. Wat. Pollut. Control 3: 19-25. Boers, P. C. M. & O. Van Hese, 1988. Phosphorus release
    from the peaty sediments of the Loosdrecht Lakes (The Netherlands). Wat. Res. 22: 355-363. Boers, P. C. M., J. Van der Does, M. Quaak, J. Van der Vlugt
    & P. Walker, 1992. Fixation of phosphorus in lake sediments using iron(III) chloride: experiences, expectations. Hydrobiologia 233/Dev. Hydrobiol. 74: 211-212.
Brinkman, A. G. & W. Van Raaphorst, 1986. De fosfaathuishouding van het Veluwemeer. Thesis University of Twente, 481 pp. (in Dutch, English summary). Buiteveld, H., 1990. UlTZlCHT-model voor berekening van
    doorzicht en extinktie. Nota 90.058, RIZA, Lelystad, The Netherlands. 33 pp. (in Dutch). Cooke, G. D., E. B. Welch, S. A. Peterson & P. R. Newroth
    (eds.), 1986. Lake and reservoir restoration. Ann Arbor Science, Butterworth Publishers, 392 pp. Cullen, P. & C. Forsberg, 1988. Experiences with reducing
    point sources of phosphorus to lakes. Hydrobiologia 170/ Dev. Hydrobiol. 48: 321-336. DiToro, D. M., J. J. Fitzpatrick & R. V. Thoman, 1971. A
    Dynamic Model of the Phytoplankton Population in the Sacramento San Joaquim Delta. Adv. Chem. Ser. 106: 131-180. American Chemical Society, Washington D.C.

Gunnison, D. & M. Alexander, 1975. Resistance and susceptibility of algae to decomposition by natural microbial communities. Limnol. Oceanogr. 20: 64-70.
Hosper, S. H. & M-L. Meijer, 1986. Control of phosphorus loading and flushing as restoration methods for Lake Veluwe, the Netherlands. Hydrobiol. Bull. 20: 183-194.
Jagtman, E., D. T. Van der Molen & S. Vermij, 1992. The influence of flushing on nutrient dynamics, composition and densities of algae and transparency in Veluwemeer, The Netherlands. Hydrobiologia 233/Dev. Hydrobiol. 74: 187-196.
Jeppesen, E., P. Kristensen, J. P. Jensen, M. Sondergaard, E. Mortensen & T. Lauridsen, 1992. Recovery resilience following a reduction in external phosphorus loading of shallow, eutrophic danish lakes: duration, regulating factors and methods for overcoming resilience. Mem. 1st. ital. Idrobiol. 48: 127-148.
Jeppesen, E., M. Sondergaard, E. Mortensen, P. Kristensen, B. Riemann, H. J. Jensen, J. P. Muller, O. Sortkjaer, J. P. Jensen, K. Christoffersen, S. Bosselmann & E. Dall, 1990. Fish manipulation as a lake restoration tool in shallow, eutrophic, temperate lakes. II Threshold levels, long-term stability and conclusions. Hydrobiologia 200-201/ Dev. Hydrobiol. 61: 219-227.
Jorgensen, S. E., 1983. Modeling the ecology processes. In G. T. Orlob (ed.), Mathematical Modeling of Water Quality: streams, lakes and reservoirs. IIASA-Wiley: 116-149.
Lijklema, L., 1986. Phosphorus accumulation in sediments and internal loading. Hydrobiol. Bull. 20: 213-224.
Los, F. J., 1991. Mathematical simulation of algae blooms by the model BLOOM II, Version 2, Documentation report. Delft Hydraulics, The Netherlands, 113 pp.
Los, F. J., 1992. Procesformuleringen DBS. T542, Waterloopkundig Laboratorium Delft, The Netherlands (in Dutch).
Los, F. J., 1993. Technisch rapport DBS. T542, Waterloopkundig Laboratorium Delft, The Netherlands (in Dutch).
Los, F. J. & J. J. Brinkman, 1988. Phytoplankton modelling by means of optimization: a 10-year experience with BLOOM II. Verh. int. Ver. Limnol. 23: 790-795.
Los, F. J., N. M. de Rooij & J. G. C. Smits, 1984. Modelling eutrophication in shallow Dutch Lakes. Verh. int. Ver. Limnol. 22: 917-923.
Meijer, M.-L., E. H. van Nes, E. H. R. R. Lammens, R. D. Gulati, M. P. Grimm, J. Backx, P. Hollebeek, E. M. Blaauw & A. W. Breukelaar, 1994. The consequences of a drastic fish stock reduction in the large and shallow Lake Wolderwijd, The Netherlands. Can we understand what happened? Hydrobiology 275-276/Dev. Hydrobiol. 94: 31-42.
Postma, L., 1988. DELWAQ Users manual, Version 3.0, Delft Hydraulics, The Netherlands, 190 pp.
Sas, H. (ed.), 1989. Lake restoration by reduction of nutrient loading: expectations, experiences, extrapolations. AcademiaVerlag, St. Augustin, Germany, 497 pp.
Smits, J. G. C. & D. T. van der Molen, 1993. Application of SWITCH, a model for sediment-water exchange of nutri-

ents, to Lake Veluwe in The Netherlands. Hydrobiologia 253/Dev. Hydrobiol. 84: 281-300.

Van der Molen, D. T. & P. C. M. Boers, 1994. Influence of internal loading on phosphorus concentration in shallow lakes before and after reduction of the external loading. Hydrobiologia 275-276/Dev. Hydrobiol. 94: 379-390.

Van der Molen, D. T. & T. H. Helmerhorst, 1991. Bodemalgen in de randmeren. H20 24: 719-724 (in Dutch with an English summary).

Vollenweider, R., 1975. Input-output models, with special reference to the phosphorus loading concept in limnology. Schweiz. Z. Hydrol. 37: 53-84.

Zevenboom, W., A. F. Post & L. R. Mur, 1982. Assessment of factors limiting growth rate of *Oscillatoria agardhii* in hypertrophic Lake Wolderwijd. Limnol. Oceanogr. 27: 39-52.

Chapter 5

Validation of water quality and ecological models

Hans Los and Herman Gerritsen, 1995

IAHR Specialist Forum on Software Validation London, September 1995

IAHR Specialist Forum
on Software Validation
London, September 1995

Validation of water quality and ecological models

Hans Los and Herman Gerritsen

(DELFT HYDRAULICS, p.o. box 177, 2600 MH Delft, The Netherlands)

1.   Model validation: a structured approach

The essence of the model validation approach we follow at DELFT HYDRAULICS is that we focus on the properties the computational model acquires during its conception and development. So, we separate the features resulting from model development from those of the model application, and structure the validation process according to the subsequent steps taken in the modelling :

- *Conceptual Model Validation* addresses the quality of the representation of the natural system or process: do the model equations represent the processes of interest, or perhaps better: which processes are described by the equations given certain parameter ranges?
- *Algorithmic implementation Validation* addresses the conversion of the Conceptual Model into procedures suitable for computation: how accurate, reliable and robust is this implementation?
- *Validation of the Software Implementation* addresses the correctness of the computer implementation. This includes aspects such as its structuring, error checking and handling, efficiency and maintainability.

While the above three address the computational model per se, the following aspect focusses on the application or use of the model:

- *Functional Validation* addresses the quality of the overall representation by the software of a given physical process or phenomenon as represented by a comprehensive set of field data or laboratory data.

We are implementing this concept in designing our test procedures and documentation. Using *objective, quantitative result assessments* whenever possible enables us to come to grips with the quality assurance of whole modelling systems. We can now address the quality of specific model features and make this quality *visible and transferable.*

Present paper

The present paper summarizes the *Functional Validation* of DELFT HYDRAULICS' ecological model DBS (DELWAQ-BLOOM II-SWITCH) on the Rijnland network system of ditches, canals and shallow and deep lakes with regulated water quantity management.

In Section 2 we present the basic characteristics of the model. Subsequent sections address the aims of the validation (Section 3), network aspects (Sections 4-6), technical approach (Section 7), and results (Sections 8-9).

In Section 10 the results are discussed and conclusions are drawn, both regarding the validation result for the selected Rijnland network and the generalization thereof, and regarding the validation approach or procedure.

2. DBS

DBS (DELWAQ-BLOOM II-SWITCH) is an integration of three individual models. It is a combination of the transport model DELWAQ and the phytoplankton model BLOOM II, to which is added the module to simulate water-bottom exchanges (SWITCH). Together, these models calculate all the biological and chemical processes occurring within the water and sediment, and the fluxes of substances occurring within the water phase and between water and sediment.

As a member of the DELWAQ family, DBS is suitable for application to a wide variety of water systems. Because the algae dynamics are a central item in DBS, the model is specifically suited to the study of systems with eutrophication problems. It can be applied to individual lakes, networks of canals and lakes, or even estuaries and seas. It has been used in several already completed and ongoing model studies, and is proposed for use in various new projects as well.

3. Aims of the Validation Study

The validation study primarily addresses the issue of whether the model is *technically* capable of calculating the water quality in a complex network system *(functionality)*. In addition, it considers the aspect of model *usability*. The main users of the DBS model for network applications are e.g. regional waterboards whose principle concern is establishing control policies for water quality management. In this respect, the model must not only be fit for use and technically robust, but must also provide its information in a manner which is helpful for the intended user.

This validation study thus addresses the *following Junctionality-related* questions:

- Is DBS technically capable of calculating water quality in network systems?
- What are the limitations of the model (if any) for application to this type of system?
- What is the correct methodology of approach for DBS modelling of networks?
- Is it possible to define a generic set of model coefficients for use with networks?

4. The Rijnland network

The Rijnland network system was selected as application case for DBS since this network satisfies the requirements on processes and field data availability that lie at the basis of any successful model validation. This includes:

1. Sufficient data available to characterize the network system for modelling purposes (e.g. hydrology, loads, boundary concentrations);
2. Sufficient water quality measurements within the network available for calibration and validation of DBS model results;
3. Maximum variability of characteristics within the network system (e.g. residence time, depth, nutrient availability) to ensure that validation study results would hold for other network systems as well.

Of the two years of data that were available, the data for 1986 was used for calibration, and that of 1989 for validation.

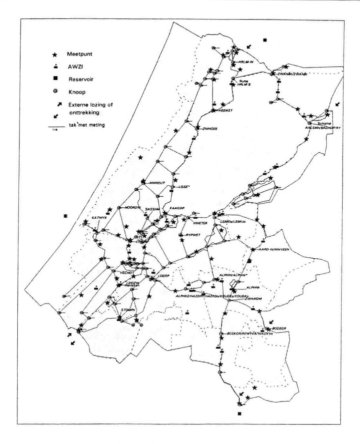

Figure 1: The Rijnland network

5.   Hydrodynamic schematization and segment aggregation

The DBS model schematization is based on that of the hydrodynamic model AAD. The latter uses three basic building blocks, namely branches (surface water bodies), nodes (water exchanges) and links (human controlled exchanges; discharges). Since DBS/DELWAQ is based on the principle of model 'elements', rather than nodes and branches, the AAD schematization cannot be used directly. Instead, the DELWAQ elements must be formed by the aggregation of branches, nodes and links.

A significant aspect of applying and calibrating the DBS model in terms of effort is often the physical schematization and aggregation of the network for the DBS application, although the basic procedure for this is well known. In addition to the basic aggregation of nodes and branches, additional aggregation of (multiple) AAD branches into DELWAQ elements is made based on:

- element residence times and related computational timesteps;
- fully mixed regions (e.g. a well mixed lake consisting of 3 AAD branches should be one DBS element);
- optimization of substance transport from external exchanges.

6.   DBS input parameters

DBS model inputs must be prepared, using field data and literature data. Measured water quality parameters, however, are usually not the same as DBS model substances. For example, measurements may include Total Nitrogen and Kjeldahl Nitrogen concentrations, while DBS model substances include ammonium, nitrate, detritus nitrogen, and refractory organic nitrogen. Thus, boundary concentrations and load values of model substances must be calculated from available measurements. The relation or mapping between the commonly measured water quality parameters and model substances requires explicit attention.

7.   The calibration procedure

A first item of attention concerns the definition of segments to be used for calibration and validation. These segments should be *representative* of the different types of regions within the network, such as lakes, canals, regions with similarities of loads, or similarities of physical characteristics, etc.

Once the calibration locations have been selected, the calibration needs to proceed in a *systematic* manner. The conservative parameter chloride should be the first to be checked and compared to water quality measurements. This serves as a check on mass balance. Significant problems with modeled chloride concentrations may indicate incorrect water flow calculations in AAD, or poor substance transport calculations, possibly due to the aggregation of AAD branches and nodes. After chloride, calculated concentrations of chlorophyll and nutrients can be compared with measurements in the following manner:

1) The most important measure for phytoplankton biomass (chlorophyll) is compared graphically with measurements.
2) The analysis of limiting factors and phytoplankton species and types is made. In general, only a qualitative analysis is possible.
3) The calculated dissolved nutrients are compared graphically with measurements. In this comparison it is important to know if a nutrient is (sometimes) limiting or not (see step 2).
4) The calculated total nutrients are compared graphically with measurements. In case of discrepancies with measurements, the comparison of individual terms (phytoplankton and dissolved species) must also be considered.
5) The calculated secchi depth and light extinction are compared graphically with measurements.

These steps should first be conducted for one single location, preferably a lake. The model results for a lake are important to check that *the internal processes* are correctly calculated. In a lake, the water quality is in general determined by the water quality processes, while in a (quickly flowing) canal, the *transport* of substances dominates the water quality.

We focus on correctly modelling the general concentration levels and the *patterns or trends* of the water quality parameters, as opposed to the absolute concentration values at all times. Concentration trends in time are critical for many substances such as chlorophyll, nutrients, etc. Comparisons of model results and data are therefore largely made graphically, as this is the most feasible and insightful method available. Comparisons are generally made for one location at a time, by looking at time series of results and data. A limited number of locations (model segments) are chosen to make these comparisons.

Criteria for evaluation

For an ecological model like DBS we found that it is still very difficult to quantify the criteria for determining if a simulation is acceptable or not acceptable. These criteria depend strongly on the specific goal(s) of the modelling study (e.g. importance of global vs. local results; or of long term mass balances and fluxes vs. instantaneous concentrations and short term mass fluxes). In our discussions with Rijnland representatives before and during the validation study the evaluation method and the criteria were discussed and mutually agreed upon.

8.   Classification of model coefficients

Based on previous (non-network) DBS studies, certain coefficients were recognized as being system dependent and thus suited for use as calibration coefficients. As part of the present validation study, a rigorous classification into three types of process coefficients in DBS was introduced:

- *system independent, fixed, coefficients*. The values of these coefficients are known from many previous DBS studies and should not be changed;
- *system dependent, fixed, coefficients*. The values of these coefficients are known, and are fixed within any one system, but they may vary from one system to another;
- *system dependent, variable, coefficients*.

For all coefficients (default) values and realistic ranges were determined. The first two types of coefficients can be named *'generic'* model coefficients.   For any new DBS application, the values can be known at the outset and can be fixed for the study.   The third type of parameter is a 'calibration' parameter, which can be varied within a certain established range during a study for the purpose of *model calibration.*   These coefficients include the phosphorus adsorption coefficient and phosphorus sedimentation, among others. Another model variable of importance is the background extinction, which in DBS is treated as a system dependent, variable coefficient.

Further calibration and model validation

By systematically following the above procedure and making adaptations in the system schematization and the *variable (calibration) coefficients,* the model was *calibrated* for the chosen calibration points and the entire test network (with the exception of segments explicitly identified as non-representative), using the available data (year 1986). At this point, there is not only a good correlation between calculated and measured water quality values at the chosen calibration points, but the full model results also allow a *thorough understanding* of the processes which control the water quality. It is this understanding which is crucial for the model to be of value as a water management tool. The data of the year 1989 were used to perform a *validation* of the model with parameter settings unchanged.

9.   Results of the Rijnland application

The DBS results for the Rijnland network give an extensive overview of the water quality and the processes controlling water quality in the system. The integrated model results for the whole system include:

Hydrologic results: These include direction of flow, quantification of inflow and outflow discharges, and the composition of the water (e.g. Waste Water Treatment Plants, district discharges, precipitation, network intakes, etc.) at different locations.

Load analysis: For total nutrients in the system, the model provides an overview of the dominant sources and sinks, and an analysis of the variability of sources and sinks throughout the year.

Water Quality results: For all regions in the network, including calibration points, model results include trends of: algae growth, algae speciation, growth limitations, oxygen concentrations, fluxes of substances from transport and/or processes, speciation of e.g. total nutrients, etc. etc.

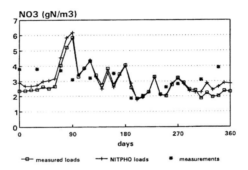

Figure 2: 1986 results for NO$_3$ at Katwijk; verification of mass balance

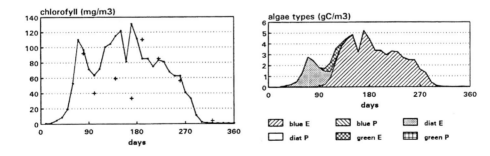

Figure 3: 1986 results for Braassemermeer elements; chlorophyll concentration and algae types

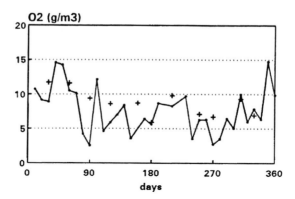

Figure 4: 1986 results for Trekvaart element; O_2 concentrations

10. Discussion and conclusion

Aims revisited

After completing the DBS application to the Rijnland test network, a reply to the original questions posed in Section 4 (Aims) can be made.

Is DBS technically capable of calculating water quality in a network system?

Based on the application of DBS to the Rijnland network and the overall good agreement between the calculated results and the measured water quality for both the calibration and validation data sets of 1986 and 1989, one can say that DBS has been successfully used in this network to calculate water quality. Specifically, the model has calculated water quality as a function of the loads to the system and the transport and processes acting on those loads, and the model can reproduce the measured water quality trends and characteristics of the system.

Based on the criteria of choosing Rijnland as a test network, it can be stated in general that DBS is capable of calculating water quality in networks.

What are the limitations of the model (if any) for application to this type of system?

A network system typically includes water bodies with a wide variety of physical and chemical characteristics. The study showed that DBS can meet the goal of a network study (i.e. give a good *global* perspective of the total system) using a finite set process equations and one primarily generic set of coefficients. Only a limited number of (calibration) coefficients are needed to calibrate the model and give an overview of the functioning of the system as a whole.

Obviously, the strength of the model in a global sense is also its limitation locally. To describe the water quality in a global perspective, concessions and compromises are being made in describing the local water quality within a specific element.

What is the correct methodology of approach for DBS modelling in networks?

The methodology presented here provides a systematic way of calibrating the model, including modifying the schematization or aggregation, as well as varying the model coefficients. This outlined systematic approach is recommended for all future applications of DBS.

Is it possible to define a generic set of model coefficients for use with networks?

A classification into three types of model coefficients for DBS was made, including (default) settings and ranges:

- *system independent, fixed, coefficients.* The values of these coefficients are known from many previous DBS studies and should not be changed;
- *system dependent, fixed, coefficients.* The values of these coefficients are known, and are fixed within any one system, but they may vary from one system to another;
- *system dependent, variable, coefficients.*

The first two types of coefficients can be named *'generic'* model coefficients. For any new DBS application, the values can be known at the outset and can be fixed for the study. *'Generic'* also implies that the coefficient values are constant for all of the model segments. The third type of parameter is a 'calibration' coefficient, which can be varied within a certain established range during a study for the purposes of model calibration. These coefficients include background light extinction, and phosphorus adsorption coefficient, among others.

To accommodate the user and to avoid errors and inconsistencies in DBS input preparations, the model now automatically sets parameter values to the recommended default values determined in the present study: overwriting is now an explicit action by the user.

Conclusion

Regarding the validation process it is concluded that the validation study has contributed to a better *insight* in and understanding of the *model functionality,* which has been made *explicit and transferable* in the form of a suggested calibration/validation approach, and a classification of model coefficients, including (default) settings and realistic parameter ranges. Integration of the latter into the model's input module has effectively *increased the model quality.*

The study has also set the conditions for *more structured and more efficient* - and therefore improved - *model applications* in the future.

The above DBS Validation Study has been fully reported elsewhere (Los et al., 1994).

References

Los, F.J., M.T. Villars and M.R.L. Ouboter. Model Validation Study DBS in networks. WL | Delft Hydraulics, Research Report, T1210, 1994.

Chapter 6

Impact of climatic fluctuations on Characeae biomass in a shallow, restored lake in The Netherlands

Winnie J. Rip, Maarten R. L. Ouboter, Hans J. Los, 2007

Hydrobiologia, Vol. 584: 415-424.

Impact of climatic fluctuations on Characeae biomass in a shallow, restored lake in The Netherlands

Winnie J. Rip • Maarten R. L. Ouboter •
Hans J. Los

Abstract External phosphorus load to a wetland with two shallow lakes in the Botshol Nature Reserve, The Netherlands, was reduced, resulting in a rapid reduction of phytoplankton biomass and turbidity, and after 4 years, explosive growth of Characeae. The clear water state was unstable, however, and the ecosystem then alternated between clear, high-vegetation and turbid, low-vegetation states. A model of water quality processes was used in conjunction with a 14-year nutrient budget for Botshol to determine if fluctuations in precipitation and nutrient load caused the ecosystem instability. The results indicate that, during wet winters when groundwater level rose above surface water level, phosphorus from runoff was stored in the lake bottom and banks. Stored phosphorus was released the following spring and summer under anaerobic sediment conditions, resulting in increased phytoplankton density and

light attenuation in the water column. During years with high net precipitation, flow from land to surface water also transported humic acids, further increasing light attenuation. In years with dry winters, the phosphorus and humic acid loads to surface water were reduced, and growth of submerged macrophytes was enhanced by clear water. Thus, the temporal pattern of precipitation and flow from land to water gave a coherent, quantitative explanation of the observed dynamics in phosphorus, phytoplankton, turbidity, and Characeae. Global warming has caused winters in The Netherlands to become warmer and wetter during the last 50 years, increasing flow from land to water of humic acids and phosphorus and, ultimately, enhancing instability of Characeae populations. In the first half of the 20th century interannual variation in precipitation was not sufficient to cause large changes in internal P flux in Botshol, and submerged macrophyte populations were stable.

Guest editors: R. D. Gulati, E. Lammens, N. De Pauw &
E. Van Donk
Shallow lakes in a changing world

W. J. Rip • M. R. L. Ouboter
Waternet (Water board Amstel Gooi en Vecht),
P.O. Box 94370, 1090 GJ Amsterdam,
The Netherlands
e-mail: winnie.rip@waternet.nl

Hans J. Los
Delft Hydraulics, P.O. Box 177, 2600 MH Delft,
The Netherlands

Keywords *Chara* ■ Runoff from land to water •
Humic acids • Lake • Light attenuation • Nutrient
loading • Phosphorus • Phytoplankton • Precipitation
• Turbidity • Climate change

Introduction

The presence of Characeae indicates a healthy aquatic ecosystem. Similar to other aquatic

123

macrophytes, *Chara* and related species support other biological components of the lake ecosystem (Timms & Moss, 1984; Carpenter & Lodge, 1986; Noordhuis et al., 2002). Submerged macrophytes also help to maintain high water transparency by a number of mechanisms (Scheffer, 1998). The benthic plants, for example, prevent resuspension of the sediment by wind or fish, compete with phytoplankton for nutrients, and offer refuge to grazing zooplankton (Moss, 1990; Scheffer et al., 1993; Van den Berg et al., 1998; Kufel & Kufel, 2002). Some species have allelopathic effects on competing epiphytes and phytoplankton (Van Donk & van de Bund, 2002). Poor light availability, due to shading by phytoplankton and epiphytes, is mostly the primary reason for the disappearance of submerged macrophytes (Philips et al., 1978). Factors other than light, such as grazing (Mitchell & Perrow, 1998) or phytotoxicity by free sulphide (Lamers et al., 1998), may also be important in determining macrophyte abundance; however, when light is insufficient, macrophytes cannot exist at all.

Shallow lakes can have alternative stable states (Scheffer et al., 1993). Most of these lakes are either rather turbid without submerged macrophytes or clear and vegetated. In some lakes repeated shifts between a clear-vegetated and a turbid state have been observed (Lake Tåkern & Lake Krankesjön) (Blindow et al., 1993), Alderfen broad (Perrow et al., 1994), Tomahawk Lagoon (Mitchell, 1998) and Botshol (Rip et al., 2005). In most lakes the shifts between clear and turbid states occur irregularly in time, suggesting that there is an infrequence external forcing, for instance by water level fluctuations or changes in external phosphorus load. Alternatively, high oxygen consumption by decomposition of macrophyte biomass accumulated during high production years can trigger the anaerobic release of phosphorus, and the resulting phytoplankton blooms limit subsequent macrophyte production (Asaeda et al., 2000). Moss (1990) suggested this intrinsic process for the regular cycles in Alderfen broad. Van Nes et al. (submitted) studied intrinsic processes triggered by submerged macrophytes that according to the "slow-fast theory (Muratori & Rinaldi, 1991) could cause regular cycles between clear and turbid states in shallow lakes. The present study examined the hypothesis that variation in precipitation and subsequent nutrient loading through runoff account for fluctuations in phosphorus, phytoplankton biomass, light attenuation, and macrophyte abundance.

A previous study (Rip et al., submitted) developed water and nutrient budgets for the entire Botshol ecosystem under the prevailing meteorological conditions. The models provided monthly estimates of nutrient loading during 1989-2002 and, ultimately, insight into the causes of interannual fluctuations in P loading. The results indicated that, during wet winters when groundwater level rose above the level of surface water in the catchment areas, water flowed from land into surface water, resulting in a rise of the P load. In dry winters, infiltration through the soil exceeded precipitation, and there was no flow from land to water, so fluctuations in P load were mainly due to the water supply.

The present study extended the nutrient budget to examine fluctuations in phosphorus loading to the two lakes of Botshol as a possible cause of instability in water clarity and abundance of submerged macrophytes. These two studies combined used an ecosystem approach, starting with P loading from catchment areas and transport to surface water, then incorporating biological processes that determine phytoplankton biomass, turbidity, and finally, biomass of Characeae.

Study area and methods

Hydrology and restoration of Botshol

The hydrology of Botshol, a nature reserve in the center of the Netherlands (52°15' N 4°26' E) is dominated by infiltration. The area is a hydrologically isolated polder. On average, the water table in Botshol drops 1.75 mm day^{-1} by water infiltrating to an adjacent polder, Groot Mijdrecht, which has water levels 4.3 m below Botshol. There is a strong gradient of infiltration within the nature reserve. Infiltration from the lakes directly bordering Groot Mijdrecht is 2.5 mm day^{-1}, while infiltration in the northwest part of Botshol is 0.1 mm day^{-1}. During spring and summer, approximately 10^6 m^3 P rich water enters Botshol to compensate for water lost through evaporation and infiltration. Prior to November 1988 three agricultural areas drained their excess water into Botshol resulting in a high nutrient load.

Up to until 1960, the lakes of the Botshol Nature Reserve were clear and dominated by Characeae. Beginning in the 1960s, water quality deteriorated due to the rise of external phosphorus input, and submerged macrophyte populations declined (Simons et al., 1994). Beginning in 1989, the external nutrient load was reduced by hydrological segregation of Botshol from the agricultural areas and by chemical stripping of 60–80% of phosphate from the water supply. The goal of restoration was to re-establish the Characeae and other submerged macrophytes. In the first 4 years after the start of phosphorus reduction, decreases were observed in phosphorus and chlorophyll *a* concentrations. Reduced phytoplankton numbers resulted in crystal clear water and explosive growth of submerged macrophytes (Rip et al., 1992; Simons et al., 1994). The dominant macrophyte species in terms of lake areal cover were *Najas marina, Fontinalis antipyretica, Chara connivens, Chara contraria, Chara hispida* and *Chara globularis*. However, this clear water state was unstable. From 1993 onwards, the ecosystem alternated between turbid water with low macrophyte cover (1993–1995, 1999–2003) and clear water with high cover of aquatic plants (1996–1998). Phosphorus concentrations in Botshol showed concomitant fluctuations (Rip et al., 2005). The composition of the fish community was determined in 1989 and 1992. The fish community was not dominated by planktivorous fish or benthivorous fish, such as bream (Abramis brama). Rather, perch, pike, and roach, species typical of clear water lakes with macrophytes, were dominant in Botshol in both surveys (Rip et al., 2005). Total fish biomass decreased over the first 4 years following the reduction of the external P-load.

Physical, chemical and biological parameters

From 1988 through 2003, samples for physical and chemical parameters (temperature, pH, conductivity, Secchi disc depth, attenuation, oxygen, chlorophyll *a*, chloride, total and dissolved P, total and

Kjeldahl N, ammonium, nitrate, silicate, and dissolved organic substances [humic acids]) were taken every 2–4 weeks in 5 subareas of Botshol, the main water supply, and Lake Vinkeveen. Phytoplankton and zooplankton densities were determined, simultaneously with the chemical samples at the two lakes (subareas I and II). In the period 1993–1999 no data for attenuation were available. Meteorological data were provided by KNMI (a national meteorological institute). Submerged macrophytes were mapped and quantified each summer 1987–2003. A transect pattern covering most of the open water and pools, watercourses and ditches was followed for each survey. Plant material was observed and collected by dredging and snorkeling from a rowing boat. Characeae species were identified according to Van Raam (2003). Detailed methods and results of the environmental and biological measurements were presented in Rip et al. (2005).

Model

The study area of Botshol was divided into five subareas. Subareas III, IV and V were catchment areas for subareas I and II, in which the two lakes were situated. Water and nutrient budgets constructed for each subarea and for the entire study area (Rip et al., submitted) were used as the basis for the ecosystem model developed in the present study.

DELWAQ-BLOOM-SWITCH (DBS), a mathematical model, was used to determine interactions between nutrient loading and transport, and other physical, chemical, and biological processes in the Botshol. The present study used the standard DBS model, which has been successfully applied to similar aquatic ecosystems (Los, 1993; Van der Molen et al., 1994). A general overview of DBS is given below. More detailed information can be found in the technical reference manual for DBS (Los, 1993). A small number of adjustments were made to model the specific situation in Botshol using measurements from 1989–1996. The accuracy of the calibrated model was validated using data from 1997–2002.

DBS modeled the cycling of carbon, oxygen, and nitrogen, phosphorus, and silicon within the

125

Botshol ecosystem, adhering to the law of conservation of mass. Nutrients in the water column could be dissolved forms, detritus components and non-detritus organic components. A portion of each nutrient pool was incorporated in the algae. Phosphorus could also be adsorbed to inorganic material (AAP). The module, SWITCH, computed the sediment-water exchange for nutrients by adsorption–desorption (Smits & Van der Molen, 1993). Nutrients could be recycled an infinite number of times with losses only through transport, chemical adsorption, denitrification, and burial in the sediment. Nutrients could also be transformed through mineralization, sedimentation, resuspension, and nitrification.

Oxygen concentration played an important role in controlling the P flux from the sediments of Botshol. A large percentage of the P from runoff was in the particulate state, resulting in P-enrichment of the sediments. Following winters that had high precipitation, anaerobic conditions in the small watercourses in spring/summer resulted in the release of the stored P from the sediments.

The BLOOM portion of DBS computed phytoplankton production based on competition among algal species (Los & Brinkman, 1988; Los, 2005). The primary state variables of the BLOOM module were different phytoplankton types. Four taxonomic groups were distinguished: Chlorophyta, Bacillariophyta, Cyanophyta, and Characeae.

BLOOM first selected the factor that was most likely to become limiting, based on environmental conditions: solar radiation, day length, water temperature, depth, background light attenuation, and nutrient concentrations. The model then selected the best-adapted phytoplankton types under those conditions. The biomass of these phytoplankton types were calculated as the net result of production, mortality, and transport during the previous time step using an optimization technique (Linear Programming).

DBS included a module called UITZICHT (Los, 1993; Buiteveld, 1995) that computed light attenuation. Total light attenuation in the water column was the sum of background attenuation and attenuation due to free floating phytoplankton, algal detritus and humic acids.

Results

Results are shown only for subarea I, but similar agreement between empirical and calculated values was found for the other subareas. Results of chloride calculations showed that the ratios of water supplies and the calculated exchange among subareas used in the water budget were accurate (Fig. 1). The interannual variation of the calculated dynamics of total phosphorus and chlorophyll were similar enough to actual concentrations to show year-to-year differences (Figs. 2, 3): low levels in 1989–1992, high levels in 1993–1995, low levels again in 1996–1998, etc. Both observed and calculated values showed a summer peak each year.

The pattern in 1998 was deviant from other years, probably due to high precipitation in this summer. The calculated values showed a sharp shift to high P and chlorophyll levels in late summer of that year.

Peaks and low points in chlorophyll *a* concentration predicted by the model corresponded well with periods of high and low levels measured in the field. In periods with high cover of Characeae, such as the summers of 1991, 1992, and 1998, predicted chlorophyll levels were higher than the observed values. This was probably due to the lack of some feedback processes in DBS. Dense *Chara* populations during those summers, for example, could have provided shelter for grazing zooplankton or could have influenced phytoplankton by allelopathic effects or increasing the sedimentation rate of phytoplankton. However, biological feedback processes were not completely lacking in the model.

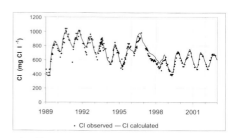

Fig. 1 Observed and calculated chloride levels at location I for 1989–2002

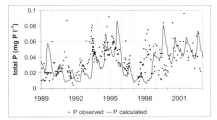

Fig. 2 Observed and calculated total P levels at location I for 1989–2002

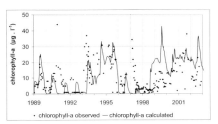

Fig. 3 Observed and calculated chlorophyll levels at location I for 1989–2002

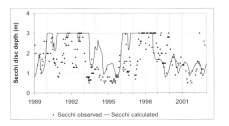

Fig. 4 Observed and calculated Secchi-disc depths at location I for 1989–2002

Phosphorus calculated to be stored in *Chara*, for example, was not available for other algae and predicted light attenuation was affected by free-floating algae, but not by the bottom dwelling *Chara*. Another example of feedback in the model concerned the relatively slow reaction of particular nutrients stored in the sediment, from decaying algae, to external forcing imposed on the lake system. The result was that the predicted remineralization flux from the sediment in a particular year was clearly affected by the previous years loadings.

Calculated and measured light attenuation were more similar during periods of high and low turbidity. Results of the model, UITZICHT, indicated that temporal variation in light attenuation was determined primarily by variation in amounts of dead (detritus) and living algae (Fig. 5). Humic acids influenced attenuation during years with high precipitation, such as 1993– 1995 and 1999–2002.

The DBS model was not very accurate in calculating Secchi disc depth, although interannual variation in turbidity was successfully predicted (Fig. 4). The generally low algal biomass in Botshol, the

lack of some feedback processes and the inaccuracy of Secchi disc depth measurements made comparison with simulations difficult. Secchi disc depths calculated for the entire years of 1991 and 1992 were at bottom depth, for example, while actual values varied.

The model's predictions of fluctuations in *Chara* biomass agreed with fluctuations in percent cover of *Chara* determined in the field (Fig. 6). Both calculated and observed levels were high, for example, in 1991, 1992, 1997, and 1998, and low in 1994, 1995, and 1999–2002. The model overestimated the biomass of *Chara* during 1993 and 1996, corresponding to overestimates of Secchi disc depth.

The runoff in dry years was about 90% reduced in comparison to wet years (Rip et al., submitted). To test the hypothesis that P loading to surface water through flow from land to water, especially after winters with high precipitation, was a major factor causing reduced cover of *Chara* in the lakes of Botshol, the model was run under the condition that the P concentration in water flow from land to surface water was reduced 90%. The P load from water inflow to maintain water level in summer remained the same. During wet years, phosphorus and chlorophyll *a* levels calculated under this condition were substantially lower than levels calculated under the standard conditions. DBS calculated a stable *Chara* population each summer.

Overall, the DBS model provided sufficient consistency between computed and measured parameters of the Botshol ecosystem to explain the dynamics of nutrient loading, phytoplankton density, light attenuation, and *Chara* populations.

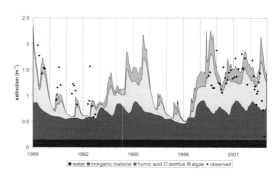

Fig. 5 Observed attenuation and calculated composition of light attenuation at location II for 1989–2002

The flow of water which is rich in phosphorus and humic acids from land to water, particularly following periods of high precipitation, is a likely explanation for the instability of the aquatic ecosystem in Botshol after the reduction of the external nutrient load.

Discussion

The main conclusion from the nutrient budget developed for Botshol was that, following nutrient reduction, P from flow from land to water acted as the primary internal nutrient source during periods when precipitation exceeded the water loss by infiltration and evaporation (Rip et al., submitted). Although similarity between calculated values and empirical measurements varied somewhat among the parameters (Figs. 1– 6), results of the calibrated DBS

model reproduced the dynamics of the Botshol ecosystem during 1989–2002 closely enough to provide a useful tool in determining causes of interannual variation.

Results of the present study showed that, during wet years, the P-rich, flow from land to water resulted in increased phosphorus levels and phytoplankton biomass in surface water of Botshol (Figs. 2, 3). In addition to high P levels, the flow from land to water contained large amounts of dissolved organic substances (humic acids) that colored the water dark brown and further reduced light penetration (Fig. 5). Consequently, light attenuation was so high during periods of high precipitation that *Chara* populations declined (Fig. 6). After dry winters, the groundwater level in terrestrial peat was below the surface water level, and the internal P source was eliminated. During the next growing season, water transparency was high, and

Fig. 6 Calculated biomass of *Chara* (kg C) as model results and field surveys (% cover) at subarea I for 1989–2002

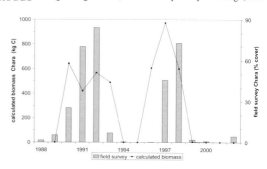

Chara populations increased. This trend was confirmed by the significant negative correlation between *Chara* cover and net precipitation during the previous 2 years ($r = -0.771$, $P < 0.01$). In a simulated situation with low precipitation, calculated *Chara* populations showed an annual cycle that was stable from year to year. Thus, variation in winter precipitation probably caused interannual changes in P flux and humic acids from terrestrial peat, which explained fluctuations in P concentration, phytoplankton biomass, and water transparency in Botshol. Hough et al. (1991) have shown similar effects of variation in precipitation on cover of submerged macrophytes.

In principle any lake with a large enough catchment area will have variation in P load from runoff that is related to variation in precipitation. However, the fluctuation in P load will not always result in a related pattern of high and low cover of submerged macrophytes. In shallow lakes with alternative states there is a range of nutrients in which either a turbid, vegetation poor or a clear, vegetation rich states can exist (Scheffer et al., 1993), sometimes called the "catastrophe fold". Although other factors, like removal of fish, can alter the ecological balance within this range, fluctuations in precipitation and runoff will only cause a switch between clear, vegetated and turbid, unvegetated states if the P content falls above or below this nutrient range. Thus, the influence of fluctuations in runoff on the cover of Characeae depends on the size of the runoff area, the P load from other sources, and the P load range where both alternative states can exist. This critical range is determined by size, depth, fetch, and sediment type of the lake (Janse, 2005). For example, the critical P load causing a switch in stable state was markedly lower in deeper lakes than in shallow lakes and the range is smaller. Botshol is relatively deep for lakes in the Netherlands, so the range of critical P load (0.8– 1.3 mg P m^{-2} year^{-1}, Rip et al., submitted) is relatively low and small and easily exceeded by small changes in P load.

Landscape and physicochemical conditions in the aquatic ecosystem of Botshol account for the large temporal variation and time lags in P flux. Subsurface flow of P-rich water was most important in subareas IV and V, which had high land/ water ratios and low infiltration rates. Although the catchment areas were not fertilized or used for agricultural purposes, phosphorus concentrations in the peat soil water were high due to mineralization and high chloride and sulfide levels (Beltman et al., 2005; Lamers et al., 1998). The mineralization of the terrestrial peat is not a 'natural' process. Botshol is below sea level and humans stabilized the landscape by influencing the water table. The water table in Botshol was maintained at a constant level by the inflow of water during the summer and outflow of water in winter. The inflowing water had high levels of P, S, and Cl. Starting in 1989, 60–80% of P was stripped from the inflowing water. The untreated sulphur and chloride levels remained high, however, and increased the internal availability of P in Botshol (Lamers et al., 1998; Beltman et al., 2005). As was found in other freshwater systems (Meyer et al., 1981), a high percentage (~85%) of P in runoff was probably particulate material. The particulate P was stored in the water bottoms and banks of the small watercourses in subarea IV and V, explaining why only a small increase in P concentration was seen in surface water during wet winters (Fig. 2). When water temperature rose in spring, the stored P was partly released from bottom sediments. In summer, however, there was an explosive P flux from the sediments due to low yearly oxygen levels. The small watercourses in subareas IV and V of Botshol became anaerobic during summer, when mineralization of organic matter in the sediment created a high oxygen demand. When P-rich water from subareas IV and V was subsequently transported to subareas I and II to make up for summer evaporation and infiltration, large amounts of P were carried into surface waters of the lakes.

Results of the present study provide a likely explanation for the decline in *Chara* populations during 1993–1995 and 1999–2002. An important condition for the growth of Characeae is light reaching the lake bottom after germination, which occurs around May. This condition was met in Botshol when P was below 0.018 mg l^{-1} (Rip et al., 2005) corresponding to periods of high macrophyte cover. When total P was between 0.018 and 0.043 mg l^{-1}, there were periods of both low and high cover of submerged macrophytes (Rip et al., submitted), supporting the hypothesis that either state can exist at intermediate nutrient levels (Scheffer, 1993). After a wet winter, however, P levels enhanced by subsurface run-off exceeded this range and stimulated growth of phytoplankton, which in turn decreased light penetration and limited growth of *Chara*.

In addition to light, other factors may influence submerged macrophyte populations, including chloride concentrations (Jeppesen et al., 1994), sediment toxicity due to sulfide (Lamers et al., 1998), grazing by herbivorous birds (Moss, 1990; Mitchell & Perrow, 1998), and accumulation of organic material of decaying submerged plants (Carpenter & Lodge, 1986; Perrow et al., 1994; Asaeda et al., 2000). The possible involvement of these factors in the observed interannual fluctuations of *Chara* in Botshol is examined in other studies (Beltman et al., 2005; Van Nes et al., submitted; Rip et al., unpublished).

The approach taken by the present study considered the ecosystem as a whole, starting with actual meteorological conditions, incorporating transport from land to surface water, as well as changes in phytoplankton and light attenuation, and finally examining effects on submerged macrophytes. In contrast, most previous studies of aquatic systems modeled either nutrient loading (Meyer et al., 1981; Mander et al., 1998) or biological processes (Janse et al., 1997; Best et al., 2001). A few previous studies modeled both, but only for average years and did not examine interannual variation (Asaeda et al., 2000; Portielje & Rijsdijk, 2003). The present study, therefore, was uniquely able to explain the role of interannual variation in precipitation as a cause of instability in an aquatic ecosystem.

Precipitation became a major factor influencing submerged macrophyte populations only after restoration efforts in Botshol were initiated. Following the reduction in external P load to very low levels, the internal P load from subsurface run-off increased in importance as a nutrient source for phytoplankton. As a high percentage of the area of Botshol is terrestrial peat, interannual variation in precipitation caused large changes in the internal P load to surface water. The high infiltration rate and

large ditch distance in much of the terrestrial peat resulted in low runoff and land-to-water P flow during dry winters (Ouboter & Rip, submitted). A side effect of the hydrological isolation of Botshol, furthermore, was an increase in chloride from 500 to 1000 mg l^{-1}, which enhanced the availability of phosphorus in peat (Beltman et al., 2005). Finally, the relatively large depths of the lakes in Botshol made submerged macrophyte populations vulnerable to even small increases in light attenuation.

External P loading to Botshol was low, and Characeae populations in Botshol did not exhibit large interannual fluctuations prior to 1960 (Simons et al., 1994). Analysis of the weather in the Netherlands during the 20th century indicated that both temperature and annual precipitation have risen since 1901, related to global changes causing a warmer and wetter climate throughout western Europe (IPCC, 2001; KNMI, 2001). Thus, all winters in which more than 500 mm of precipitation were recorded in De Bilt, a site in the Netherlands close to Botshol, came after 1960. Before that time, interannual variation in precipitation was not sufficient to cause large changes in internal P flux in Botshol, and submerged macrophyte populations were stable (Simons et al., 1994).

The first 4 years of the Botshol restoration project (1988–1992) coincided with four dry winters, so the initial change to the clear water state and increased *Chara* populations were due to the combination of reduced external P load and low precipitation. The pre-restoration P load to the lakes was high and reduction of external P load was necessary to allow growth of submerged macrophytes. Now that the external P load is not a limiting factor, additional measures are needed to decrease internal P flux in Botshol during wet winters to sustain stable populations of submerged macrophytes. In wet years, internal P flux could be reduced by allowing the water level to fluctuate more naturally. Desiccation of peat during dry periods, for example, would reduce P in runoff, due to immobilization of P by oxidation of Fe (Lucassen et al., 2005). Another approach would be to reverse terrestrialization by removing large areas of terrestrial peat 1 m below water level. This would reduce the surface area where nutrient-rich groundwater can flow into surface water. Finally, the creation of larger areas of surface water in the subareas with the highest runoff would raise oxygen levels, decrease reduction of sulphate to sulphide, and reduce P release from sediment. Results of the present study demonstrated that the future stability of macrophyte populations in Botshol will depend on improved management of internal P flux.

Acknowledgements The authors thank the Society for the Preservation of Nature in the Netherlands for permission to perform the research in the Botshol nature reserve, and Jos van Gils, Arjen Markus and Paul Grashoff for discussions and assistance during the development of the models of Botshol. We are grateful to Boudewijn Beltman and Val Gerard for improving the manuscript.

References

Asaeda, T., V. K. Trung & J. Manatunge, 2000. Modeling the effects of macrophyte growth and decomposition on the nutrient budget in Shallow Lakes. Aquatic Botany 68: 217–237.

Beltman, B., W. J. Rip & A. Bak, 2005. Nutrient release after water quality restoration measures. A phytometer assessment in the Botshol wetlands, The Netherlands. Wetlands Ecology and Management 13: 577–585.

Best, E. P. H., C. P. Buzzelli, S. M. Bartell, R. L. Wetzel, W. A. Boyd, R. D. Doyle & K. R. Campbell, 2001. Modeling submersed macrophyte growth in relation to underwater light climate: modeling approaches and application potential. Hydrobiologia 444: 43–70.

Blindow, I., G. Andersson, A. Hargeby & S. Johansson, 1993. Long-term pattern of alternative stable states in two shallow eutrophic lakes. Freshwater Biology 30: 243–252.

Buiteveld, H., 1995. A model for calculation of diffuse light attenuation (PAR) and Sechi depth. Netherlands Journal of Aquatic Ecology 29: 55–65.

Carpenter, S. R. & D. M. Lodge, 1986. Effects of submersed macrophytes on ecosystem processes. Aquatic Botany 26: 341–370.

Hough, R. A., T. E. Allenson & D. D. Dion, 1991. The response of macrophyte communities to drought-induced reduction of nutrient loading in a chain of lakes. Aquatic Botany 41: 299–308.

IPCC, 2001. Climate Change 2001: The Scientific Basis. Cambridge University Press, 892 pp.

Janse, J. H., E. van Donk & T. Aldenberg, 1997. A model study on the stability of the macrophyte-dominated clear water state as affected by biological factors. Water Research 32: 2696–2706.

Janse, J. H., 2005. Model studies on the eutrophication of shallow lakes and ditches. University of Wageningen. Thesis, 376 pp.

Jeppesen, E., M. Sondergaard, E. Kanstrup & B. Petersen, 1994. Does the impact of nutrients on the biological structure and function of brackish and freshwater differ? Hydrobiologia 276: 15–30.

KNMI, 2001. De toestand van het klimaat in Nederland 2000 (in Dutch).

Kufel, L. & I. Kufel, 2002. *Chara* beds acting as nutrient sinks in shallow lakes—A review. Aquatic Botany 72: 249–260.

Lamers, L. P. M., H. B. M. Tomassen & J. G. M. Roelofs, 1998. Sulphate-induced eutrophication and phytotoxicity in freshwater wetlands. Environmental Science and Technology 32: 199–205.

Los, F. J. & J. J. Brinkman, 1988. Phytoplankton modelling by means of optimization: a 10-year experience with Bloom II. Verhandlungen Internationale Verein. Limnologie 23: 790–795.

Los F. J., 1993. DBS. Tech. Rep. T542, Delft Hydraulics, The Netherlands (in Dutch).

Los, F. J., 2005. An algal biomass prediction model. In Daniel, P. Loucks & Eelco van Beek (eds), Water Resources Systems Planning and Management - an introduction to methods, models and applications. UNESCO, 2005.

Lucassen, C. H. E. T., A. J. P. Smolders & J. G. M. Roelofs, 2005. Effects of temporary desiccation on the mobility of phosphorus and metals in sulphur-rich fens: differential responses of sediments and consequences for water table management. Wetlands Ecology and Management 13: 135–148.

Mander, Ü., A. Kull, V. Tamm, V. Kusemets & R. Karjus, 1998. Impact of climatic fluctuations and land use change on runoff and nutrient losses in rural landscapes. Landscape and Urban planning 41: 229-238.

Meyer, J. L., G. E. Likens & J. Sloane, 1981. Phosphorus, nitrogen and organic carbon flux in a headwater stream. Archiv fu¨r Hydrobiologie 91: 28–44.

Mitchell, S. F. & M. R. Perrow, 1998. Interactions Between Grazing Birds and Macrophytes. In Jeppesen, E., Ma. Sondergaard, Mo. Sondergaard & K. Christoffersen (eds), The structuring Role of Submerged Macrophytes in Lakes. Springer Verlag, New York, 175–195.

Moss, B., 1990. Engineering and biological approaches to the restoration from eutrophication in which aquatic plant communities are important components. Hydrobiologia 275(276): 367–377.

Mutatori, S. & S. Rinaldi, 1991. A separation condition for existence of limit cycles in slow-fast systems. Applied Mathematical Modelling 15: 312–318.

Noordhuis, R., D. T. Van der Molen & M. S. Van den Berg, 2002. Response of herbivorous water-birds to the return of *Chara* in Lake Veluwemeer, The Netherlands. Aquatic Botany 72: 349–367.

Perrow, M. R., B. Moss & J. Stansfield, 1994. Trophic interactions in a shallow lake following a reduction in nutrient loading: A long term study. Hydrobiologia 275: 43–52.

Philips,G.L.,D.Eminson&B.Moss, 1978.A mechanism to account for macrophyte decline in progressively eutrophicated freshwaters. Aquatic Botany 4: 103–126.

Portielje, R. & R. E. Rijsdijk, 2003. Stochastic modeling of nutrient loading and lake ecosystem response in relation to submerged macrophytes and benthivorous fish. Freshwater Biology 48: 741–755.

Rip, W. J., K. Everards & A. Houwers, 1992. Restoration of Botshol (The Netherlands) by reduction of external nutrients load: The effects on physico-chemical conditions, plankton and sessile diatoms. Hydrobiological Bulletin 25: 275–286.

Rip, W. J., M. Ouboter, E. H. van Nes & B. Beltman, 2005. Oscillation of a shallow lakes ecosystem upon reduction in external phosphorus load. Archiv fü̈r Hydrobiologie 164: 387–409.

Scheffer, M., 1998. Ecology of shallow lakes. Population and Community Biology series, Kluwer Academic Publishers, 357 pp.

Scheffer, M., S. H. Hosper, M. L. Meijer, B. Moss & E. Jeppesen, 1993. Alternative equilibria in shallow lakes. Trends in Ecology and Evolution 8: 275–279.

Simons, J., M. Ohm, R. Daalder, P. Boers & W. J. Rip, 1994. Restoration of Botshol (The Netherlands) by reduction of external nutrient load: recovery of a characean community, dominated by Chara connivens. Hydrobiologia 275/276: 243–253.

Smits, J. G. C. & D. T. van der Molen, 1993. Application of SWITCH, a model for sediment-water exchange of

nutrients, to Lake Veluwe in the Netherlands. Hydrobiologia 253: 281–300.

Timms, R. M. & B. Moss, 1984. Prevention of growth of potentially dense phytoplankton populations by zooplankton grazing in the presence of zooplanktivorous fish in a shallow wetland ecosystem. Limnology and Oceanography 29: 472–486.

Van den Berg, M. S., H. Coops, M.-L. Meijer, M. Scheffer & J. Simons, 1998. Clear water associated with a dense Chara vegetation in the shallow and turbid Lake Veluwemeer, The Netherlands. In Jeppesen, E., Ma. Sondergaard, Mo. Sondergaard & K. Christoffersen (eds), The Structuring Role of Submerged Macrophytes in Lakes. Ecological Studies. Springer Verlag 131: 339–352.

Van der Molen, D. T., F. J. Los, L. van Ballegooijen & M. P. van der Vat, 1994. Mathematical modelling as a tool for management in eutrophication control of shallow lakes. Hydrobiologia 275/276: 479–492.

Van Donk, E. & W. J. van de Bund, 2002. Impact of submersed macrophytes including charophytes on phyto- and zooplankton communities: allelopathy versus other mechanisms. Aquatic Botany 72: 261–274.

Van Raam J. C., 2003. Handboek Kranswieren. Chara boek, Hilversum (in Dutch).

Chapter 7

Trend Analysis of Eutrophication in Dutch
Coastal Waters for 1976 through 1994 Using
a Mathematical Model

F.J. Los, M. Bokhorst, I. de Vries and R.J.
Jansen, 1997

Extended version of:
Los, F.J., and M. Bokhorst, Trend analysis
Dutch coastal zone. In: New Challenges for
North Sea Research, Zentrum for Meeres-
und Klimaforschung, University of Hamburg,
1997: 161-175.

TREND ANALYSIS OF EUTROPHICATION IN DUTCH COASTAL WATERS FOR 1976 THROUGH 1994 USING A MATHEMATICAL MODEL

F.J. LOS[1], M. BOKHORST[1], I. DE VRIES[2] AND R.J. JANSEN[2]

[1]) *Delft Hydraulics / WL, Marine and Coastal Management, p.o. box 177, Rotterdamseweg 185, 2600 MH Delft, The Netherlands*

[2]) *National Institute for Coastal and Marine Management / RIKZ, p.o. box 20907, 2500 EX The Hague, The Netherlands*

Abstract

In the last decades a number of measures have been taken to reduce the nutrient discharges to the western European surface waters. As a result the phosphorus concentrations of several major European rivers, the Rhine in particular, have decreased. Locally this has resulted in a noticeable reduction of algal biomasses in general and of blue green algae in particular. The anthropogenic phosphorus discharges into the Dutch part of the North Sea have also declined recently. However the impacts on the North Sea ecosystem have so far been limited. Time series analysis reveals a significant decreasing trend for dissolved inorganic phosphorus (40%) whereas chlorophyll, an indicator for total algal biomass, shows a small and mostly not statistically significant decreasing trend (De Vries, et al., 1998). To explain these observations the National Institute for Coastal and Marine Management and the North Sea Directorate of the Dutch Ministry of Public Works have committed Delft Hydraulics to reproduce the observed trends, the spatial gradients and long term seasonal patterns for the Dutch coastal zone for a 20 year period from 1975 through 1994 with a detailed coupled physical-ecological model. The model results indicate a small effect of decreasing phosphorus but an important role of both nitrogen and light climate for primary production.

Overview of the modelling instrument

About ten years ago the first Dutch eutrophication models were developed for various parts of the North Sea. In this first generation of models processes were described by rather simple equations. Transports were based on 2D hydrodynamic calculations. As a compromise between computational performance and accuracy a uniform grid size of 16x16 km was selected (the so called GENO grid). For this application to the southern Bight of the North Sea this model includes 1395 computational elements. With the water quality model, named DYNAMO, it was possible to obtain reasonable results for total algal biomass (expressed in chlorophyll concentrations) and nutrients. With this model it was not possible to compute the composition of the phytoplankton bloom, variations in internal characteristics (stoichiometry) such as nutrient to biomass ratios and steep gradients due to temporal or spatial variations.

Algal blooms usually consist of various species of phytoplankton belonging to different taxonomic or functional groups such as diatoms, micro flagellates and dinoflagellates. They have different requirements for resources (nutrients; light) and they have different ecological properties. Some species are considered to be objectionable for various reasons. Among these are *Phaeocystis*, which causes foam on the beaches and various species of dinoflagellates, which among others may cause diuretic shell fish poisoning. To deal with these phenomena it is necessary to distinguish different types of phytoplankton in a model.

For Dutch fresh water systems the development of eutrophication models had started about ten years earlier, in the late 1970s. Much of this work was performed at Delft Hydraulics for both the national and for several local governmental authorities. These experiences were formalised in the DELWAQ-BLOOM-SWITCH (DBS) model (Delft Hydraulics, 1992). This model and some of its predecessors have been successfully applied to many different systems with great differences in physical and chemical conditions. They range in depth from 1 to 30 m, in extinction from 0.5 to 10 m-1, in residence times from about 1 day to several years. Nutrients conditions vary from meso- to hyper eutrophic, chlorophyll peaks range from 1 to 500 μg/l, occasionally even more. Compared with DYNAMO, many processes are described in greater detail in this model. The performance of this fresh water modelling instrument was thoroughly validated against numerous sets of observations (Delft Hydraulics, 1991; Delft Hydraulics, 1994)

Phytoplankton within DBS is computed by the module BLOOM II. It is based upon the principle of competition between different species. The basic variables of this module are called types. A type represents the physiological state of a species under strong conditions of limitation. Usually a distinction is made between three different types: an N-type for nitrogen limitation, a P-type for phosphorus limitation and an E-type for light energy limitation. The solution algorithm of the model considers all potentially limiting factors and first selects the one which is most likely to become limiting. It then selects the best adapted type for the prevailing conditions. The suitability of a type (its fitness) is determined by the ratio of its requirement and its growth rate. This means that a type can become dominant either because it needs a comparatively small amount of a limiting resource (it is efficient) or because it grows rapidly (it is opportunistic). The algorithm considers then the next potentially limiting factor and again selects the best adapted phytoplankton type. This procedure is repeated until it is impossible to select a new pair of a type and limiting factor without violating (i.e. over-exhausting) some limiting factor. Thus the model seeks the optimum solution consisting of n types and n limiting factors. As a further refinement BLOOM takes the existing biomasses of all phytoplankton types into account. These are the result of production, loss processes and transport during the previous time-steps. So the optimisation algorithm does not start from scratch. As they represent different stages of the same species, the transition of one type to another is a rapid process with a characteristic time-step in the order of a day. Transitions between different species is a much slower process as it depends on mortality and net growth rates. It is interesting that the principle just described, by which each phytoplankton type maximises its own benefit, effectively means that the total net production of the phytoplankton community is maximised. Several modules from this fresh water model system were gradually included in the salt water model to obtain a greater accuracy and new output variables such as a further distinction of the algal community into major functional groups or even species. The marine version of BLOOM, called North Sea BLOOM, considers diatoms, flagellates, dinoflagellates and *Phaeocystis*.

Calibration procedure

Methodology   Selecting a methodology for the calibration of a complex model for a complex system is not trivial. First of all the results for about half a dozen of different substances have to be considered. Second seasonal and regional variations must be taken into account. For the trend analysis project chlorophyll (as an indicator for total biomass), dissolved nutrients (nitrate, ortho-phosphorus, silicate), the distinction between different groups of phytoplankton (only qualitatively) and the extinction of light (little data) have been evaluated. To check the hydrodynamics also observations for the salinity have been used. Model outputs were compared to observations for about 20 different stations arranged in several transects. The locations are shown in Figure 1. For some of these stations yearly seasonal data are available for 1974 through 1993. For most stations some years are lacking, but in general data availability is unusually large. For the calibration of previous versions of North Sea BLOOM data for 1985 were used. In recent years some changes were observed due to the aforementioned reduction in phosphorus discharges. Moreover in the international political arena, 1990 is a milestone. Thus 1990 was selected for the calibration of the application of North Sea BLOOM.

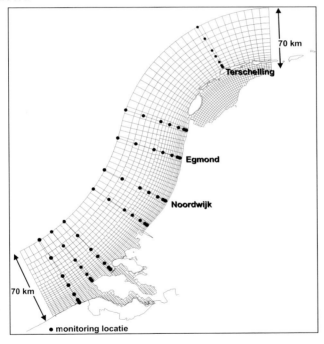

Figure 1 Grid layout and monitoring locations.

Hydrodynamics   Size, arrangement and transport between computational elements was derived from a 3-D hydrodynamic model of the Dutch coast. From these a 2D vertically averaged schematisation was determined for the water quality model. The water quality model consists of over 2000 elements with varying sizes of about 1x1 km near the coast up to a few square kilometres at the model boundaries. With the underlying hydrodynamic model eleven representative situations were computed for different combinations of wind speed and direction. An attempt was made to simulate actual transports by interpolation between the eleven flow fields. However, computations with BLOOM showed that the results lacked realism. For instance at the Noordwijk 10 station spring values of nitrate were computed for the calibration year 1990 that were clearly outside the range that was observed over the entire 20 year period. Instead a different approach was adopted. The residual flows computed for the average south western winds were used, but with a dispersive correction factor for other wind directions. So if in a particular period during a particular year the wind is for instance north west, the horizontal dispersion perpendicular to the coast is increased. This approach proved to be quite successful in reproducing observations.

Nutrient loading   Data for nutrient loading for the entire 20 year period were collected by the National Institute of Coastal and Marine Management. The most important source of nutrients is the river Rhine. For the calibration of the model the 1990 measurements were used to estimate the nutrient loading.

Sea boundaries   The model area has open boundaries to the south, west and north (See Figure 1). Since the prevailing transport direction along the Dutch coast is from south west to north east, the southern boundary is by far the most important. Concentration values here were derived from the measurements at the nearby Appelzak transect. Values for elements for which there are no measurements, were obtained by fitting an exponential curve through the observations. Little variations are observed along the western boundary. Therefore concentration values here were obtained by simply taking the average of all observations at all stations 70 kilometres off the coast. Concentrations at the northern boundary were obtained from measurements at the Rottum transact. Notice that this is by far the least important boundary due to the prevailing direction of the currents. Most of the model substances are measured directly and the observed data were directly used as boundary concentrations. For other substances such as the biomasses of different types of phytoplankton some conversion functions were needed as there are no direct measurements. These functions are based on the experience obtained in numerous studies for both fresh and salt water systems.

Model forcing   Data for the solar radiation, wind speed and wind direction were obtained from the Royal Dutch Meteorological Institute. Data for the water temperature were provided by the National Institute of Coastal and Marine Management. As there are little variations within the model area, the same forcing is used for all computational elements.

Light attenuation   Light is an important limiting factor for phytoplankton in Dutch coastal waters. Determining adequate values for the light attenuation factor is therefore important for the model performance. Originally the total contribution of inorganic material was computed as the sum of the background extinction coefficient and the extinction due to suspended matter, which is computed by a sub-model. Seasonal variations are (simply) computed using a cosine function. This approach proved to be adequate in reproducing the observed suspended matter data. It was, however, impossible to compute correct values for the extinction contribution due to non algal material using these suspended values and a single, uniform conversion coefficient. In particular in regions with large fresh water discharges the extinction coefficient was under estimated. Hence the model would compute algal biomasses exceeding the observations. Using a larger specific coefficient for suspended matter would improve the model's results near the coast, however at the expense of off-shore regions where the inorganic fraction of the extinction coefficient would be too large. Thus an additional term was introduced into the equation to compute the extinction based upon the local salinity values. This was previously proposed for the Eastern Scheldt (Peeters et al., 1991). The underlying assumption is that fresh water contains various substances such as humic acids contributing to the extinction. The coefficients by which the salinity and suspended matter are multiplied have been determined by calibration using observations on the extinction coefficient from the EUZOUT project for 1987 through 1990 (Peeters et al., 1991).

Water quality and ecological processes   At the beginning of the project a calibrated North Sea BLOOM version existed for the GENO grid. The set of processes and their corresponding coefficient values were used as the starting point for the model calibration. In general the appropriate value for a model coefficient must always be regarded in relation to the process formulations adopted by that model. In other words the coefficient values will always compensate some of the inaccuracies of a model. So when the moderately refined GENO grid was replaced by the much more detailed grid used in the trend analysis study it was conceivable that many changes to the coefficients for the water quality processes would be necessary. Fortunately preliminary results showed that only some minor changes were necessary.

Light dependency   So far it was assumed that the light dependency of growth was best described by an optimum curve as is often observed in the laboratory. Hence growth declined at intensities exceeding the optimum value (photoinhibition). From a further analysis of the mixing patterns and light intensities it became obvious that inhibition will be a rare phenomenon in the Dutch coastal waters considered here: in general phytoplankters simply do not spend a sufficient amount of time at high intensities. Therefore the light dependency curves have been adapted, maintaining the initial part, but now assuming the growth efficiency to remain equal to 1.0 at the optimum and all higher intensities. Due to this modification the average growth rates increase, which is particularly important at locations where light limitations frequently occur. With the old curves considering photo-inhibition, the maximum biomasses in relatively shallow locations near the coast could not be reproduced as they proved to be considerably below the observations.

Maximum growth rates   In relation to the changes in the light response curves some minor adjustments were also made to the maximum growth rates of the species groups considered in North Sea BLOOM. In general, maximum growth rates were somewhat decreased. This is true in particular for dinoflagellates. Literature sources indicate that a further reduction might still be necessary, but within the current model set-up this would practically eliminate this group from the output.

Stoichiometry   The combination of the growth rates and the stoichiometric coefficients determine which type will dominate under each set of conditions. To improve the selection of types, some minor adjustments with respect to the previously used coefficient values have been made. These modifications have no significant effect on the total biomass, only on its composition.

139

Mortality and sedimentation Previously diatoms were the only algae group which were considered to settle. Observations indicate that in general, algal biomasses often decline rapidly immediately following a bloom. There is some debate on the mechanisms, however. It is often assumed that the grazing pressure increases (external cause). As an alternative assumption the vitality of phytoplankters might decline under nutrient stress due to a degradation of their fitness (internal cause). Perhaps both mechanisms are important. What ever the mechanism, the result is a rapid decline in biomass and an enhanced production of detritus. This effect can be easily included in the model by a change in the net sedimentation rate of nutrient limited types. During the calibration procedure a settling rate 0.5 m/day has been established. This value is used for all groups except dinoflagellates, as sedimentation is probably insignificant for this group with active buoyancy regulation. Compared to previous model versions concentration peaks become steeper, first of all due to the enhanced sedimentation rate when nutrients get depleted but secondly because a larger fraction of the nutrients is now mineralised at the bottom rather than in the water column. Since degradation processes proceed at a lower rate in the bottom compared to the water phase, there is shift towards the end of the season in the recycling of nutrients. As a result, biomasses during summer tend to be lower in nutrient limited locations, but distinct autumn peaks are now a more common phenomenon in the model. In general this behaviour compares better with measurements.

CALIBRATION RESULTS

There are numerous ways to consider the results of a complicated model such as North Sea BLOOM because variations in both space and time for a large number of substances and fluxes might be considered. Here three types of outputs are presented:

1. Times series of some important substances at two representative locations,
2. Gradients perpendicular to the coast for two substances at one transect,
3. Geographical plots for some important substances at three representative times.

Time series A large number of potential locations and years were available for the calibration of the model. Only the model results for 1990 have been used during the calibration and a careful comparison between model outputs and observations has been made for all locations monitored during this particular year. The substances that we considered have been mentioned in "Methodology". The 1990 model results are presented against the mean and median of the observations over the period 1975-1993. To give an indication of the temporal variations the 16.6% and 83.3% quantils are shown.

Figure 2 shows the time series for the location Noordwijk 10, where nutrient levels are so high that light energy is the most important limiting factor. Both the calculated concentration of about 20 µg/l and calculated timing of the chlorophyll spring peak agrees well with the measurements. Computed summer values tend to be high in comparison to the observations. Computed levels of nitrate start at about 0.60 mg/l and achieve a maximum value in March of about 1.0 mg/l, after which they decline rapidly due to uptake by phytoplankton. They remain low during the growing season until August and rise to about 0.60 mg/l by the end of the year. The agreement with the observations is excellent. Computed and observed ortho phosphate levels agree well, but in contrast to nitrate (and silicate) the model tends to under estimate the summer values. The main reason for this discrepancy is most likely the lack of inorganic phosphate adsorption and desorption in the present model version (See "Conclusions and future developments"). Computed and observed levels for silicate are very similar to those for nitrate. The main difference is the period at which silicate is virtually depleted due to uptake by diatoms: a period of about 100 days (from May to August).

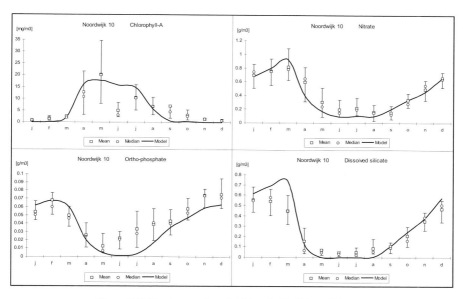

Figure 2 Calibration result station Noordwijk 10 km offshore.

Figure 3 shows the results for the location Terschelling 4, where all three macro nutrients get depleted during the growing season. The computed chlorophyll peak is of the same magnitude (20 µg/l) and occurs earlier compared to Noordwijk 10. Notice that a clear autumn peak is computed in August In general the computed levels agree well with the observations. Computed and observed nitrate levels are about 40 percent below those at Noordwijk 10. As a result this nutrient is now limiting for a period of almost 100 days. Computations reflect measurements accurately. There is considerable variation in the measured phosphate levels at his station. In winter computed and measured levels agree well. The observed summer levels are under estimated by the model. There is an excellent agreement between computed and measured levels of silicate at this station.

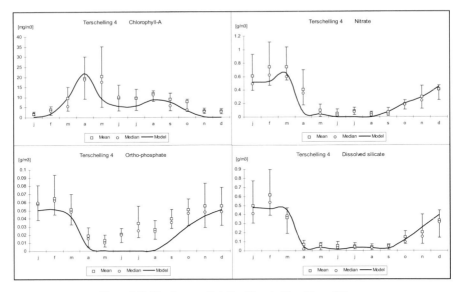

Figure 3 Calibration result station Terschelling 4 km offshore.

Gradients perpendicular to the coast   With the Dutch coastal waters the concentrations of various substances may vary an order of magnitude over a distance of about 50 km. This is due to several factors: (1) large nutrient discharges from the rivers Rhine and Meuse which remain in a 'plume' along the coast, (2) a large increase in depth from about 5 m adjacent to the coast up to about 20 m at 10 km off shore and (3) high suspended matter concentrations and hence a high turbidity at locations near the coast. The regions at which these phenomena manifest themselves are not identical, but do show considerable overlap. The resulting gradients lead to variations in concentrations over distances of only a few km. To simulate these by a model, the grid size must be of the same order of magnitude, which is the case in the coastal waters model. As an example of the model performance the results for chlorophyll and ortho phosphorus at the Egmond transect in May 1990 are presented (Figure 4). Together with the model results the average, maximum and minimum value of the observations from 1975 up to 1993 are shown. It may be concluded that the model results are close to the averages of the observations for both substances. This is even true for the first few km offshore, where the level of variation in the measurements is relatively large.

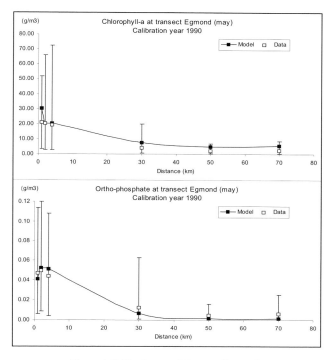

Figure 4 Calibration result transect Egmond.

Geographical images   The model results for individual locations or for a transect at a single time-step, compare well with the observations and show that a broad range of conditions can be adequately reproduced by the model. To illustrate the spatial behaviour of the model a number of characteristic images for chlorophyll, the main limiting factors and the species groups will be presented. The time steps selected for presentation are begin April during the spring bloom, begin June in the midst of summer and begin August during the autumn bloom.

Spring peak (Figure 5)   Chlorophyll levels are relatively high during the spring bloom varying from 4 up to 50 µg/l in some regions. Light energy is by far the most important limiting factor; there is no nitrogen limitation, and phosphorus limitation is confined to a small region. A complex interplay of depth, transport and suspended matter leads to a biomass pattern in Dutch coastal waters where biomass levels are highest adjacent to the coast and in some regions between 20 and 40 km off-shore. Concentrations are comparatively low at locations 10 km offshore because here the background extinctions are generally high (suspended matter and fresh water) and the depths are already in the order of 20 meters. The model biomass mainly consists of diatoms and *Phaeocystis*; flagellates and dinoflagellates are virtually absent.

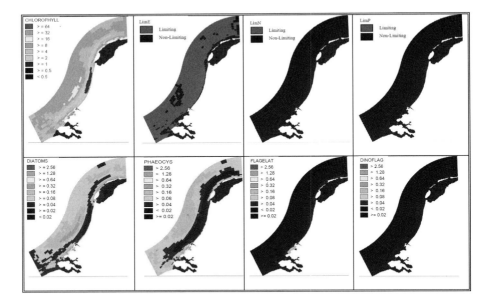

Figure 5 Simulated spatial distribution of chlorophyll-*a*, limiting factors and species composition begin April.

Summer situation (Figure 6)   Two months later during the summer the situation has changed drastically. Light limitation is now confined to areas with high suspended matter concentrations. These are the erosion areas in front of the coast of Zeeland (relatively low nutrient concentrations) and the plume of the river Rhine (high nutrient concentrations). Everywhere else nitrogen and phosphorus are limiting, usually simultaneously. For chlorophyll, the overall picture is considerably different compared to begin April. In areas where light is limiting, concentrations are generally higher because the solar radiation has increased. In areas where nutrients have become limiting concentrations are significantly lower than begin April. This is true in particular where the distance to the Rhine outflow larger. Thus chlorophyll decreases as the distance from the coast increases. The changes in the pattern of limiting factors also lead to changes in the composition of the phytoplankton. Diatom concentrations have declined, but still this group is present nearly everywhere. *Phaeocystis* is out competed in many regions by flagellates due to phosphorus limitation and is now confined to a small range of 20 km immediately off shore. Flagellates are now the most abundant group, though not in the south, where light is the main limiting factor. Here dinoflagellates start to increase.

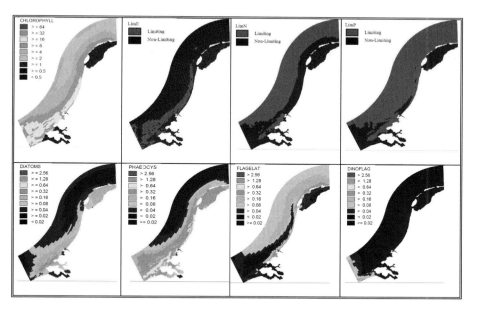

Figure 6 Simulated spatial distribution of chlorophyll-*a*, limiting factors and species composition begin June

Autumn situation (Figure 7)   During the autumn, light limitation is restricted to more or less the same areas compared to the summer situation. The coverage of areas with nitrogen limitation has slightly increased. There is less phosphate limitation in the coastal zone and no phosphate limitation outside the 50 km zone. The chlorophyll pattern is more or less the same compared to the summer situation, and the levels are of the same magnitude. The species composition has changed drastically. The group of dinoflagellates has become dominant as they have a relatively high growth rate at higher temperatures and are well adapted to nitrogen limitation. Flagellates remain in a small strip before the coast of Holland and the Wadden Sea where there is still phosphate limitation. Diatoms only appear in light limited zones like the plume of the river Rhine.

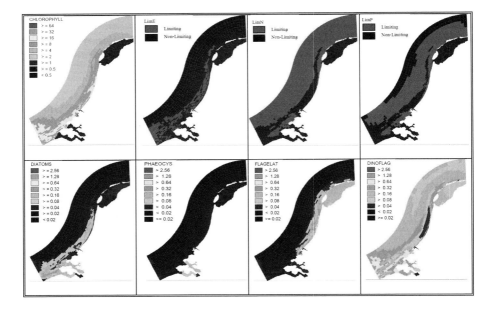

Figure 7 Simulated spatial distribution of chlorophyll-*a*, limiting factors and species composition begin August.

Trend analysis

After calibration, the model was used to simulate a 20 year period from 1975 up to 1994. The simulation takes the variability of nutrient discharges, irradiation and water temperature into account. Natural variations and/or possible trends in transport, boundary conditions and suspended matter concentrations were not taken into account. In combination with measurements, the simulation results were analysed to determine the changes in system behaviour over time: the actual trend analysis (de Vries et al., 1997). The simulation can also be regarded as a model validation because none of the model coefficients where changed. To give an impression of the model behaviour Terschelling 4 km offshore was chosen, a location which is relatively sensitive towards changes in (anthropogenic) nutrient discharges by the river Rhine. At this location a small downward trend in both measured and computed chlorophyll concentrations is detected as a result of a decrease in nutrient discharges starting mid 80's. Figure 8 shows slightly increasing concentrations of ortho-phosphate from 1975 up to 1980, a stabilisation at rather high levels, followed by a decrease. Every year ortho-phosphate is depleted. The period in which ortho-phosphate is depleted has become somewhat longer during recent years. Measured and simulated nitrate and silicate agree well. There is no observed or simulated trend. Depletion of both nutrients occurred every summer.

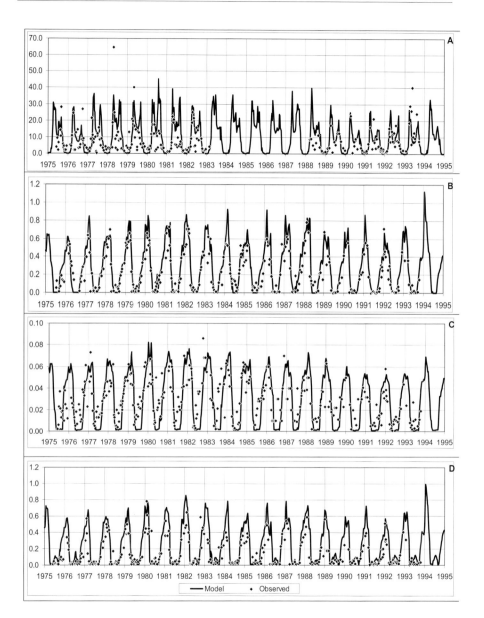

Figure 8   Validation result for 2D BLOOM/GEM application: station Terschelling 4, 1975-1994 (A: chlorophyll-a in μg.l⁻¹, B: nitrate in mg.l⁻¹, C: ortho-phosphate in mg.l⁻¹, D: dissolved silicate in mg.l⁻¹)

Conclusions and future developments

Replacing the DYNAMO model by BLOOM and high resolution transport modelling resulted in a huge step forward in modelling eutrophication in the Dutch coastal zone. The model provides an adequate reproduction of a twenty year period of field observations and reproduces steep gradients perpendicular to the coast. Still further improvement of the biological processes in the model is foreseen in the near future.

Biological processes   The model provides a rather reliable reproduction of observed chlorophyll levels. The modelled changes in species composition is however less reliable because of insufficient knowledge of certain (groups of) species. In co-operation with the Dutch Institute for Research of the Sea (NIOZ) competition experiments are being carried out to establish BLOOM coefficients for various algal species, thus improving the capability of the model to compute species composition. During summer, the computed chlorophyll levels are high compared to in-situ measurements. This is likely caused by grazing of zooplankton, a process not included in the model. The model will be extended with dynamical grazing. The modelling of sorption kinetics of phosphate and the sediment-water exchange of nutrients is somewhat simplistic. A more elaborate model of the sediment and inorganic kinetics of phosphate is operational and will be applied in the near future.

Trend analysis   The anthropogenic phosphorus discharges into the Dutch part of the North Sea have declined recently. However the impacts on the North Sea ecosystem have so far been limited. Chlorophyll levels, an important target parameter of Dutch policy, are hardly affected. Only in phosphate limited areas, for example near the coast of Terschelling, chlorophyll levels declined slightly. The nitrogen discharges have not declined over the past twenty years. A reduction seems desirable because algal growth is frequently limited due to the depletion of nitrogen.

References

Delft Hydraulics (1991). Mathematical simulation of algae blooms by the model BLOOM II. *Report T68.*

Delft Hydraulics (1992). Proces formuleringen DBS. Report T542 (inDutch).

Delft Hydraulics (1994). Model validation study DBS in networks. *Report T1210.*

De Vries, I., R.N.M. Duin, J.C.H. Peeters, F.J. Los, M. Bokhorst, R.W.P.M. Laane (1998). Patterns and trends in nutrients and phytoplankton in Dutch coastal waters: comparison of time series analysis, ecological model simulation and mesocosm experiments. ICES Journal of Marine Science, Vol. 55: 620-634.

Peeters, J.C.H., H.A. Haas, and L. Peperzak (1991). Eutrofiering, primaire produktie en zuurstofhuishouding in de Noordzee. *National Institute of Coastal and Marine Management. Report GWAO-91.083 (in Dutch)*

Chapter 8

GEM: a generic ecological model for estuaries and coastal waters

Anouk N. Blauw, Hans F. J. Los, Marinus Bokhorst, Paul L. A. Erftemeijer, 2009

Hydrobiologia DOI 10.1007/s10750-008-9575-x

GEM: a generic ecological model for estuaries and coastal waters

Anouk N. Blauw • Hans F. J. Los • Marinus
Bokhorst • Paul L. A. Erftemeijer

Received: 17 March 2008/Revised: 18 August 2008/Accepted: 24 August 2008
©Springer Science+Business Media B.V. 2008

Abstract The set-up, application and validation of a generic ecological model (GEM) for estuaries and coastal waters is presented. This model is a comprehensive ecological model of the bottom of the foodweb, consisting of a set of modules, representing specific water quality processes and primary production that can be combined with any transport model to create a dedicated model for a specific ecosystem. GEM links different physical, chemical and ecological model components into one generic and flexible modelling tool that allows for variable sized, curvilinear grids to accomodate both the requirements for local accuracy while maintaining a relatively short model run-time. The GEM model describes the behaviour of nutrients, organic matter and primary producers in estuaries and coastal waters, incorporating dynamic process modules for dissolved oxygen, nutrients and phytoplankton. GEM integrates the best aspects of existing Dutch estuarine models that were mostly dedicated to only one type of ecosystem, geographic area or subset of processes. Particular strengths of GEM include its generic applicability and the integration and interaction of biological, chemical and physical processes into one predictive tool. The model offers flexibility in choosing which processes to include, and the ability to integrate results from different processes modelled simultaneously with different temporal resolutions. The generic applicability of the model is illustrated using a number of representative examples from case studies in which the GEM model was successfully applied. Validation of these examples was carried out using the 'cost function' to compare model results with field observations. The validation results demonstrated consistent accuracy of the GEM model for various key parameters in both spatial dimensions (horizontally and vertically) as well as temporal dimensions (seasonally and across years) for a variety of water systems without the need for major reparameterisation.

Keywords Generic ecological model • GEM •
Nutrients • Phytoplankton modelling •
Validation • Water quality

Handling editor: P. Viaroli

A. N. Blauw • H. F. J. Los • P. L. A. Erftemeijer
Deltares (formerly WL | Delft Hydraulics), PO Box 177, 2600
MH Delft, The Netherlands
e-mail: paul.erftemeijer@deltares.nl

M. Bokhorst
Centre for Water Management (formerly National Institute for
Coastal and Marine Management/RIKZ), Zuiderwagenplein
2, 8224 AD Lelystad, The Netherlands

Introduction

Growing stresses from conflicting human demands, anthropogenic impacts and climate change on coastal and estuarine environments (UNEP, 2006; Airoldi &

151

Beck, 2007), along with requirements from recent international legislative agreements (e.g. EU Water Framework Directive) to control and reduce undesirable ecosystem changes (Devlin et al., 2007), demand greater knowledge and understanding of the dynamics and driving forces of these complex water systems. Three-dimensional ecological models have the capacity to provide consistent distributions and dynamics of the lower trophic levels on their regional, annual and decadal scales, which cannot be derived to this degree of coverage by field monitoring observations (Moll & Radach, 2003).

Estuarine and coastal waters pose a challenge to modellers, in terms of physics, biogeochemistry and ecology. Substantial river discharges and relatively shallow nearshore waters often result in large fluctuations and strong spatial gradients in salinity, suspended matter concentrations, nutrient concentrations and algal biomass in such water systems. These characteristics, along with complex benthic–pelagic interactions and light attentuation issues, have proven difficult to replicate in models (Radach & Moll, 2006), in particular with regard to scales of temporal and spatial resolution required to simulate the possible impacts of future conditions, including management scenarios.

Over the past decades, a relatively large number of models have been developed for simulating nutrient cycles, primary production and ecosystem functioning in Dutch estuarine and coastal waters. Examples include MANS (Los et al., 1994), North Sea BLOOM (Los & Bokhorst, 1997; Los & Wijsman, 2007), SMOES (Klepper et al., 1994; Scholten & Van der Tol, 1994), MOSES (Soetaert & Herman, 1995a, b; Soetaert et al., 1994), ECOWASP (Brinkman, 1993) and ERSEM (Baretta et al., 1995). In addition, various model applications simulating aspects of ecosystem functioning of the North Sea (and adjacent Dutch coastal waters) have been developed at WL | Delft Hydraulics in response to specific management questions, often in close cooperation with the National Institute for Coastal and Marine Management (De Vries et al., 1998; Los & Bokhorst, 1997; De Kok et al., 1995, 2001; Gerritsen et al., 2001; Delhez et al., 2004). These latter models typically have a relatively high spatial resolution, especially in the coastal zone, compared to most other North Sea models (Moll & Radach, 2003), but they traditionally use relatively simple (if any) model formulations for

food web interactions and organisms other than phytoplankton. This has been in contrast with most ecological models, which are usually detailed in ecological parameterization for specific ecological processes but lack spatial resolution in the underlying hydrodynamics and are mostly developed for application in only one geographic area. The ECOWASP model (Brinkman, 1993), for example, simulates the population ecology of mussels in the Wadden Sea at the level of size classes and year classes. The spatial resolution of such models, however, has been very low, ranging from 6 segments in the Wadden Sea (ECOWASP) to large ICES boxes, including the entire cross-shore gradient in Dutch coastal waters in only one model segment (ERSEM) (Baretta et al., 1995).

All of these models have proven useful tools in scientific research of estuarine ecosystems and for site-specific scenario studies of management strategies (Moll & Radach, 2003). Most were, however, developed for a specific region, focussed on a specific area of expertise, differed markedly in model complexity and level of temporal and spatial resolution, and served different objectives. Typically, such individual process-oriented ecological models perform well for the particular water system for which they were developed, but when applied to other systems, their performance tends to be poor even after reparameterization (Fitz et al., 1996).

Therefore, the Dutch National Institute for Coastal and Marine Management (RIKZ) initiated the development of a Generic Ecological Model for estuaries (GEM), an integrated model that includes physical, chemical and ecological processes at a sufficient level of detail and in a consistent way. The goal of the generic ecological model was to integrate the best aspects of the existing (Dutch) models and expertise that are dedicated to one ecosystem or only a subset of relevant processes into an integrated coherent model that allows for general application to different coastal and estuarine systems.

The resulting GEM model has been applied for over a decade in a range of different consultancies and studies by WL | Delft Hydraulics, but the model set-up has not yet been scientifically published. A recent audit of the GEM model by an independent panel of international scientific experts (including Alain Menesguen, Paul Tett and William Silvert) strongly encouraged publication of the model (in particular, the approach for phytoplankton) and its promising results (Van de Wolfshaar, 2007).

The present paper describes the background, model set-up, application and validation of GEM. Our main objectives were to integrate biological and physical processes in a simulation of basic ecosystem dynamics for generic application to estuarine and coastal waters. The generic applicability of the model is illustrated using a number of representative examples from four selected case studies encompassing different spatial and temporal dimensions and a variety of different water systems in which the GEM model was successfully applied and validated.

Description of the 'gem' model

Modelling environment

GEM consists of a subset of process formulations from the process library of DELWAQ: the program for modelling water quality and aquatic ecology in the Delft3D modelling suite (WL | Delft Hydraulics, 2003). Delft3D-WAQ, Delft3D-ECO, Delft3D-SED and DBS are other subsets from the same process library that partly overlap with GEM. DELWAQ uses a finite grid approach. Sources and sinks of variables due to processes in the water are included in the advection–diffusion equation. A large selection of numerical schemes is available to solve the transport part in the advection–diffusion equation below. Processes (P) are all simulated with an explicit numerical scheme.[1]

$$\frac{\partial C}{\partial t} = -u\frac{\partial C}{\partial x} - v\frac{\partial C}{\partial y} - w\frac{\partial C}{\partial z} + \frac{\partial}{\partial x}(D_x\frac{\partial C}{\partial x})$$
$$+ \frac{\partial}{\partial y}(D_y\frac{\partial C}{\partial y}) + \frac{\partial}{\partial z}(D_z\frac{\partial C}{\partial z}) + S + P$$

where

C: concentration (g m^{-3})
u, v, w: components of the velocity vector (m s^{-1})

D_x, D_y, D_z: components of the dispersion tensor (m^2 s^{-1})
x, y, z: coordinates in three spatial dimensions (m)
S: source and sinks of mass due to loads and boundaries
P: sources and sinks of mass due to processes
t: time (s)

The advection and diffusion fluxes between grid cells are usually derived from a hydrodynamic model (e.g. Delft3D-FLOW) for the same model area. DELWAQ has been used successfully for many different types of applications, including the simulation of dredging plumes, thermal discharges and various water pollution studies (Van Gils et al., 1993; Van der Molen et al., 1994; Ouboter et al., 1998; Van Gils, 1998).

Not all processes incorporated need the same level of detail with respect to time step and grid size. DELWAQ enables the use of different time steps and grids for different processes. The model will then aggregate and de-aggregate the input and output parameters of the processes. One can, for example, use a different time step for transport (tide resolving) and water quality processes, which in general show less steep gradients in both space and time. This way a considerable reduction of simulation time can be achieved with only a limited loss in accuracy.

The set-up of a GEM model application for a specific area and period comprises, besides the GEM set of processes and parameter setting, input for schematisation and transport, loadings, boundaries, forcings and initial conditions. GEM can be combined with any hydrodynamic model and additional processes to create a dedicated model application for a specific ecosystem.

Model set-up

GEM comprises a set of process formulations quantifying the P term of the advection–diffusion equation. The set of process formulations is dedicated to modelling the nutrient cycling and primary production in coastal and estuarine systems. The approach for the development of GEM started from an existing ecological model at WL | Delft Hydraulics, which proved useful for the simulation of eutrophication in Dutch coastal waters (e.g. Los & Bokhorst, 1997; De Vries et al., 1998). In a series of

---

[1] Numerical schemes used in the four case studies described in this paper were as follows: implicit upwind scheme with an iterative solver (Case 1, Veerse Meer); horizontal: FCT Scheme, vertical: implicit in time and central discretisation (Case 2, North Sea); flux correct transport (FCT) method (Case 3, Venice Lagoon); and horizontal upwind scheme, vertical: implicit in time and central discretisation (Case 4, Sea of Marmara).

research and consultancy projects, partly in cooperation with other Dutch institutes for marine research, GEM has been further elaborated and improved. The resulting GEM includes default parameter settings that have been calibrated for the North Sea and that have proven to be applicable for a range of other coastal ecosystems as well.

GEM simulates the nutrient cycles of nitrogen (N), phosphorus (P) and silicate (Si). For the dissolved inorganic state of these nutrients, the following state variables are included in the model: nitrate (NO_3) representing the sum of nitrite and nitrate, ammonia (NH_4), ortho-phosphate (PO_4) and dissolved silicate (Si). Four phytoplankton species groups are simulated: diatoms, flagellates, dinoflagellates and Phaeocystis. Dead particulate organic matter in water is included as separate variables for particulate organic carbon (POC), particulate organic nitrogen (PON), particulate organic phosphorus (POP) and opal silicate (POSi). Similarly, four organic matter variables are defined in the sediment (POC_S, PON_S, POP_S and $POSi_S$). Additional model variables are dissolved oxygen (O_2) and salinity (SAL). Suspended matter concentrations and water temperature are forcing parameters in the model.

Processes included in GEM

Figure 1 gives a schematic overview of the variables and processes incorporated in the present set-up of GEM. GEM includes the following processes:

- phytoplankton processes: primary production, respiration and mortality
- extinction of light
- decomposition of particulate organic matter in water and sediment
- nitrification and denitrification
- reaeration
- settling
- burial
- filterfeeder processes: grazing, excretion and respiration

The above-mentioned processes in GEM are described in more detail below. For some processes, alternative (more detailed) formulations are available and additional processes are available as well. These are used for specific model applications that require more detailed process formulations, for example, for

nutrients in the sediment, phosphate adsorption, salinity effects on mortality and decomposition of organic matter or benthic algae. In this paper, however, we focus on the set of formulations that is most commonly used. Equations used in GEM, state variables and process parameter (default) settings, and phytoplankton parameter values in the BLOOM module are presented in Tables 1–3, respectively.

Phytoplankton processes

In GEM, the phytoplankton module, BLOOM, simulates primary production, respiration and mortality of phytoplankton. This module allows for the modelling of species competition and adaptation of phytoplankton to limiting nutrients or light (Los et al., 1984; Los & Brinkman, 1988; Van der Molen et al., 1994; Los & Bokhorst, 1997; Los, 2005). For the simulation of species competition, four species groups are defined in GEM: diatoms, flagellates, dinoflagellates and Phaeocystis. BLOOM can also be used for other species groups, including fresh water species and benthic species, but these are generally not used in GEM. Within each of these groups, three phenotypes are defined to account for adaptation to changing environmental conditions:

- energy types, with relatively high growth rates, low mortality rates and high N/C and P/C ratio
- nitrogen types, with typically lower internal N/C ratio, lower maximum growth rates, higher mortality rates, higher settling velocities and higher chlorophyll content
- phosphorus types, with typically lower internal P/C ratio, lower maximum growth rates, higher mortality rates, lower settling velocities and lower chlorophyll-a content

The different phenotypes of a species group are modelled as separate variables with different parameter settings, e.g. growth rates, settling velocities and respiration rates. When conditions in the water change, biomass of one phenotype can be instantaneously converted into another phenotype of the same species group, representing rapid adaptation of individual algal cells. Since the phytoplankton types represent different phenotypes of the same species, the transition of one type to another occurs at the time scale of a cell division, which is in the order of a day. Due to this characteristic time scale (i.e. Chapter 5 of

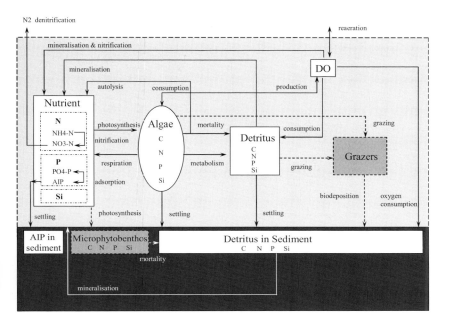

Fig. 1 Schematic overview of all state variables and processes included in the ecological model instrument GEM. State variables in grey and processes indicated by dashed lines are optional and have not been included in the North Sea modelling applications. AIP is 'adsorbed inorganicphosphorus'

Harris, 1986), the simulation time step for the BLOOM phytoplankton processes is usually chosen to be 24 h. Nutrient fluxes associated with switches between types are dealt with in the following way: If the nutrient content requires a larger amount of nutrients than previously stored in the phytoplankton, it is taken up from the dissolved fraction in the water. If, in contrast, the nutrient content declines, the extraneous amount previously stored in the phytoplankton is released into the dissolved fraction in the water. This is similar to autolysis. Switch-associated uptake of one nutrient and release of another may occur simultaneously, i.e. when a P-type is replaced by an N-type. A shift in species composition due to changing environmental conditions is a slower process, involving changing dominance of species groups. This shift in species composition is restricted by growth and mortality rates. BLOOM allows for simultaneous co-existence of different phenotypes and/or species groups, representing a continuous variable stoichiometry restricted only by the limits set for the individual types. Notice that the stoichiometry of the types is fixed, but the relative proportion varies (see Los et al. (2008) for a more detailed description of BLOOM). BLOOM was originally developed as a steady-state model that is now being applied in a dynamic setting. In a typical simulation, transport and a small number of processes are simulated with a short time step (i.e. 30 min of less), while the majority of processes are simulated using a much longer time step of 24 h. Consistency and mass conservation issues due to the different time steps and approaches are resolved by the numerical solver of the mode. Comparative simulations have shown that the results of simulations with a 30 min time step for all processes are similar to those with

Table 1 Equations in GEM

A. Balance equations for state variables:

$$\frac{dNO_3}{dt} = nit - den - upt_N * (1 - f_{am})$$ (A.1)

$$\frac{dNH_4}{dt} = dec_{PON} - nit + dec_{PON_S} + rsp_{N,G} + f_{aut} * mor_N - upt_N * f_{am}$$ (A.2)

$$\frac{dPO_4}{dt} = dec_{POP} + dec_{POP_S} + rsp_{P,G} + f_{aut} * mor_P - upt_P$$ (A.3)

$$\frac{dSi}{dt} = dec_{POSi} + dec_{POSi_S} + f_{aut} * mor_{Si} - upt_{Si}$$ (A.4)

$$\frac{dALG_i}{dt} = gro_i - mrt_i - sed_i - grz_i$$ (A.5)

$$\frac{dO_2}{dt} = rea + (gro_C - dec_{POC} - dec_{POCs}) * s_o - nit * s_{no}$$ (A.6)

$$\frac{dPOX}{dt} = mor_X * (1 - f_{aut}) - sed_{POX} - dec_{POX} - grz_{POX} + exc_{POX}$$ (A.7)

$$\frac{dPOXS}{dt} = (sed_{POX} - dec_{POXs} - bur_{POXs} + exc_{POXs}) * Z$$ (A.8)

where:

i = algae type 1 – 12

X = element carbon, nitrogen, phosphorus and silicate

$$upt_X = \sum_{i=1}^{n} (gro_i * s_{X,i})$$ (A.9)

$$mor_X = \sum_{i=1}^{n} (mrt_i * s_{X,i})$$ (A.10)

$$f_{am} = \frac{MIN(NH_4, upt_N * \Delta t)}{upt_N * \Delta t}$$ (A.11)

Table 1 continued

B. Phytoplankton processes

$$gro_i = \frac{ALG_{i,new} - ALG_i}{\Delta t} + mrt_i \tag{B.1}$$

$$mrt_i = m_i * ALG_i \tag{B.2}$$

$$p_i = p_{i,0} * (T - kt_{p,i}) \tag{B.3}$$

$$r_i = r_{i,0} * kt_{r,i}^T \tag{B.4}$$

$$pg_i = p_i + r_i \tag{B.5}$$

$$m_i = m_{i,0} * kt_{m,i}^T \tag{B.6}$$

Optimisation: find set of new concentrations of ALG_i with maximum:

$$\sum_{i=1}^{n} (pg_i * le_i - r_i) * ALG_i \tag{B.7}$$

satisfying the following constraints:

$$\sum_{i=1}^{3} ALG_{i,new} \leq \sum_{i=1}^{3} (ALG_i) e^{(pg_i * le_i - r_i)*\Delta t} \qquad \text{(growth constraint per species group)} \tag{B.8}$$

$$\sum_{i=1}^{3} ALG_{i,new} \geq \sum_{i=1}^{3} ALG_i * e^{-m_i * \Delta t} \qquad \text{(mortality constraint per species group)} \tag{B.9}$$

$$\sum_{i=1}^{n} (s_{N,i} * ALG_{i,new}) \leq \sum_{i=1}^{n} (s_{N,i} * ALG_i) + NO_3 + NH_4 \qquad \text{(nutrient constraint, nitrogen)} \tag{B.10}$$

$$\sum_{i=1}^{n} (s_{P,i} * ALG_{i,new}) \leq \sum_{i=1}^{n} (s_{P,i} * ALG_i) + PO_4 \qquad \text{(nutrient constraint, phosphorus)} \tag{B.11}$$

$$\sum_{i=1}^{n} (s_{Si,i} * ALG_{i,new}) \leq \sum_{i=1}^{n} (s_{Si,i} * ALG_i) + Si \qquad \text{(nutrient constraint, silicate)} \tag{B.12}$$

$$k_{min,i} \leq k_d \leq k_{max,i} \qquad \text{(light constraint)} \tag{B.13}$$

where:

$$k_{max,i} = f_{table}(le_{cr}) \tag{B.14}$$

$$le_{cr} = \frac{m_i + r_i}{pg_i} \tag{B.15}$$

Extinction of light

$$k_d = k_b + k_{SPM} + k_{POM} + k_{ALG} + k_{HUM} \tag{C.1}$$

$$k_{POM} = e_{POC} * POC \tag{C.2}$$

$$k_{ALG} = \sum_{i=1}^{n} (e_{ALG_i} * ALG_i) \tag{C.3}$$

$$k_{SPM} = e_{SPM} * SPM \tag{C.4}$$

$$k_{HUM} = e_{HUM,0} * (1 - \frac{SAL}{SAL_b}) \tag{C.5}$$

Table 1 continued

D. Decomposition of organic matter

$$dec_{POX} = (f_{T,dec} * k_{decL,POX} + (k_{decH,POX} - k_{decL,POX}) * f_{nut}) * POX \qquad (X = C, N, P) \qquad (D.1)$$

$$dec_{POSi} = (f_{T,dec} * k_{decL,POSi} + (k_{decH,POSi} - k_{decL,POSi}) * \frac{(POSi/POC) - sl_{Si}}{su_{Si} - sl_{Si}}) * POSi \qquad (D.2)$$

$$\frac{dec_{POXS} = k_{dec,POXs} * f_{T,dec} * POX_S}{Z} \qquad (D.3)$$

where:

$$f_{T,pr} = kt_{pr}^{(T-20)} \qquad \text{where: pr = dec, den, nit} \qquad (D.4)$$

$$f_{nut} = MIN(\frac{PON/POC - sl_N}{su_N - sl_N}, \frac{POP/POC - sl_P}{su_P - sl_P}) \qquad (D.5)$$

E. Nitrification and denitrification

$$nit = k_{nit} * NH_4 * f_{T,nit} \qquad (E.1)$$

$$den = k_{den} * NO_3 * f_{T,den} \qquad (E.2)$$

F. Reaeration

$$rea = (0.3 + 0.028 * W^2) * \frac{(O_{2,eq} - O_2)}{Z} * k_{rea} \qquad (F.1)$$

G. Sedimentation

$$sed_Y = \frac{v_Y * Y}{Z} \qquad \text{where: Y = ALG}i, POX \qquad (G.1)$$

H. Burial

$$\frac{bur_{POX_S} = b * POXS}{Z} \qquad (H.1)$$

Table 1 continued

I. Filterfeeder processes

$$grz_i = up * ALG_i \tag{I.1}$$

$$grz_{POX} = up * POX \tag{I.2}$$

$$rsp_{X,G} = (ass2 * rg_{G,20} + G_i * rm_{G,20}) * f_{T,grz} * s_{X,G} \tag{I.3}$$

$$exc_{POX} = exc_X * (1 - f_{sed}) \tag{I.4}$$

$$exc_{POXs} = exc_X * f_{sed} \tag{I.5}$$

where:

$$exc_X = eg_X + ass1_X - \frac{ass4 * s_{X,G}}{\Delta t} - ass2 * rg_{G,20} * f_{T,grz} * s_{X,G} \tag{I.6}$$

$$up = G * f_{T,grz} * MIN(k_{flt,20} \frac{c_f}{c_f + k_f}, k_{upt,20} \cdot \frac{1}{c_f}) \tag{I.7}$$

$$c_f = \sum_{i=1}^{n}(ALG_i) + POC \tag{I.8}$$

$$eg_X = grz_{POX} * f_{eg} + \sum_{i=1}^{n}(grz_i * f_{eg} * s_{X,i}) \tag{I.9}$$

$$ass1_X = grz_{POX} + \sum_{i=1}^{n}(grz_i * s_{X,i}) - eg_X \tag{I.10}$$

$$ass2 = MIN(ass1_C, \frac{ass1_N}{s_{N,G}}, \frac{ass1_P}{s_{P,G}}) \tag{I.11}$$

$$\tag{I.12}$$

$$ass3 = ass2 * (1 - rg_{G,20} * e^{ktg*(T-20)}) \tag{I.13}$$

$$ass4 = MIN(ass3 * \Delta t, G_c - G * (1 - rm_{G,20} * f_{T,grz} * \Delta t))$$

$$G_c = MAX(G_i, G * (1 + g_{G,20} * f_{T,grz} * \Delta t)) \qquad \text{if } G_i > G \tag{I.14}$$

$$G_c = MIN(G_i, G * (1 - m_{G,20} * f_{T,grz} * \Delta t)) \qquad \text{if } G_i < G \tag{I.15}$$

$$G_{new} = MIN(G_c, G * (1 - rm_{G,20} * f_{T,grz} * \Delta t) + ass3 * \Delta t \tag{I.16}$$

$$f_{T,grz} = e^{ktg(T-20)} \tag{I.17}$$

variable time steps (see Los et al., 2008 for further details).

BLOOM assumes that fast-growing (less efficient) phytoplankton species dominate in a situation where resources (light, nutrients) are abundant, while slow-growing but efficient phytoplankton species gain dominance when resources become limited. This assumption is based on the theory on k- and r-strategies (e.g. Reynolds et al., 1983; Harris, 1986). Linear programming is used as an optimisation technique to determine the species composition that is best adapted to prevailing environmental

Table 2. Explanation of symbols

Symbol	Description	value	unit
	State variables		
NO_3	nitrate		$gN.m^{-3}$
NH_4	ammonium		$gN.m^{-3}$
PO_4	ortho-phosphate		$gP.m^{-3}$
Si	dissolved silicate		$gSi.m^{-3}$
O_2	dissolved oxygen		$gO_2.m^{-3}$
ALG_1	diatoms energy type: diat-E		$gC.m^{-3}$
ALG_2	diatoms nitrogen type: diat-N		$gC.m^{-3}$
ALG_3	diatoms phosphorus type: diat-P		$gC.m^{-3}$
ALG_4	flagellates energy type: flag-E		$gC.m^{-3}$
ALG_5	flagellates nitrogen type: flag-N		$gC.m^{-3}$
ALG_6	flagellates phosphorus type: flag-P		$gC.m^{-3}$
ALG_7	dinoflagellates energy type: dino-E		$gC.m^{-3}$
ALG_8	dinoflagellates nitrogen type: dino-N		$gC.m^{-3}$
ALG_9	dinoflagellates phosphorus type: dino-P		$gC.m^{-3}$
ALG_{10}	*Phaeocystis* energy type: *Phaeo*-E		$gC.m^{-3}$
ALG_{11}	*Phaeocystis* nitrogen type: *Phaeo*-N		$gC.m^{-3}$
ALG_{12}	*Phaeocystis* phosphorus type: *Phaeo*-P		$gC.m^{-3}$
POC	particulate organic carbon		$gC.m^{-3}$
PON	particulate organic nitrogen		$gN.m^{-3}$
POP	particulate organic phosphorus		$gP.m^{-3}$
POSi	particulate organic silicate		$gSi.m^{-3}$
POC_S	particulate organic carbon in the sediment		$gC.m^{-2}$
PON_S	particulate organic nitrogen in the sediment		$gN.m^{-2}$
POP_S	particulate organic phosphorus in the sediment		$gP.m^{-2}$
$POSi_S$	particulate organic silicate in the sediment		$gSi.m^{-2}$
SAL	Salinity		ppt
	Fluxes		
sed	settling		$gX.m^{-3}.d^{-1}$
mrt	phytoplankton mortality		$gC.m^{-3}.d^{-1}$
gro	net phytoplankton growth		$gC.m^{-3}.d^{-1}$
mor_X	formation of dead organic matter by phytoplankton mortality		$gX.m^{-3}.d^{-1}$
upt_X	uptake of nutrients by phytoplankton growth		$gX.m^{-3}.d^{-1}$
dec_X	decomposition of dead particulate organic matter		$gX.m^{-3}.d^{-1}$
bur_X	burial		$gX.m^{-3}.d^{-1}$
den	denitrification		$gN.m^{-3}.d^{-1}$
nit	nitrification		$gN.m^{-3}.d^{-1}$
rea	rearation		$gO_2.m^{-3}.d^{-1}$
grz_X	grazing by filterfeeders		$gX.m^{-3}.d^{-1}$
rsp_X	respiration by filterfeeders		$gX.m^{-3}.d^{-1}$
exc_X	excretion of organic matter by filterfeeders		$gX.m^{-3}.d^{-1}$

Table 2. Continued

Symbol	Description	value	unit
Δt	time step for processes (different from that of transport)	1	d
$ALG_{i,new}$	concentration of algae type i at the end of the time step	**	$gC.m^{-3}$
$ass1_X$	nett uptake flux of food of nutrient X (C, N or P)	**	$gX.m^{-3}.d^{-1}$
$ass2$	nett uptake flux corrected for stoichiometry filterfeeders	**	$gC.m^{-3}.d^{-1}$
$ass3$	maximum assimilation flux corrected for growth respiration	**	$gC.m^{-3}.d^{-1}$
$ass4$	effectively assimilated organic matter during time step	**	$gC.m^{-3}$
b	burial rate	0.0025	d^{-1}
c_f	available food concentration	**	$gC \cdot m^{-3}$
Cl	chloride concentration	***	$g.m^{-3}$
e_{ALGi}	specific extinction of algae type i	*	$m^2.gC^{-1}$
eg_X	egestion flux by grazers of nutrient X (C, N or P)	**	$gX.m^{-3}.d^{-1}$
$e_{HUM,0}$	extinction due to humic substances in pure fresh water	0.97	m^{-1}
e_{POC}	specific extinction of particulate dead organic matter	0.1	$m^2.gC^{-1}$
e_{SPM}	specific extinction of inorganic suspended matter	0.025	$m^2.g^{-1}$
f_{am}	fraction ammonium in nitrogen uptake	**	-
f_{aut}	autolysis fraction of mortality	0.3	-
f_{eg}	fraction of uptake that is egested by grazers	0.5	-
f_{nut}	function for relative nutrient availability	**	-
f_{sed}	fraction of excretion by grazers released to sediment	1	-
f_{table}	tabulated function relating k_d to growth efficiency, converted from the tabulated function of growth efficiency and light	**	m^{-1}
$f_{T,pr}$	temperature function for process pr = dec, den, nit	**	-
$f_{T,grz}$	temperature function for grazing processes	**	-
G	grazer biomass	**	$gC.m^{-3}$
G_c	feasible grazer biomass at maximum growth or mortality rates	**	$gC.m^{-3}$
$g_{G,20}$	maximum growth rate grazers at 20 °C	0.2	d^{-1}
G_i	imposed grazer biomass by forcing function	***	$gC.m^{-3}$
G_{new}	effective grazer biomass at the end of the time step	**	$gC.m^{-3}$
K_{ALG}	total extinction due to phytoplankton	**	m^{-1}
k_b	background extinction	0.08	m^{-1}
k_d	total extinction coefficient	**	m^{-1}
$k_{decL, POC}$	minimum decomposition rate for POC at 20°C	0.12	d^{-1}
$k_{decL, PON}$	minimum decomposition rate for PON at 20°C	0.08	d^{-1}
$k_{decL, POP}$	minimum decomposition rate for POP at 20°C	0.08	d^{-1}
$k_{decL, POSi}$	minimum decomposition rate for POSi at 20°C	0.04	d^{-1}
$k_{dec,POCS}$	decomposition rate for POC in sediment at 20 °C	0.015	d^{-1}
$k_{dec,PONS}$	decomposition rate for PON in sediment at 20 °C	0.015	d^{-1}
$k_{dec,POPS}$	decomposition rate for POP in sediment at 20 °C	0.025	d^{-1}
$k_{dec,POSiS}$	decomposition rate for POSi in sediment at 20 °C	0.008	d^{-1}
$k_{decH, POC}$	maximum decomposition rate for POC at 20°C	0.18	d^{-1}
$k_{decH, PON}$	maximum decomposition rate for PON at 20°C	0.18	d^{-1}
$k_{decH, POP}$	maximum decomposition rate for POP at 20°C	0.18	d^{-1}
$k_{decH, POSi}$	maximum decomposition rate for POSi at 20°C	0.08	d^{-1}
k_{den}	denitrification rate	0.003	d^{-1}
k_f	half saturation constant for grazing	0.1	$gC \cdot m^{-3}$
$k_{flt,20}$	maximum rate of filtration by mussels	0.05	$m^3 \cdot gC^{-1} \cdot d^{-1}$
k_{HUM}	extinction due to humic substances from fresh water input	**	m^{-1}
$k_{max,I}$	maximum extinction where the net growth of algae type i is positive; above this level self shading limits growth	**	m^{-1}
$k_{min,I}$	minimum extinction where the net growth of algae type i is positive; below this level photo-inhibition limits growth	0	m^{-1}

Table 2. Continued

k_{nit}	nitrification rate	0.07	d^{-1}
k_{POM}	extinction of dead particulate organic matter	**	$m^2.gC^{-1}$
k_{rea}	reaeration rate	4	d^{-1}
k_{SPM}	extinction of inorganic suspended matter	**	$m^2.gC^{-1}$
kt_{dec}	temperature coefficient for decomposition of POX and POX_S, except for opal silicate in sediment	1.11	-
$kt_{dec,POSiS}$	temperature coefficient for dissolution of opal silicate in sediment	1.047	-
kt_{den}	temperature coefficient for denitrification	1.11	-
kt_G	temperature coefficient for grazing processes	0.04	-
$kt_{m,i}$	temperature coefficient for mortality of algae type i	*	-
kt_{nit}	temperature coefficient for nitrification	1.06	-
$kt_{p,i}$	temperature coefficient for phytoplankton growth	*	°C
$kt_{r,i}$	temperature coefficient for phytoplankton respiration	1.07	-
$k_{upt,20}$	maximum food uptake rate by grazers	0.1	$gC \cdot gC^{-1} \cdot d^{-1}$
le_{cr}	critical light efficiency where phytoplankton growth just balances losses	**	-
le_i	growth efficiency of algae type i, tabulated function of light	**	-
$m_{i,0}$	mortality rate for algae type i at 0 °C	*	d^{-1}
$m_{G,20}$	maximum mortality rate of grazers at 20 °C	0.2	d^{-1}
m_i	mortality rate for algae type i	**	d^{-1}
n	number of algae types in calculation	12	-
$O_{2,eq}$	saturation concentration of oxygen	**	$gO_2.m^{-3}$
$p_{i,0}$	maximal net growth rate of algae type i at 0 °C	*	d^{-1}
pg_i	maximal gross growth rate algae type i	**	d^{-1}
p_i	maximal net growth rate for algae type i	**	d^{-1}
$r_{i,0}$	maintenance respiration rate for algae type i at 0 °C	0.06	d^{-1}
$rg_{G,20}$	growth respiration fraction of grazers at 20 °C	0.2	-
$rm_{G,20}$	maintenance respiration rate of grazers at 20 °C	0.005	d^{-1}
r_i	maintenance respiration rate for algae type i	**	d^{-1}
SAL_b	background salinity	34.97	ppt
sl_N	lower limit stoichiometric constant PON	0.1	$gN.gC^{-1}$
sl_P	lower limit stoichiometric constant POP	0.01	$gP.gC^{-1}$
sl_{Si}	lower limit stoichiometric constant POSi	0.01	$gSi.gC^{-1}$
$s_{N,G}$	stochiometry of nutrient N in grazers	.1818	$gN.gC^{-1}$
s_{NO}	oxygen nitrogen ratio in NO_3	4.571	$gO_2.gN^{-1}$
s_O	oxygen carbon ratio in detritus	2.67	$gO_2.gC^{-1}$
$s_{P,G}$	stochiometry of nutrient P in grazers	.0263	$gP.gC^{-1}$
SPM	(suspended) inorganic matter concentration	***	$gDM.m^{-3}$
su_N	upper limit stoichiometric constant PON	0.15	$gN.gC^{-1}$
su_P	upper limit stoichiometric constant POP	0.015	$gP.gC^{-1}$
su_{Si}	upper limit stoichiometric constant POSi	0.005	$gSi.gC^{-1}$
$s_{X,i}$	stochiometry of nutrient X in algae type i	*	$gX.gC^{-1}$
T	water temperature	***	°C
up	uptake rate of organic matter by grazers	**	d^{-1}
v_{ALGi}	settling velocity of algae type i	*	$m.d^{-1}$
v_{POX}	settling velocity of particulate dead organic matter	1.5	$m.d^{-1}$
W	wind velocity	***	$m.s^{-1}$
Z	water depth	***	m

*	in Table 3
**	calculated in GEM
***	model forcing

Table 3 Parameter values in the algal module (BLOOM) of the GEM model

ALGi	eALGi	$s_{N,i}$	$s_{P,i}$	$s_{Si,i}$	$s_{Chl,i}$	$p_{i,0}$	$kt_{p,i}$	$m_{i,0}$	ktm,i	V_{ALGi}
diat-E	0.24	0.255	0.032	0.447	0.053	0.083	-1.75	0.07	1.072	0.5
diat-N	0.21	0.07	0.012	0.283	0.01	0.066	-2	0.08	1.085	1
diat-P	0.21	0.105	0.01	0.152	0.01	0.066	-2	0.08	1.085	1
flag-E	0.25	0.2	0.02	0	0.023	0.09	-1	0.07	1.072	0
flag-N	0.225	0.078	0.01	0	0.007	0.075	-1	0.08	1.085	0.5
flag-P	0.225	0.113	0.007	0	0.007	0.075	-1	0.08	1.085	0.5
dino-E	0.2	0.163	0.017	0	0.023	0.132	5.5	0.075	1.072	0
dino-N	0.175	0.064	0.011	0	0.007	0.113	4.75	0.08	1.085	0
dino-P	0.175	0.071	0.01	0	0.007	0.112	4.75	0.08	1.085	0
Phaeo-E	0.45	0.188	0.023	0	0.023	0.084	-3.25	0.07	1.072	0
Phaeo-N	0.413	0.075	0.014	0	0.007	0.078	-3	0.08	1.085	0.5
Phaeo-P	0.413	0.104	0.011	0	0.007	0.078	-3	0.08	1.085	0.5

conditions. In accordance with the aforementioned principle, the suitability of a type (its fitness) is determined by the ratio of its requirement and its growth rate. It can be shown mathematically that the principle by which each phytoplankton type maximises its own benefit effectively means that the total net production of the phytoplankton community is maximised. This makes it possible to use the computationally efficient Linear Programming technique (Danzig, 1963) to compute the phytoplankton biomasses according to the competition rules formulated for the module. Mathematically, this means that the optimisation process finds the combination of phytoplankton types that maximise the net growth subject to the following set of constraints.

* growth constraint: the biomass increase of any of the species groups cannot exceed the maximum net growth rate (production minus respiration) at actual temperature and light intensity. The relation between light intensity and growth efficiency (as a fraction of the maximum net growth rate) is included in the model as tabulated P–E curves, based on laboratory studies.

* mortality constraint: the mortality rate of any of the species groups cannot exceed the maximum mortality rate at actual temperature and salinity.

* light constraints: the total extinction of light by phytoplankton cannot exceed the threshold level where the light intensity becomes insufficient to maintain further net growth. The growth response to varying light intensities is implemented as tabulated data, based on laboratory studies. The tabulated response curve is converted to account for variable depth and day length.

* nutrient constraints: the total uptake of each of the nutrients (N, P, Si) must not exceed the

availability. The total available amount of a nutrient is defined as the sum of dissolved inorganic nutrient plus the amount of nutrients in phytoplankton.

Due to mortality, phytoplankton biomass is released partly as dead particulate organic matter and partly as inorganic nutrients in the water column, the latter representing autolysis. Different kinds of freshwater and marine phytoplankton can be modelled simultaneously. The uptake of nitrogen is done preferably as ammonium ($NH_4^?$). When ammonium is depleted, the remainder of nitrogen uptake is done as nitrate. Most models require an explicit term to delimit the growth rate, forcing it to zero when a nutrient gets depleted. BLOOM does not require such a term since the Linear Programming procedure automatically stops the uptake when the concentration becomes zero. The growth rate in BLOOM declines step-wise as a nutrient gets depleted. When the E-types dominate, the net growth rate is high; then, it declines abruptly when a nutrient limited type takes over, and finally, it is put to zero when the nutrient is depleted.

Extinction of light

Light is simulated as total photosynthetic active radiation (PAR). Extinction of light by substances in the water is modelled as an exponential decrease of light intensity with depth according to the Lambert–Beer formula. The extinction coefficient is calculated as the sum of the extinction by inorganic suspended matter, particulate organic matter, phytoplankton (self-shading), dissolved humic substances (approximated by salinity) and background extinction. Each of the substances, including the phytoplankton species, is characterised by a specific extinction coefficient.

Decomposition of dead particulate organic matter

The decomposition rate of dead particulate organic matter (POM) in water is dependent on the nutrient stoichiometry of detritus, since bacteria need a supply of nutrients that is proportional to the supply of organic carbon. The decomposition rate of POM is highest for POM with a high nutrient content (expressed as N/C and P/C ratios). Therefore, if the nutrient content is above a threshold value, a high decomposition rate is applied in the model. If the nutrient content is below another threshold, a low decomposition rate is applied. If the nutrient content is in between the two thresholds, the decomposition rate is linearly interpolated. The nutrient content of the most limiting nutrient (N or P) determines the overall decomposition rate for POC, PON and POP. This approach was first applied in the freshwater model DBS as described by Van der Molen et al. (1994).

Particulate organic matter in the sediment is formed upon settling of phytoplankton and dead particulate organic matter. In the model, decomposition rate is affected only by temperature, which is assumed to be lower in the sediment than in the water. Remineralised inorganic nutrients are released back into the water column. In most GEM applications, the sediment is not included as a separate layer in the model grid. This means that only settled organic matter (C, N, P, Si) is simulated as a model variable, but nutrient concentrations and oxygen in porewater are not. Instead, remineralised nutrients are released directly back into the water column. The advantage of this simplification is the reduction of the number of grid cells and simulation time. The disadvantage is that processes that are in reality strongly non-linear, such as the release of ortho-phosphate from anaerobic sediments, cannot be taken into account properly. This disadvantage is particularly apparent in shallow coastal areas. In such shallow (often eutrophic) systems, a more sophisticated sediment module may have to be applied, which includes reduction & oxidation as well as P adsorption and desorption processes in the sediment. This model improvement for shallow systems is currently being developed and tested.

Nitrification and denitrification

Nitrification and denitrification are modelled as simple first order processes, corrected for temperature and oxygen concentration in the water column. Nitrification rates decrease when the oxygen concentrations in the water drop below a critical level.

Reaeration

The difference between the actual oxygen concentration in the water and the saturation concentrations is reduced by the reaeration rate. The saturation concentration of oxygen is a function of temperature and salinity of the water. The reaeration rate in most

GEM applications is a function of wind speed and water depth.

Settling

Settling in GEM is generally modelled as a net settling process with a settling velocity that is constant in time, and independent from turbulence and bottom shear stress. In 3D simulations, the effect of turbulence is accounted for by vertical dispersion between the water layers. Optionally, resuspension and the effect of bottom shear stress can be included. For each phytoplankton species and phenotype, a separate settling velocity is specified. For POM, one settling velocity is specified that is applied for all nutrient fractions (POC, PON, POP and $POSi_S$).

Burial

Particulate organic matter in the sediment is removed from the active sediment layer through burial to deeper sediment layers. These deeper sediment layers are not included in the model, and so in effect, the burial is a sink from the nutrient cycle in the model. The burial rate is a constant fraction of the particulate organic matter in the sediment throughout the year. The burial rate is a calibration parameter. It does not only represent the actual burial process, which is generally unknown for marine ecosystems, but also all other unknowns in the mass balance, including uncertainties on loading and transport across open boundaries.

Filterfeeder processes

Primary consumers are not simulated as real state variables in GEM. Instead, grazer biomass can be imposed to the model as a forcing condition (based on field data). The model then simulates the grazing effects of the imposed filterfeeder biomass. Five types of filterfeeders can be defined, both pelagic and benthic. The imposed biomass of primary consumers is adjusted by the model if availability of food (algae and detritus) is insufficient to support the imposed filterfeeder biomass, according to the specified filtering rates and metabolic coefficients. In order to ensure conservation of mass balance, nutrient concentrations are corrected in accordance with these adjustments in filterfeeder biomass. In case the filterfeeder module is applied for scenario simulations, a decision needs to be made whether or not to adjust its biomass function. This is usually done on the basis of expert knowledge taking observed grazer biomasses in various comparable water systems into account This approach, originally developed by Van der Molen et al. (1994), is considered as a first step in dynamic modelling of primary consumers.

At high densities of phytoplankton and detritus, the uptake of food is determined by the uptake rate. At low densities of phytoplankton and detritus, the uptake of food is determined by the filtration rate. Part of the uptake is egested as detritus. The egestion rate is a constant fraction of the uptake rate per type of food. The egestion fraction (f_{eg}) may be different for each phytoplankton or detritus type. For benthic filterfeeders, the detritus is excreted to the sediment, and for pelagic filterfeeders, it is excreted to the water column. The composition of grazers as C:N:P ratio is constant over time. Therefore, the assimilated food should have the same C:N:P ratio as the grazers. The part of the uptake of the non-limiting nutrients that cannot be assimilated is excreted as detritus. Part of the food uptake is lost due to growth respiration. The remaining part is available for biomass increase of the filterfeeders. The realised biomass increase (or decrease) is given by the forcing function, but is constrained by maximum growth, maximum mortality and food availability. The food that is not used for biomass increase and organic matter associated with biomass decrease is all released as detritus. Uptake, filtration, growth, maintenance respiration, growth respiration and mortality are all affected by temperature. The temperature coefficients can be chosen differently for each process, but generally, the same temperature coefficient is used for each process.

Applicability of the 'gem' model

The GEM model has been successfully applied and validated in several projects in a range of different estuarine and coastal water systems. This section presents a number of selected case study simulations to illustrate the generic applicability of the model. It would be beyond the scope of the present paper to describe each of these case studies in great detail. They are introduced here as representative examples to demonstrate the success with which GEM was repeatedly applied in a range of different water systems and geographical areas without the need for continuous remaking of (new) models for each different system, site or objective, or for reparameterization of process parameters and state variables.

The 'goodness-of-fit' between these model results and associated field measurements was calculated using the 'cost function' (OSPAR, 1998; Radach & Moll, 2006). The cost function is a mathematical function that gives a non-dimensional number, which is indicative of the 'goodness of fit' between two sets of data, in this case, model results and observations. It can be defined as the sum of the absolute deviations of the model values from the observations, normalised by the deviations of the observations for the chosen temporal and spatial range. Thus, it is a standardised, relative mean error. Cost function results for GEM applications have been calculated as

$$C_x = \frac{\sum |M_{x,t} - D_{x,t}| / n}{sd_x} * ((1-c) + c(1-r_x))$$

where C_x is the normalised deviation per station, annual value, $M_{x,t}$ is mean value of the model results per station per month,[2] $D_{x,t}$ is mean value of the in situ data per station per month, sd_x is standard deviation of the annual mean based on the monthly means of the in situ data ($df = 11$), n is 12 months, c is 0.5 and r_x is the correlation over time between $M_{x,t}$ and $D_{x,t}$ (OSPAR, 1998). The validation results were classified according to the following rating criteria (Radach & Moll, 2006) for the cost functions (cf):

Rating	Condition	
Very good	0<cf ≤ 1	Standard deviations
Good	1<cf ≤ 2	Standard deviations
Reasonable	2<cf ≤ 3	Standard deviations
Poor	3 >cf	Standard deviations

Case 1—Veerse Meer

The 'Veerse Meer' is a brackish coastal lake in the southwestern part of the Netherlands. This former

[2] For the Marmara Case study, in situ data were only available for two 60-day periods in a year, which were compared with corresponding 60-day periods of the model results.

estuary has been transformed into a non-tidal brackish lake, which is characterised by highly eutrophic conditions and seasonal blooms of the macroalga Ulva sp. (Malta & Verschuure, 1997). The GEM model was used to study the effects of various water management alternatives on the water quality and ecology of the lake, in particular, the effects of water level changes and flushing scenarios (Kernkamp et al., 2002). Figure 2 presents the results of GEM model baseline simulations for the 'Veerse Meer' against monthly field observations (DONAR database, 1 station) for nutrient concentrations (nitrate, ammonium, phosphate and silicate), chlorophyll-a and dissolved oxygen for the period 1995–1999. Cost function results ranged from 0.235 to 2.330 (Table 4), indicating that these model results can objectively be classified as 'very good' for nitrate and ammonium, 'good' to 'very good' for phosphate and silicate, and 'reasonable' to 'good' for chlorophyll-a. These results demonstrate that the GEM model produced consistently good and acceptable results, in this case, for a range of different key parameters. Similarly, good results for a range of key parameters were obtained at other stations and in other case studies without the need for recalibration or reparameterisation.

Case 2—North Sea

The North Sea can be characterised as a coastal shelf sea with relatively shallow (10–50 m) coastal waters. Substantial river discharges result in large fluctuations and strong temporal and spatial gradients in salinity, suspended matter concentrations, nutrient concentrations and algal biomass. Over the years, GEM was applied to the North Sea system in a range of studies, including simulations to evaluate the impact of a proposed land reclamation scheme (Nolte et al., 2005), and as a screening tool for the Water Framework Directive (WFD) to demonstrate the potential impacts of reductions in nutrient loads (Blauw & Los, 2004; Los & Wijsman, 2007). Most recently, BLOOM/GEM applications to the southern North Sea were subjected to a detailed three-dimensional model validation (Los et al., 2008). Here, we present the results of a trend analysis of chlorophyll-a concentrations in the North Sea over a 24-year period (1975–1998) predicted by the GEM model and

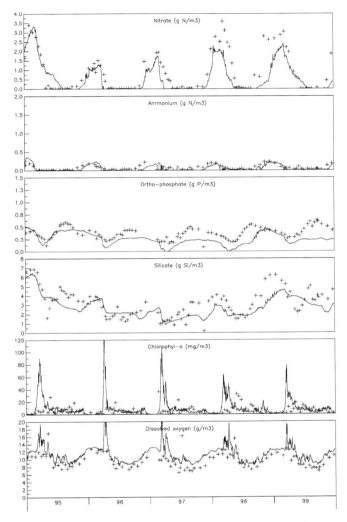

Fig. 2 Results of GEM baseline simulations for the "Veerse Meer" (surface water layers and bottom water layers) against actual field observations for nutrient concentrations (nitrate, ammonium, phosphate and silicate—all in mg l^{-1}), chlorophyll-a (in µg l^{-1}) and oxygen (in mg l^{-1}) for the period 1995–1999 (station: Soelekerkepolder)

compared with field observations (DONAR database, monthly means measured over the same 24-year period) at six monitoring stations (Fig. 3 and Table 5). Figure 3 shows the results for the station

Terschelling 4 km. The goodness-of-fit between model prediction and field measurements in the 1970s is not as perfect as in the 1980s and 1990s, and does not adequately reproduce their interannual

167

Table 4 Cost function results for the Veerse Meer case (5 parameters, surface water, 1995–1999)

Parameter	Year				
	1995	1996	1997	1998	1999
Chl-a	2.298	2.094	2.330	2.124	1.860
Si	1.026	0.809	1.179	0.778	1.006
PO₄	0.393	0.864	1.257	1.389	1.090
NO₃	0.265	0.441	0.480	0.249	0.235
NH₄	0.698	0.536	0.512	0.329	0.470

variability. This may be related to the following two issues: [1] all model predictions were based on simulations with the same hydrodynamics (single, representative day) and same suspended sediment forcings (steady state, cosine-transformed for seasonality) (as described in Los & Bokhorst, 1997), which may have reduced interannual variability in the model results; [2] there has been a shift from light limitation in the 1970s to nutrient (P) limitation in the late 1980s and 1990s (see www.waterbase.nl). Apparently, the model performs better under P-limited conditions given the stoichiometric settings chosen in the model set-up. Cost function results for individual years and stations ranged from 0.287 to 1.348 (Table 5), indicating that the model results for chlorophyll-a remained 'good' (one station) to 'very

good' (other stations) throughout all of these years. In order to illustrate that it is also possible to make reasonable predictions of the phytoplankton species composition using GEM (Los & Blaas, in prep.), an example of detailed phytoplankton results is presented in Fig. 4, showing phytoplankton species composition as modelled in GEM versus field observations. The results of the 24-year analysis of chlorophyll-a demonstrate that the GEM model produces consistently good and acceptable results for this key parameter over many consecutive years without the need for recalibration or reparameterisation.

Case 3—Venice Lagoon

The Lagoon of Venice is a very shallow (average depth 1.1 m), saline, subtropical, semi-enclosed estuarine lagoon system bordering the Italian city of Venice along the Mediterranean coast. The lagoon system covers an area of approximately 550 km² and is characterised by eutrophic conditions and seasonal blooms of the macroalga Ulva sp.. Over the past two decades, the lagoon has gone through substantial changes in ecological quality and water quality as a result of changes in nutrient loads, sediment

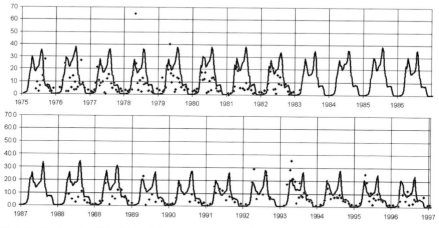

Fig. 3 Results of a trend analysis of chlorophyll-a concentrations (in μg l⁻¹) in the North Sea over a 24-year period (1975–1998) predicted by the GEM model in comparison with field observations (station: Terschelling 4 km)

Table 5 Cost function results for the North Sea case (chlorophyll-a, 24 years, 6 stations)

Year	Station					
	Noordwijk 4	Noordwijk 10	Noordwijk 20	Noordwijk 70	Terschelling 4	Terschelling 10
1975	0.314	0.353	0.714	1.347	0.751	0.744
1976	0.315	0.347	0.704	1.348	0.903	0.730
1977	0.309	0.348	0.699	1.341	0.775	0.726
1978	0.311	0.366	0.710	1.340	0.718	0.723
1979	0.321	0.353	0.714	1.343	0.869	0.703
1980	0.321	0.347	0.712	1.344	0.923	0.715
1981	0.319	0.359	0.710	1.343	0.889	0.711
1982	0.318	0.359	0.703	1.342	0.743	0.731
1983	0.305	0.352	0.688	1.341	0.656	0.727
1984	0.312	0.366	0.710	1.344	0.829	0.690
1985	0.313	0.356	0.708	1.344	0.872	0.697
1986	0.305	0.351	0.691	1.343	0.719	0.725
1987	0.301	0.360	0.680	1.336	0.621	0.673
1988	0.315	0.359	0.693	1.341	0.641	0.720
1989	0.301	0.345	0.661	1.341	0.600	0.698
1990	0.304	0.355	0.644	1.336	0.441	0.660
1991	0.305	0.376	0.656	1.335	0.514	0.667
1992	0.308	0.410	0.653	1.332	0.431	0.693
1993	0.299	0.381	0.664	1.333	0.521	0.658
1994	0.304	0.347	0.615	1.335	0.423	0.693
1995	0.311	0.409	0.668	1.333	0.456	0.690
1996	0.290	0.420	0.645	1.329	0.402	0.641
1997	0.297	0.425	0.622	1.327	0.395	0.684
1998	0.287	0.403	0.634	1.330	0.417	0.638

resuspension and anthropogenic perturbations (Sfriso et al., 2003). Large and lagoon-wide blooms of Ulva have not been observed anymore in the Lagoon of Venice over the last 15 years. The GEM model was used to study the potential effects of closure of mobile storm protection gates and reduction of nutrient loads on the environment, water quality and ecology of Venice Lagoon (Boon et al., 2006). Figure 5 presents the spatial results (horizontal dimension) of GEM model baseline simulations for chlorophyll-a concentrations (seasonal mean) for the entire Venice Lagoon in comparison with measured field data (Arresto campaign, 23 stations). This figure illustrates that the model was reasonably good in representing observed spatial variability in chloro-phyll-a concentrations. Cost function results (based on the comparison of model results with monthly field measurements from the period June 1988 to June 1989) for chlorophyll-a, nitrate, ammonium and phosphate (Table 6) indicate that, in total, 78% of model results for these key parameters for all of 26 different stations can objectively be classified as 'good' to 'very good' and a further 13% as reasonable. Best results were obtained for nitrate and chlorophyll-a, with 92% and 73% of all model predictions falling in the categories 'good' and 'very good', respectively. The results of this analysis demonstrate that the GEM model produces consistently good and acceptable results for various key parameters throughout the horizontal (spatial) dimension without the need for recalibration or reparameterisation.

Case 4—Sea of Marmara

The Sea of Marmara is a deep (> 2,000 m), oligotrophic, temperate, stratified coastal sea in Turkey. As

169

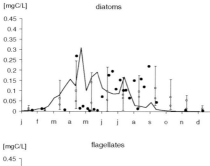

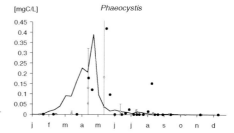

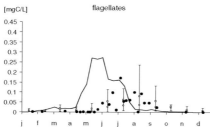

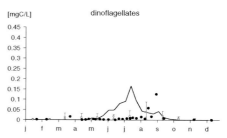

Fig. 4 Validation result for GEM algal species composition (1998 simulation) at station Noordwijk 10 km (m gC l⁻¹). Closed circles are measurements for 1998; bars indicate 90 percentile of measurements for the years 1991–2003, with open circles representing median and open squares representing mean values

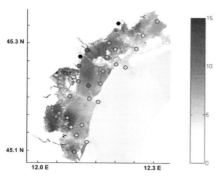

Fig. 5 Spatial results of GEM baseline simulation of chloro-phyll-a in μg l⁻¹ (depth averaged) for Venice Lagoon showing seasonal mean values (April–September) of model results for summer 1988 in comparison with field measurements (Arresto Degrada monitoring data for April–September 1988). Marker colours of field data (circles) refer to the same scale as model output

part of a study for an environmental master plan and investment strategy for the Marmara Sea Basin, the GEM model was applied to simulate baseline water quality conditions and the implications of a number of predefined management scenarios (Smits, 2006). Figure 6 presents the results of an analysis of chlorophyll-a concentrations at different water depths (upper 60 m) predicted by the GEM model (baseline 2003) for four different seasons, in comparison with in situ measurements at station MD56 (MEMPIS campaign). This figure illustrates that the model was reasonably good in representing observed seasonal variability in stratification patterns for chlorophyll-a. For the calculation of cost functions, model results for two 60-day periods (baseline 2003) were compared with available field observations (MEMPIS campaign, June and September 2005, 8 stations aggregated). Cost function results for nutrient concentrations (nitrate, phosphate and silicate), dissolved oxygen and chlorophyll-a at five different water depths (Table 7) indicate that, in total, 60% of model results can objectively be classified as 'good' to 'very

Table 6 Cost function results for the Venice Lagoon case (4 parameters, 26 locations)

Location	Parameter			
	Chl-a	NO$_3$	NH$_4$	PO$_4$
Arresto 1	0.799	0.946	1.389	1.582
Arresto 2	3.479	0.690	1.460	1.130
Arresto 3	1.423	2.250	8.096	1.289
Arresto 4	2.041	1.943	3.666	1.578
Arresto 5	2.600	1.119	1.711	1.506
Arresto 6	1.019	0.339	1.186	1.252
Arresto 7	0.748	0.271	1.027	0.731
Arresto 8	0.441	1.155	1.376	1.571
Arresto 9	0.705	0.512	1.692	1.629
Arresto 10	0.504	0.927	1.342	1.782
Arresto 11	0.729	0.904	1.568	1.456
Arresto 12	1.303	0.602	1.628	2.331
Arresto 13	0.900	1.240	1.287	1.595
Arresto 14	2.024	0.546	2.137	1.332
Arresto 15	1.112	0.446	1.665	1.172
Arresto 16	3.285	0.656	2.813	3.364
Arresto 17	1.322	0.985	3.011	2.531
Arresto 18	0.813	1.093	1.842	1.882
Arresto 19	1.137	2.231	3.100	4.069
Arresto 20	1.058	1.472	4.779	2.805
Arresto 22	0.835	0.459	0.847	0.865
Arresto 23	3.209	0.936	2.335	1.893
Arresto 24	0.810	0.311	0.999	0.678
Arresto 25	0.607	0.496	0.724	1.235
Arresto 26	0.823	0.567	1.154	1.088
Arresto 27	2.007	0.638	2.596	1.520
Percentages	Very good	Good	Reasonable	Poor
Chl-a	46	27	15	12
NO$_3$	69	23	8	0
NH$_4$	12	54	15	19
PO$_4$	12	69	12	8

good' and a further 12% as reasonable. GEM model results for chlorophyll-a, nitrate and dissolved oxygen in the deeper layers (40 and 50 m) compared poorly with in situ data, but the cost function results were 'reasonable' to 'good' for silicate and 'very good' for phosphate at these depths. The overall results of this analysis demonstrate that the GEM model produces consistently good and acceptable results for various key parameters throughout the vertical dimension (depth) without the need for recalibration or reparameterisation.

Discussion

The present paper describes the set-up, application and validation of a GEM for estuaries and coastal waters. Since its development in the mid-1990s, the GEM model has been applied successfully in a range of consultancy and scenario studies, which have formed the basis for several major policy and management decisions and infrastructural developments in coastal zones in the Netherlands and worldwide.

Fig. 6 Chlorophyll-a concentrations (in µg l⁻¹) at different water depths in the Sea of Marmara as predicted by the GEM model (baseline 2003) for four different seasons in comparison with field observations from 2005 (station: MD56)

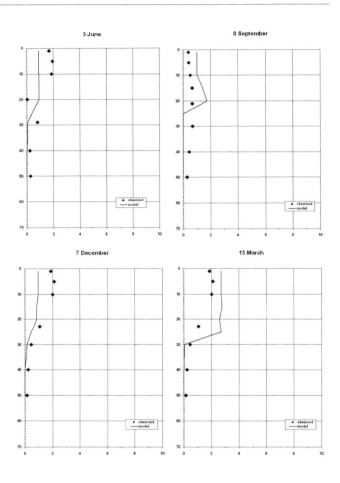

Table 7 Cost function results for the Sea of Marmara case (5 parameters, 5 depths)

Parameter	Depth				
	10	20	30	40	50
Chl-a	0.408	3.136	1.821	4.618	8.235
Si	2.374	0.649	1.453	1.736	2.033
PO₄	1.412	0.468	0.499	0.955	0.910
NO₃	1.483	1.264	2.961	4.780	5.792
Oxygen	0.141	0.421	1.988	4.704	7.888

Using a number of representative examples from case studies in which the GEM model was applied, the generic applicability of the GEM model was evaluated. Validation of the examples using the 'cost function' revealed consistent accuracy of the GEM model for various key parameters (Case 1), for both horizontal (Case 3) and vertical (Case 4) spatial dimensions, as well as temporal dimensions (seasonally and across years, Case 2) for a variety of estuarine and coastal water systems without the need for substantial reparameterisation. In addition, it is

172

also possible to make reasonable predictions of the phytoplankton species composition using GEM (Fig. 4), which is further elaborated in another forthcoming paper (Los & Blaas, in prep.).

In each of the four applications of GEM presented in this paper, model implementation for a new area usually required careful checking and improving physical forcings (loads, suspended matter, hydrodynamic and hydrological boundary conditions), which are typically area-specific. In addition, it was sometimes necessary to make minor adjustments to a few of the process parameters (in particular, to the closure terms denitrification and burial), resulting in minor differences in parameter settings between the four case studies. Dentrification rates applied in case studies 1–4 were 0.003, 0, 0.05 and 0.2, respectively, in the sediment and 0, 0.2, 0.02 and 0.15, respectively, in the water column. Burial rates were only applied in the 2D cases of North Sea (0.003 day^{-1}) and Venice Lagoon (0.025 day^{-1}). In the 3D cases (Veerse Meer and Sea of Marmara), settling rates for particulate variables were calibrated instead of burial, since settling in such deep permanently stratified systems has a similar effect as burial. Most other (esp. biological) model parameters, in particular, the settings for all of the key state variables for algae, were identical between the four case studies (Table 3). Table 2 shows the parameter settings used for the southern North Sea model, which is representative for the other model applications. Depending on the objectives of a particular study, the model setup can be extended with new processes specific to the particular site or application. For example, the phytoplankton group Phaeocystis was replaced by Ulva in the Venice Lagoon and Veerse Meer case studies, microphytobenthos was added in the Veerse Meer case study and pH-dependent phosphorus adsorption was included in the Sea of Marmara and Veerse Meer case studies. For a more detailed description of the calibration and validation procedure for GEM & BLOOM applications, see Los et al. (2008). A recent sensitivity analysis of GEM (Salacinska, 2008) using the Moris method (Saltelli et al., 2004), revealed that the Dutch coastal zone application of the GEM model is particularly sensitive to Chl:C ratios and light-related parameters (excl. extinction by POM).

While in most cases GEM produced good to very good results, the goodness of fit between model results and field measurements was rather poor for some parameters in deeper layers (40 and 50 m) of the oligotrophic Sea of Marmara and in very shallow (intertidal) areas of the Lagoon of Venice. For the Sea of Marmara, only a limited set of field observations was available for validation (due to low sampling frequency) and there was a mismatch between the timing of boundary forcings of the model (meteorological and hydrological data of 2003) and available field observations (2005), and so these results should be interpreted with some caution. For the Lagoon of Venice, poor model performance in very shallow areas of the lagoon (with extensive tidal flats) might be related to the periodic occurrence of anoxia. The formulations used for the sediment sub-model in GEM are not suited to account for such conditions.

Besides marine and brackish water systems, close relatives to the GEM model have also been applied successfully in a number of freshwater systems, such as the Dutch 'IJsselmeer' (Ibelings et al., 2003) and 'Veluwemeer' (Van der Molen et al., 1994) as well as 'Laguna de Bay' in the Philippines (Nolte & Chua, 2003). For these freshwater applications, alternative algal parameters (not presented here) were used to reflect the different algal community in freshwater environments. With these additional (freshwater) algal parameters incorporated into GEM, the model is currently being applied in a study that compares (simultaneously) different freshwater and marine flushing scenarios for the Dutch 'Volkerak-Zoommeer'.

Limitations of the GEM model in its present state include application to very shallow dynamic environments with extensive tidal flats and periodic anoxia, due to inappropriate formulations used for the sediment sub-model in GEM. Application of GEM in deep and stratified oligotropic seas has been a challenge, producing poor results for some parameters at greater depths as shown in the Marmara case study. Performance of GEM in the Marmara case, however, was highly dependent on the accuracy of the underlying hydrodynamic model, which did not perfectly reproduce vertical stratification patterns. Since phytoplanktonic growth in GEM depends linearly on water temperature, while respiration and mortality follow the usual Arrhenius law, the ratio between growth and loss processes varies as a function of the temperature. During applications to tropical waters with higher temperatures compared to those for which the model was originally developed, we have found that loss rates could get unrealistically high relative to growth rates. As a practical solution, the temperature functions for all growth and loss processes had to be shifted in the model, while keeping the same relative proportion between them. Temperature coefficients of processes of the grazer module were modified in a similar way.

It should be stressed that these limitations are primarily and inherently related to the generic character of GEM (remaining unchanged in its settings and parameterisation between different case studies). Generally, such limitations can be overcome by further modifications to a specific model application, resulting in a more dedicated model. However, such dedicated models, while successful in their specific applications, tend to produce unsatisfactory results when applied (unchanged) in other case studies. Without compromising on its generic applicability, these issues and limitations are currently being addressed in further improvements to the GEM model.

Modelling of primary consumer processes remains a challenge to date. Existing grazer modules, such as ERSEM (and others), are currently being tested and validated for possible application and incorporation within the GEM model to allow for the modelling of higher trophic levels. Other developments within GEM that are currently underway include an improved module for sediment processes, including key geochemical processes (such as P adsorption and desorption) and more complex sedimentation and resuspension formulations.

The current set-up of the GEM-model, including the explicit choices of which processes to include, allows for an optimal balance among ecological (process) resolution, model grid resolution and model run-time.[3] Inclusion of microbial loops (incl. pico-plankton etc.) and more detailed benthic processes (incl. improved mortality terms, differentiating complex burial processes), as is often done in various other (Dutch) ecological models, may yield greater insight into specific ecological processes, which can be useful when the model study has a more fundamental scientific focus. However, this invariably leads to exponential increases in model run-time, as

a result of which the spatial resolution of the model is usually compromised. In most model applications designed to answer practical management questions, however, greater spatial resolution is usually required, while simplified closure terms for some of these more specific microbial and benthic processes often suffice, especially since field data on such processes are often lacking (Los et al., 2008).

Acknowledgments The authors would like to acknowledge the cooperation from all the people who participated in the development and various application studies of GEM, including Hanneke Baretta-Bekker, Marcel van der Tol, Peter Herman, Rene Jansenf, Piet Ruardij, Bert Brinkman, Johan de Kok, Johannes Smits, Leo Postma, Jan van Beek, Arno Nolte, Jos van Gils, Meinte Blaas, Xavier Desmit, Johan Boon and Tjitte Nauta. The development and validation of GEM was made possible by financial support of the National Institute of Coastal and Marine Management (RIKZ). Permission to make use of the monitoring data sets on Dutch waters (DONAR: www.waterbase.nl), Venice Lagoon (ARRESTO campaign) and Sea of Marmara (MEMPIS campaign) is kindly acknowledged. The Netherlands Ministries of Transport & Public Works and Economic Affairs are thanked for their financial support within the framework of the WL | Delft Hydraulics R&D Programme 2007, which enabled us to prepare this paper.

References

Airoldi, L. & M. W. Beck, 2007. Loss, status and trends for coastal marine habitats of Europe. Oceanography and Marine Biology—Annual Review 45: 345–405.

Baretta, J. W., W. Ebenhoh & P. Ruardij, 1995. The European Regional Seas Ecosystem model (ERSEM), a complex marine ecosystem model. Netherlands Journal of Sea Research 33: 233–246.

Blauw, A. N. & F. J. Los, 2004. Analysis of the response of phytoplankton indicators in Dutch coastal waters to nutrient reduction scenarios. WL | Delft Hydraulics, technical report Z3844, December 2004, 145 pp.

Boon, J. G., X. Desmit & M. Blaas, 2006. Ecological model of the Lagoon of Venice. Part II: Set-up, calibration and validation. WL | Delft Hydraulics, technical report Z3733, May 2006, 232 pp.

Brinkman, A. G., 1993. Biological processes in the Ecowasp ecosystem model. Institute for Forestry and Nature Research (IBN-DLO), IBN research report 93/6, ISSN: 0928–6896.

Danzig, G. B., 1963. Linear programming and extensions. Princeton University Press, Princeton, NJ.

De Kok, J. M., C. de Valk, J. H. T. M. Van Kester, E. de Goede & R. E. Uittenbogaard, 2001. Salinity and temperature stratification in the Rhine plume. Estuarine, Coastal and Shelf Science 53: 467–475.

De Kok, J. M., R. Salden, I. D. M. Roozendaal, P. Blokland & J. Lander, 1995. Transport paths of suspended matter along the Dutch coast. In Brebbia, C. A., L. Traversoni &

---

[3] Model run times in the four case studies ranged from approximately 1 h (Venice and Veerse Meer) to 11 h (North Sea) and 13 h (Sea of Marmara) per simulation.

L. C. Wrobel (eds), Computer modelling of seas and coastal regions II. Computational Mechanics Publications, Southampton: 75–86.

Delhez, E. J. M., P. Damm, E. D. De Goede, J. M. De Kok, F. Dumas, H. Gerritsen, J. E. Jones, J. Ozer, T. Pohmann, P. S.Rasch, M.D. Skogen&R. Proctor, 2004. Variability of shelf-seas hydrodynamic models: lessons from the NOMADS2 project. Journal of Marine Systems 45: 39–53.

Devlin, M., M. Best & D. Haynes, 2007. Implementation of the Water Framework Directive in European marine waters. Marine Pollution Bulletin 55: 1–2.

De Vries, I., R. N. D. Duin, J. C. H. Peeters, F. J. Los, M. Bokhorst & R. W. P. M. Laane, 1998. Patterns and trends in nutrients and phytoplankton in Dutch coastal waters: comparison of time-series analysis, ecological model simulation and mesocosm experiments. ICES Journal of Marine Science 55: 620–634.

Fitz, H. C., E. B. DeBellevue, R. Costanza, R. Boumans, T. Maxwell, L. Wainger & F. H. Sklar, 1996. Development of a general ecosystem model for a range of scales and ecosystems. Ecological Modelling 88: 263–295.

Gerritsen, H., J. G. Boon, T. Van der Kaaij & R. J. Vos, 2001. Integrated modelling of suspended matter in the North Sea. Estuarine, Coastal and Shelf Science 53: 581–594.

Harris, G. P., 1986. Phytoplankton ecology. Chapman and Hall, London.

Ibelings, B. W., M. Vonk, H. F. J. Los, D. T. Van der Molen & W. M. Mooij, 2003. Fuzzy modelling of cyanobacterial surface waterblooms: validation with NOAA-AVHRR satellite images. Ecological Applications 13: 1456–1472.

Kernkamp, H., G. Boot & A. Nolte, 2002. Onderzoek naar de toekomstige waterkwaliteit en ecologie van het Veerse Meer. Studie naar het effect van het doorlaatmiddel en aanvullende maatregelen. Deel 1: Opzet en kalibratie hydrodynamisch en waterkwaliteitsmodel. WL | Delft Hydraulics, technical report Z3304, November 2002, 81 pp.

Klepper, O., W. M. Van der Tol, H. Scholten & P. M. J. Herman, 1994. SMOES: a simulation model for the Oo-sterschelde ecosystem. Part 1: Description and uncertainty analysis. Hydrobiologia 282/283: 453–474.

Los, F. J., 2005. An algal biomass prediction model. In Loucks, D. P. & E. Van Beek (eds), Water resources systems planning and management—an introduction to methods, models and applications. UNESCO, pp. 408–416.

Los, F. J. & M. Bokhorst, 1997. Trend analysis Dutch coastal zone. In: New challenges for North Sea research, Zentrum fur Meeres- und Klimaforschung, University of Hamburg: 161–175.

Los, F. J. & J. J. Brinkman, 1988. Phytoplankton modelling by means of optimization: a 10-year experience with BLOOM II. Verhandlungen der Internationalen Vereini-gung für Theoretische und Angewandte Limnologie 23: 790–795.

Los, F. J., N. M. de Rooij & J. G. C. Smits, 1984. Modelling eutrophication in shallow Dutch lakes. Verhandlungen der Internationalen Vereinigung für Theoretische und Ange-wandte Limnologie 22: 917–923.

Los, F. J., R. Jansen & S. Cramer, 1994. MANS eutrophication modelling system. National Institute for Coastal and Marine Management/RIKZ, The Hague: 137.

Los, F. J., M. T. Villars, & M. W. M. Van der Tol, 2008. A 3-dimensional primary production model (BLOOM/GEM) and its applications to the (southern) North Sea (coupled physical–chemical–ecological model). Journal of Marine Systems. doi:101016/j.jmarsys.2008.01.002.

Los, F. J. & J. W. M. Wijsman, 2007. Application of a validated primary production model (BLOOM) as a screening tool for marine, coastal and transitional waters. Journal of Marine Systems 64: 201–215.

Malta, E. J. & J. M. Verschuure, 1997. Effects of environmental variables on between-year variation of Ulva growth and biomass in a eutrophic brackish lake. Journal of Sea Research 38: 71–84.

Moll, A. & G. Radach, 2003. Review of three-dimensional ecological modelling related to the North Sea shelf system—Part 1: models and their results. Progress in Oceanography 57: 175–217.

Nolte, A., P. Boderie, & J. van Beek, 2005. Impacts of Ma-asvlakte 2 on the Wadden Sea and North Sea coastal zone. Track 1: Detailed modelling research. Part III: Nutrients and Primary Production. WL | Delft Hydraulics, technical report Z3945, November 2005, 175 pp.

Nolte, A. & G. Chua, 2003. Evaluation and future prospects of the Laguna de Bay Water Quality Model. Technical report for the ''Sustainable Development of the Laguna de Bay Environment'' project, WL | Delft Hydraulics, June 2003.

OSPAR, 1998. Report of the ASMO modelling workshop on eutrophication issues, 5–8 November 1996, The Hague, The Netherlands. OSPAR Commission Report, Netherlands Institute for Coastal and Marine Management/ RIKZ, The Hague, The Netherlands.

Ouboter, M. R. L., B. T. M. Van Eck, J. A. G. Van Gils, J. P. Sweerts & M. T. Villars, 1998. Water quality modelling of the western Scheldt estuary. Hydrobiologia 366: 129–142.

Radach, G. & A. Moll, 2006. Review of three-dimensional ecological modelling related to the North Sea shelf system. Part II: Model validation and data needs. Oceanography and Marine Biology—Annual Review 44: 1–60.

Reynolds, C. S., S. W. Wiseman, B. M. Godfrey & C. But-terwick, 1983. Some effects of artificial mixing on the dynamics of phytoplankton populations in large limnetic enclosures. Journal of Plankton Research 5: 203–234.

Salacinska, K., 2008. Sensitivity analysis of the 2D application of the Generic Ecological Model to the North Sea. MSc thesis, Delft University of Technology, Delft, The Netherlands, 84 pp.

Saltelli, A., S. Tarantola, F. Campolongo & M. Ratto, 2004. Sensitivity analysis in practice: a guide to assessing scientific models. Halsted Press, New York, NY, 2004.

Scholten, H. & W. M. Van der Tol, 1994. SMOES: a simulation model for the Oosterschelde ecosystem. Part 2: Calibration and validation. Hydrobiologia 282/283: 453–474.

Sfriso, A., C. Facca & P. F. Ghetti, 2003. Temporal and spatial changes of macroalgae and phytoplanton in a Mediterranean coastal area: the Venice Lagoon as a case study. Marine Environmental Research 56: 617–636.

Smits, J., 2006. Environmental master plan and investment strategy for the Marmara Sea Basin. Water quality modelling of the Sea of Marmara: Model development and scenario simulations. WL | Delft Hydraulics, technical report Z3804.50, October 2006, 167 pp.

Soetaert, K. & P. M. J. Herman, 1995a. Carbon flows in the Westerschelde estuary (the Netherlands) evaluated by means of an ecosystem model (MOSES). Hydrobiologia 311: 247–266.

Soetaert, K. & P. M. J. Herman, 1995b. Nitrogen dynamics in the Westerschelde estuary (SW Netherlands) estimated by means of the ecosystem model MOSES. Hydrobiologia 311: 225–246.

Soetaert, K., P. M. J. Herman & J. Kromkamp, 1994. Living in the twilight: estimating net phytoplankton growth in the Westerschelde estuary (The Netherlands) by means of an ecosystem model (MOSES). Journal of Plankton Research 16: 1277–1301.

United Nations Environment Programme (UNEP), 2006. Marine and coastal ecosystems and human well-being: a synthesis report based on the findings of the millennium ecosystem assessment. UNEP, 76 pp, Nairobi (see http://www.MAweb.org), ISBN: 92-807-2679-X.

Van der Molen, D. T., F. J. Los & M. Van der Tol, 1994. Mathematical modelling as a tool for management in eutrophication control of shallow lakes. Hydrobiologia 275/276: 479–492.

Van de Wolfshaar, K. E., 2007. Beoordeling Generiek Eco-logisch Model, GEM. Report summarising the outcome of an international audit of the GEM model. WL | Delft Hydraulics, technical report Z4267, January 2007, 47 pp.

Van Gils, J., 1998. The SOBEK processes editor: a flexible tool for tailor-made water quality modelling. In Babovic, V. & L.C. Larsen (eds), Proceedings of the Third International Conference on Hydroinformatics, Copenhagen, Denmark. Balkema, Rotterdam, pp. 591–595.

Van Gils, J. A. G., M. R. L. Ouboter & N. M. De Rooij, 1993. Modelling of water sediment quality in the Scheldt Estuary. Netherlands Journal of Aquatic Ecology 27: 257– 265.

WL | Delft Hydraulics, 2003. Delft3D-WAQ users manual. WL | Delft Hydraulics, Delft, The Netherlands.

176

Chapter 9

A 3-dimensional primary production model (BLOOM/GEM) and its applications to the (southern) North Sea (coupled physical–chemical–ecological model)

F.J. Los, M.T. Villars, M.W.M. Van der Tol, 2008.

Journal of Marine Systems, 74: 259-294.
doi:101016/j.jmarsys.2008.01.002.

A 3-dimensional primary production model (BLOOM/GEM) and its applications to the (southern) North Sea (coupled physical–chemical–ecological model)

F.J. Los [a,*], M.T. Villars [a], M.W.M. Van der Tol [b]

[a] WL | Delft Hydraulics, P.O Box 177, 2600 MH Delft, The Netherlands

[b] National Institute for Coastal and Marine Management/RIKZ, Kortenaerkade 1, 2518 AX Den Haag, The Netherlands

Received 21 April 2006; received in revised form 21 November 2007; accepted 11 January 2008

Abstract

This paper presents the ecological modelling instrument BLOOM/GEM and several applications to the southern North Sea, including a 3-dimensional model validation. The current instrument and its predecessors have been used since the 1980s for evaluating the ecological status of the North Sea and potential effect of management strategies. The main modelled processes of nutrient cycling, oxygen dynamics and primary production are described, as well as the external forcings required for the ecological model (hydrodynamics, suspended sediments and river loads). In the development of the BLOOM/GEM modelling instrument, the explicit choice of processes to include (the 'ecological' resolution) has been balanced with the need for high spatial resolution in the model applications for the Dutch coastal zone. The calibration and validation of the BLOOM/GEM modelling instrument as well as the model applications have mainly drawn on the extensive dataset available for the Dutch coastal waters (http://www.waterbase.nl). A specific 3-dimensional model application to the North Sea is described including the model validation results based on the use of an objective cost function for a number of different substances. Plotted model results showing seasonal as well as regional variations and spatial gradients for many substances at several stations give additional support to the validation. As such, the model is well suited to support many management decisions, related to e.g. the OSPAR convention and the European Water Framework Directive and the construction of infrastructural works.

Keywords: Ecological modeling; Primary production; North Sea; Model validation

1. Introduction

In the Netherlands, the generic modelling instrument BLOOM/GEM (Generic Ecological Model) is regularly applied to assess ecological quality of the Dutch coastal waters and the southern North Sea, as well as to evaluate potential effects of e.g. new coastal infrastructure projects, and new national and international policies (OSPAR, Water Framework Directive). A model application requires a hydrodynamics calculation, which is then coupled (off-line) to the water quality–ecological modelling instrument BLOOM/GEM.

The generic modelling instrument, originally developed at WL | Delft Hydraulics, has evolved over the past 20 years. In the early stages of the development of GEM,

* Corresponding author. Tel: +3115 285 8549; fax: +3115 285 8582.
  E-mail address: hans.los@deltares.nl (F.J. Los).

0924-7963/$ - see front matter © 2008 Elsevier B.V. All rights reserved.
doi:10.1016/j.jmarsys.2008.01.002

WL | Delft Hydraulics has been advised by other Dutch institutes with respect to the general model set-up (National Institute for Coastal and Marine Management/ RIKZ, National Institute for Ocean Research/NIOZ, and National Institute for Ecologic Research/NIOO, Alterra). Some specific formulations provided by these institutes are available in GEM (see also Section 3.2). This paper gives a general description of BLOOM/GEM and specific applications to the Dutch coastal zone and southern North Sea, focusing on the extensive validation conducted using available data from project-related and regular monitoring.

The BLOOM/GEM North Sea application is one example of several existing European large-scale 3D ecosystem models applied to the North Sea and parts of the Northwest European Continental Shelf. Others include ERSEM (Baretta et al., 1995), NORWECOM (Skogen and Soiland, 1998), POL3dERSEM (Allen et al., 2001), COHERENS (Luyten et al., 1999), ECOHAM (Moll, 1997, 1998), Elise (Ménesguen et al., 1995), MIRO (Lancelot et al., 1995, 2000, 2005; Lacroix et al., 2007) and ECOSMO (Schrum etal., 2006a,b). Moll and Radach (2003) and Radach and Moll (2006) have written two extensive papers on the description and comparison of a number of these existing eco-hydrodynamical models, in which they also provide many additional references. They deal both with the model structure and with the validation, and include a review of model inter-comparisons which have been performed during the last decade. Lacroix et al. (2007) in their description ofthe MIRO model give a more comprehensive, but also more recent description of some of the North Sea models.

There are several reasons for the existence of so many different models. A somewhat trivial explanation is that authorities in different countries want to have national modelling systems. From a scientific point of view it is obvious that the driving physical and ecological forces vary considerably across the North Sea. These differences are reflected by the models which are being developed with respect to the area included, the size of the computational elements (grid cells), the level of detail of the water quality and ecological processes and the amount of data used for the validation of these models.

Models covering the entire North Sea area while using computational elements small enough to describe coastal gradients which also include a large number of potentially relevant ecological functional groups and processes, would require simulation times in the order of several weeks for a one year period. To keep models manageable, modelers make specific choices. Thus the models developed in countries with a long shoreline generally cover a large area but use a coarse grid. Among these are NORWECON and POL3dERSEM. Other models with a

similar physical schematization are ECOHAM and CSM-NZB (Los and Bokhorst, 1997). In contrast, models developed for regions with a short shoreline characterized by strong physical gradients tend to cover a much smaller area but also use a more refined grid. Elise, MIRO and DCM-NZB (Los and Bokhorst, 1997) are typical representatives of these kinds of models. There is a tendency for the regional models to also describe the ecology in greater detail but obvious exceptions are POL3dERSEM and CSM-NZB, both of which describe a number of processes at a similar level of detail as the aforementioned regional models.

In the Dutch coastal zone, observed gradients of nutrients, algal biomass and suspended matter are very steep; even a concentration difference by a factor of 10 over a distance of only 10 km is not uncommon (http:// www.waterbase.nl). In addition, the Rhine-Meuse river system is the major fresh water source of the North Sea. Its impact is not confined to the Dutch coastal zone, but extends much further i.e. to the German Bight. The Dutch continental waters stretch as far as 400 km north of the Wadden Islands and include important regions such as the Oystergrounds and Dogger Bank. Observed horizontal gradients in these waters are less pronounced compared to the coastal zone. In order to consider both the coastal waters and off shore regions, previously two different schematizations were developed: CSM-NZB covering a wide area including the Dutch coastal zone with a coarse 18×18kmgrid and DCM-NZB including onlythecoastal zoneuptill 70kmusing a grid sizein the order of 1×1 km. In both cases the same ecological modules were applied (North Sea BLOOM; see also Table 1). In recent years, these two model applications have been merged into a single one using a variable grid size with a minimum of about 1×1 km in the coastal zone and a maximum of about 20×20 km in the most north westerly part of the model domain. In order to match the rounded shape of the Dutch coastline the model uses curve-linear elements. The ecological processes are based on those of its predecessors.

Comparing the present North Sea application of BLOOM/GEM to these other models, it should be noted that the modelled area is intermediate: for instance NORWECON and POL3dERSEM consider a larger area, but Elise and MIRO consider a much smaller area. All other models use rectangular grids; no other model uses a curve-linear grid. The obvious advantage is that the grid can be better adapted to spatial gradients using a fine resolution where necessary and a much coarser resolution elsewhere. Hence the spatial resolution in the coastal zone of BLOOM/GEM is considerably more refined than in any other model (approximately 1×1 km)

Table 1
The evolution of the BLOOM/GEM ecological modelling for the North sea

Project name/Year	Area	Schematisation	Hydrodynamics	No. hor. elements	Wind	River loads	Ecology processes
MANS/1989–1993	N.S.	GENO	2D tidal ave.	1395	uniform SW 4.5 m s^{-1}	1985	DYNAMO
MANS/1989–1993	N.S.	GENO+ NZSTRAT	2D+1Dv tidal average.	1395+3	uniform SW 4.5 m s^{-1}+actual	1985– DYNAMO	ECOLUMN/ BLOOM
MANS/1989–1993	N.S.	Zunowak	2D tidal ave.	approx. 12,000	uniform SW 4.5 m s^{-1}	1990	NZBLOOM
KSENOS/1993–1995	Dutch CZ	Kuststrook	3D tidal ave*.	2153	daily average 1990; interpolated hydrodynamics, wind dependent dispersion	1990	NZBLOOM
KSENOS/1993–1995	N.S.	CSM	2D tidal ave.	3915	uniform SW 7 m s^{-1}	1990	NZBLOOM
Impact high river discharge/1996	N.S.	CSM	2D tidal ave.	3915	uniform SW 7 m s^{-1}	1990	NZBLOOM
Trend analysis KSENOS/1996–1997	Dutch CZ	Kuststrook 1	3D* tidal ave.	2153	daily average 1990; interpolated hydrodynamics, wind dependent dispersion	1975–1994	NZBLOOM
WL Profile/1998	Dutch CZ	Kuststrook 2	3D* tidal ave.	3150	uniform SW 7 m s^{-1}	1990	NZBLOOM
Application of GEM to Dutch coastal waters/1997–1998	Dutch CZ	Kuststrook 3	3D* tidal ave.	1154	uniform SW 7 m s^{-1}	1990	BLOOM/ GEM
GEM coastal zone and indicators/1999	Dutch CZ	Kuststrook 3	3D* tidal ave.	1154	uniform SW 7 m s^{-1}	1975–1996	BLOOM/ GEM
Rotterdam Harbour extension/1999	Dutch CZ	Kuststrook 4	3D* actual tides 4 layers	2457	uniform SW 7 m s^{-1}	1990	BLOOM/ GEM
Combined impact management measures/1999	Dutch CZ	Kuststrook 4	3D* actual tides 4 layers	2457	uniform SW 7 m s^{-1}	1990	BLOOM/ GEM
Screening impacts airport North Sea	Dutch CZ	Kuststrook 2	3D* tidal ave. 10 layers	3150	uniform SW 7 m s^{-1}	1990	BLOOM/ GEM
Flyland: future airport North Sea/2000–2002	N.S.	Zunogrof and Zunofine	3D* actual tides 10 layers	4350	historic wind field	1989–1998	BLOOM/ GEM
Further validation North Sea model/2003–2004	N.S.	Zunogrof	3D actual tides 10 layers	4350	historic wind field	1989	BLOOM/ GEM
Application of GEM for Water Framework Directive/2004	N.S.	Zunogrof	3D* actual tides 10 layers	4350	historic wind field	1989, 1998, average year	BLOOM/ GEM
EIA Rotterdam Harbour extension/2005	N.S.	Zuno-DD	3D* actual tides 10 layers	10892	historic wind field	1989–1998	BLOOM/ GEM
EIA Sand mining Rotterdam Harbour extension/ 2005-2006	N.S.	Zunogrof	3D* actual tides 10 layers	4350	historic wind field: 1989, 1995–2003	1989–2003	BLOOM/ GEM

N.S.: North Sea.

Dutch CZ: Dutch coastal zone.

CSM: Continental Shelf Model Schematisation.

GENO: Schematisation of the southern North Sea (1395 elements).

ZUNOWAK: Schematisation of the southern North Sea (approx. 12,000 elements).

Kuststrook: Various schematisations of the Dutch coastal zone (approx. 70 km wide).

Zunogrof: Schematisation southern North Sea (4350 horizontal elements, 10 layers).

Zunofine: Schematisation southern North Sea (app. 12,000 horizontal elements, 10 layers).

Zuno-DD: Special 10 domain version of Zunogrof (refined in coastal areas; approx. 40,000 horizontal elements, 10 layers; aggregated to 10892 elements for BLOOM/GEM).

NZBLOOM: DELWAQ–BLOOM model.

GEM: Generic Ecological Model.

3D*: In these calculations, 3D hydrodynamics were calculated and then aggregated to 2D for the ecological model.

3D: Both 3D hydrodynamics and ecological processes were calculated.

while in the central North Sea the resolution is similar to those of the other non-regional models.

With respect to the level of detail of the ecological modules the phytoplankton module (BLOOM) is the most extensive as it includes several functional groups and types. ERSEM also considers several functional groups, but these are defined with respect to the trophic web while types in BLOOM are primarily defined with respect to resource competition. MIRO and BLOOM are the only two phytoplankton modules in which the nuisance species *Phaeocystis* is explicitly considered. BLOOM/GEM includes complete cycles for all three nutrients (N, P and Si) and for oxygen, which is not always the case in the other models. It also includes a simple benthic module for organic nutrients but not nearly as complex as the one included in ERSEM. From the review by Moll and Radach (2003) it is obvious that there is little consistency in the way existing models deal with grazing by zooplankton and zoobenthos. Some models do not include grazing at all, some use a forcing function approach, and some models include grazers as state variables. Although the BLOOM/GEM model code includes a number of grazing algorithms, including the one from ERSEM, zooplankton grazing is not explicitly taken into account in the standard version of BLOOM/GEM presented in this paper.

In summary, it can be stated that the main purpose is to demonstrate that BLOOM/GEM is sufficiently well validated according to objective criteria to simulate primary production and nutrients in the southern North Sea area under historic as well as under future conditions. This requires a specific level of spatial and ecological resolution.

2. Modelling instrument BLOOM/GEM

BLOOM/GEM is a generic ecological modelling instrument that can be applied to any water systems (fresh, transitional or coastal water) to calculate the primary production, chlorophyll-a concentration and phytoplankton species composition.We define a 'generic model' here as a computer code comprising a consistent set of formulations of processes that together describe (part of) ecosystem functioning which can be applied to those water bodies that satisfy those ecosystem characteristics. BLOOM/GEM is part of the Delft3D integrated modelling system of WL | Delft Hydraulics, which includes separate modules for hydrodynamics as well as for waves, morphology, and suspended sediments. In this chapter the general set upof BLOOM/GEM is described. The specific application of this generic model to the North Sea is described in the next chapter.

The ecological modelling instrument BLOOM/GEM has two main tasks:

1. It calculates the transport of model substances (state variables) in the water column as a function of advective and dispersive transport (provided by a hydrodynamic model, such as Delft3D-FLOW (Lesser et al., 2004; WL | Delft Hydraulics, 2005), Telemac (EDF-DER, 1998, 2000) or others).

2. It calculates the *water quality and ecological processes* affecting the concentrations of the state variables. These processes are defined as 'reactions' that causes one or more state variables of the model to appear, to disappear or to change into another state. Within the advection-dispersion equation (below), the processes are included in the source and sink term (S).

2.1. Transport of substances

The transport of dissolved or suspended matter in a fluid is commonly described by the advection-dispersion equation in Cartesian coordinates (Crank, 1975):

$$\frac{\partial C}{\partial t} = -u \frac{\partial C}{\partial x} - v \frac{\partial C}{\partial y} - w \frac{\partial C}{\partial z} + \frac{\partial}{\partial x}(D_x \frac{\partial C}{\partial x})$$

$$+ \frac{\partial}{\partial y}(D_y \frac{\partial C}{\partial y}) + \frac{\partial}{\partial z}(D_z \frac{\partial C}{\partial z}) + S + P$$

where:

C	concentration (kg m^{-3})
u,v,w	components of the velocity vector (m s^{-1})
D_x, D_y, D_z	components of the dispersion tensor (m^2 s^{-1})
x,y,z	coordinates in three spatial dimensions (m)
S	source or sink of mass due to physical, chemical and biological processes (kg m^{-3} s^{-1})
t	time (s)

In BLOOM/GEM, this equation is the basis for transport of substances. This equation states that the change of the concentration in time is caused by advective transport due to translation with the velocity vector (u,v,w) and by diffusive and/or dispersive transport, plus addition or extraction of mass (source/sink).

The source and sink term 'S' on the right hand side represents waste loads as well as various (kinetic) water quality and ecological processes. The coefficients terms in this equation are the three components of the velocity vector, the dispersion coefficients D_x, D_y and D_z as well as the source and sink term S. In order to solve the equation, these coefficient terms must be known. The structure of the source/sink term S must also be known. The reactions or other processes which represent S can be 0-order, first-order, or second-order reactions. At the open boundaries of the model domain, suitable boundary conditions for C need to be prescribed. These area dependent conditions determine the uniqueness and physical correctness of the solution of Eq. (1).

The advection-dispersion Eq. (1) can be solved analytically for relatively simple problems. For most practical applications, numerical simulation techniques are required due to the complexity of the geometry, the coefficients and the boundary conditions. A number of options for the numerical solution scheme are provided by BLOOM/GEM. In view of these practical applications, BLOOM/GEM has been designed with the aim of flexibility:

- it supports 1D, 2D and 3D model schematisations with a complex and irregular geometry;
- it is equipped with different finite difference solution methods, for both stationary and time varying problems; and
- the input of the geometry and the model coefficients can be done in various ways and the programme supports interfacing with other hydrodynamic models through files.

In addition to transport, concentrations of substances are determined by various physical, chemical and biological reactions, which are referred to as 'water quality and ecological processes'.

2.2. Water quality and ecological processes

Included in BLOOM/GEM are physical, biological and/or chemical reactions that cause one or more state variables of the model to appear, to disappear or to change into another state variable. These processes are related to algae growth and mortality, mineralization of organic matter, nutrient uptake and release, oxygen production and consumption. The BLOOM/GEM modelling instrument considers three nutrient cycles: nitrogen, phosphorus and silica. The carbon cycle is partially modelled, and a mass-balance of organic carbon is made. The model assumes that the availability of inorganic carbon for uptake by algae is unlimited. Furthermore, different groups of algae are considered, either phytoplankton (e.g. diatoms, flagellates, dinoflagellates, *Phaeocystis*) or macroalgae (*Ulva* 'attached' or *Ulva* 'suspended'), oxygen, suspended detritus, and inorganic particulate matter. Light availability for phytoplankton growth is calculated based on the light irradiance and extinction, due to suspended sediment as well as phytoplankton and other organic matter. Different

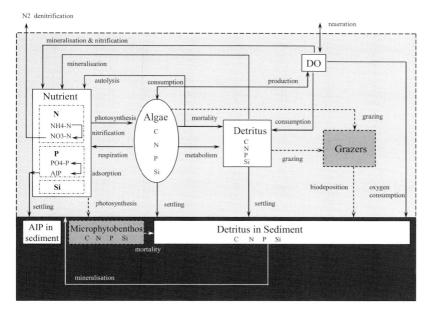

Fig. 1. Schematic overview of all state variables and processes included in the ecological model instrument BLOOM/GEM. State variables in grey and processes indicated by dashed lines are optional and have not been included in the North Sea modelling applications. AIP is 'adsorbed inorganic phosphorus'.

formulations are available for characterisation of grazers, microphytobenthos, bottom sediment and sediment-water exchange. The formulations can range from simple functions (e.g. grazing) to fully dynamical processes (e.g. algae growth and mortality). The full set of possible processes is illustrated schematically in Fig. 1 and the most standardly used ones are described briefly below. Depending on the objective of the application, a decision as to which ones to include must be made for each model application. In addition it should be noted that not all processes have been equally tested and validated (e.g. microphytobenthos). More detailed description of the processes, including model formulations and process coefficients are given in WL | Delft Hydraulics (2002, 2003a) and Los and Wijsman (2007).

2.2.1. Nutrient cycling processes

There is a closed mass balance of all nutrients (P, N and Si) and the processes included for the three nutrients are generally similar. Processes include the uptake of inorganic nutrients by algae and by benthic algae (microphytobenthos). Algae mortality produces detritus and inorganic nutrients (via autolysis). Mineralization of detritus in the water column and in the bottom sediment produces inorganic nutrients, in water and sediment, respectively. Different sediment water formulations are available to define how inorganic nutrients in the sediment are released back into the water column (see 2.2.5 sediment–water interaction, below). Nutrients in detritus can be sedimented and resuspended, in a similar manner to inorganic suspended sediment. Nutrients in benthic detritus (sediment) can also become buried, and essentially removed from the system. Specific processes for nitrogen are nitrification and denitrification. Specific processes for phosphorus are sedimentation of adsorbed inorganic phosphorus (AIP), creating a pool of AIP in the sediment, and adsorption/desorption of orthophosphate. There is no explicit microbial loop – the mineralization of algae and detritus is modelled with decay coefficients, which vary with the stoichiometry of the dead organic matter (WL | delft Hydraulics, 2003a).

2.2.2. *Phytoplankton growth and mortality processes*

BLOOM/GEM calculates the growth of algae as a function of nutrient and light conditions. Phytoplankton mortality and grazing produces detritus and releases inorganic nutrients (autolysis) in the water column, while mortality of benthic algae produces benthic detritus and releases inorganic nutrients (autolysis) ineither the bottom sediment or directly into the overlying water depending on the sediment formulation chosen. Benthic algae can also be buried and removed from the system if there is significant sedimentation. Pelagic algae can sediment to the bottom and benthic algae can be resuspended into the water column. Zooplankton and benthic grazers can be explicitly modelled or may be represented by a grazing function or as part of the total ('natural') mortality (Loucks and Van Beek, 2005; Los and Wijsman, 2007).

2.2.3. *Oxygen related processes*

BLOOM/GEM solves the mass balance for oxygen, in which several oxygen producing and oxygen consuming processes are considered. Oxygen is produced by algae (primary production) and is consumed by algal respiration, by mineralization of detritus and other organic material (in the water column and bottom sediment), and by nitrification. Exchange of oxygen with the atmosphere (e.g. reaeration) can result in either a gain or loss of oxygen in the water column (WL | delft Hydraulics, 2003a).

2.2.4. *Light attenuation*

Since primary production is strongly influenced by light availability and can even become limited if there is too little light, the correct calculation of light conditions in the water column is essential. The availability of light is a function of the solar irradiation within a certain wave length range (photosynthetically available radiation: PAR) and of the extinction due to absorption and scattering of the light inside the water column. The extinction of light underwater is described by the Lambert-Beer law, the empirical relationship between the absorption of light and the concentration of absorbing species in water:

$$I_z = I_0 * e^{-kz}$$

In this equation I_z is the underwater light intensity at depth z, I_0 is the surface irradiance, k is the extinction coefficient, which can be related to the absorption and scattering properties of the water constituents.

In BLOOM/GEM, the light extinction coefficient (k) is calculated with an empirical model as the sum of the extinction by inorganic suspended solids, labile organic matter, phytoplankton and other organic substances which is taken as a function of the amount of fresh water. Each of the substances, including different phytoplankton species, is characterized by a specific extinction coefficient (Los and Bokhorst, 1997; Van Gils and Tatman, 2003). A detailed description of the equation used in BLOOM/GEM is presented by Los and Wijsman (2007). Inorganic suspended sediment can be modelled as a state variable (Delft3D-WAQ, WL | Delft Hydraulics, 2003a), or concentrations canbeprovidedas aforcing function. Recently, methods for obtaining suspended sediment concentrations from remote sensing data are being developed (WL | Delft Hydraulics, 2003b).

2.2.5. Sediment-water interaction

BLOOM/GEM provides several options for modelling bottom sediment processes and sediment-water interactions. These vary in complexity and ease of use. Using the simplest description, the sediment serves as a temporary storage of detritus and adsorbed phosphate just below the lowest water segments. The sediment pool can accumulate inorganic and organic particulate matter due to sedimentation from the water column. As the organic matter mineralises in the bottom, related nutrients are released directly back into the overlying water column. In the event of erosion, the bottom particulate matter is resuspended back into the water column. Burial or more generically 'inactivation' permanently removes nutrients from the sediment (WL | Delft Hydraulics, 2002; See also Fig. 1). This approach is often sufficiently accurate, but proves to be inadequate in shallow areas in which a more intense exchange of material between water and sediment prevails. For these type of systems, more complicated modules have been designed. Among these are a four layer biochemical model called SWITCH (Smits and Van der Molen, 1993), GEMSED a special version of the benthic nutrient model in ERSEM (Ruardij and Van Raaphorst, 1995), and a new, n-layer model (WL | Delft Hydraulics, 2002). Potentially these models provide a more accurate description, but they are complex and in many cases there is a lack of data from sediments to validate them.

2.2.6. BLOOM concept of optimisation

The total algal biomass usually consists of various species of phytoplankton belonging to different taxonomic or functional groups such as diatoms, flagellates, green algae and cyanobacteria, commonly referred to as blue-green algae. This is true for both fresh-water as well as for marine systems. They have different requirements for resources (e.g. nutrients, light) and they have different ecological properties. Some species are considered to be objectionable due to their effect on the turbidity of the water, the formation of scums or the production of toxins. For example, *Oscillatoria* can achieve very high biomass levels in shallow lakes causing a very low transparency (Zevenboom et al., 1982; Berger and Bij de Vaate, 1983; Berger 1984), and *Microcystis* is notorious for the formation of scums and has been reported to produce toxins that are harmful to animals (e.g. cattle) and men (Chorus et al., 2000; Atkins et al., 2001). In the marine environment, *Phaeocystis* is responsible for foam on beaches (Lancelot et al., 1987) and mass mortality of shellfish due to the settlement of a bloom in sheltered areas and subsequent depletion of oxygen (Rogers and Lockwood 1990). To deal with these phenomena, it is necessary to distinguish between different types of phytoplankton in a model.

The phytoplankton module BLOOM is based upon the principle of competition between different species (Los et al., 1984; Los and Brinkman, 1988; Los, 1991; Van der Molen et al., 1994; Loucks and Van Beek, 2005; Los and Wijsman, 2007). Two basic units are distinguished by the model: The first unit is that of functional groups which we refer to as 'species groups' or simply 'species'. A model species may be equivalent to a biological species, or it can be representative for larger taxonomic units, which are supposed to have similar ecological characteristics. The second unit is that of 'types'. Model species usually consist of several types. A type represents the physiological state of the model species under various possible conditions of limitation. Model species are usually divided into three different types: an N-type representing the ecophysiological condition of a species under nitrogen limitation, a P-type for phosphorus limitation and an E-type, representing the state of a species under low light conditions.

The solution algorithm of the model considers all potentially limiting factors in terms of the available amounts and the requirements by each type of phytoplankton, i.e. the equations are solved under optimisation constraints. The optimisation algorithm selects the resource that is most likely to become limiting and the best adapted type under the prevailing conditions. The suitability of a type (its fitness) is determined by the ratio of its requirement and its growth rate. This means that a type can either become dominant because it needs a comparatively small amount of a limiting resource (it is efficient) or because it grows rapidly (it is opportunistic).

Subsequently, the algorithm considers the next potentially limiting factor and again selects the best adapted phytoplankton type. This procedure is repeated until it is impossible to select a new pair of phytoplankton type and limiting factor without violating (i.e. over-exhausting) one of the resources. Thus the model seeks the optimum solution consisting of n types and n limiting factors.

As a further refinement, BLOOM takes the existing biomasses of all phytoplankton types into account. Because the phytoplankton types represent different phenotypes of the same species, the transition of one type to another occurs at the time scale of a cell division, which is in the order of one day. Due to this characteristic time scale (i.e. Chapter 5 of Harris, 1986), the simulation time step for the BLOOM phytoplankton processes is usually chosen to be 24 h. A transition between different species is a much slower process as it depends on mortality and net growth rates of different species. In the model, therefore, transitions between types can occur with each BLOOM time step, which is thus more rapid than the transition between species, which may require weeks. It can be shown mathematically that the principle by which each phyto-plankton type maximizes its own benefit, effectively means that the total net production of the phytoplankton community is maximized. This makes it possible to use the computationally efficient Linear Programming technique (Danzig, 1963) to compute the phytoplankton biomasses according to the competition rules formulated for the module.

Note: The BLOOM/GEM model allows a 'variable timestep' to be used for efficient computer run time. For transport, typically a time step between 1 and 30 min is required depending on the grid and numerical method that is being used. For calculation of ecological pro-cesses within BLOOM/GEM, a time step of 24 h is usually short enough to provide sufficiently accurate results for most biological and chemical processes. Two exceptions are the calculation of the primary production rate and the dissolved oxygen concentration, both of which are calculated with the same time step as transport (i.e. in the order of minutes) to account for the diurnal cycle in the light intensity. To this purpose, the measured daily solar radiation is distributed over the known day length according to a predetermined cosine function. Thus the diurnal cycle of dissolved oxygen is simulated at a 30 min time step. The result for primary production is integrated over 24 h to generate the daily growth rate of phytoplankton taking other fluxes (i.e. mortality, respira-tion and nutrient uptake) into account. Hence in a typical simulation, transport and a small number of processes are simulated with a short time step (i.e. 30 min or less) while the majority of the processes are simulated using a much longer time step of 24 h. Comparative simulations have shown that results of simulations with a 30 min time step for all processes are similar to those with the variable time-steps.

The number and the characteristics of the phytoplank-ton species are inputs to the model. Data for about 20 different marine and fresh-water species have been collected over the years based on literature, laboratory experiments (Zevenboom and Mur, 1981; Zevenboom et al., 1983; Zevenboom and Mur, 1984; Post et al., 1985; Jahnke, 1989; Riegman et al., 1992, 1996; Riegman, 1996) and previous model applications. Depending on the problem and the water system being modelled (specific model application to either fresh, transitional or coastal water), sometimes only major groups are included such as diatoms, greens and blue-greens, and sometimes indivi-dual genera are modelled such as *Ulva* or *Phaeocystis*. For each species and species type, there is a different factor for converting biomass to chlorophyll-a concentra-tion. This factor is variable depending on the limiting factor (light; nutrients) and ranges from 0.0067 to

0.0533 mg Chla per mg C. A complete overview is given in Los and Wijsman (2007).

3. Model application to the North Sea: calibration and validation

3.1. Concepts of validation

3.1.1. Definitions

Although model validation receives much attention in the literature (e.g. Los andGerritsen, 1995; Scholten etal., 2000; Refsgaard and Henriksen, 2004), the concept of validation and calibration is ill-defined, and is interpreted differently by different modellers. Often validation is considered as proving that the model behaves like reality for the processes of interest. Mankin et al. (1975) postu-lates that *"a valid model has no behaviour which does not correspond to system behaviour and that a useful model predicts some system behaviour correctly. Since no model is perfect, available ecosystem models may be described as invalid but useful models. "* Following this concept, this means that in fact all "validated" ecosystem models that are presented in literature are invalid. Therefore it makes more sense to focus validation procedures on the use-fulness of the model for processes of interest rather than at the disqualification of the model. Clearly, if it turns out that the model is invalid and not useful at all, the model should not be used for system analysis or predictions. The criteria for the 'usefulness' of a model depend strongly on the specific processes and goal(s) of the modelling study (e.g. importance of global vs. local results; or of long term mass balances and fluxes vs. instantaneous concentrations and short term mass fluxes). It is crucial to have an agreement beforehand as to the specific goal(s) of a model application and the choice of focus in order to conduct a successful study.

Based on the work of Schlesinger et al. (1979), Refsgaard and Henriksen (2004) propose the following terminology for calibration and validation:

Model calibration: The procedure of adjustment of parameter values of a model to reproduce the response of reality within the range of accuracy specified in the performance criteria.

Model validation: Substantiation that a model within its domain of applicability possesses a satisfactory range of accuracy consistent with the intended appli-cation of the model.

In this sense, validation means testing the usefulness of the model related to specific objectives, as postulated by Mankin et al. (1975), rather than proving that the model is good or false. In fact this terminology describes the approach we have followed in the development of the BLOOM/GEM North Sea model during the last two decades. A similar definition by Lynch and Davies (1995) describes validation of a computational model as the process of formulating and substantiating explicit claims about the applicability and accuracy of computational results, with reference to the intended purposes of the model as well as to the natural system it represents. One component of model validation regards the correctness of coding, also referred to as model verification: checking that the mathematical equations are being solved numerically correctly. This is done in many ecological models by budget calculations, where conservation of mass is taken as confirming that the numerical methods and coding are correct.

3.1.2. Cost functions

For many modellers, validation is performed by measuring the 'goodness' of the simulations. This is often done by graphically presenting data for observed and simulated parameters and visually assessing the goodness of fit. Even if Box-Whisker plots, data ranges or other similar statistics are used when graphically presenting the performance of the model vs. observations, the assessment and comparison remains subjective.

One objective method that can be used is that of the 'cost function'. The cost function is a mathematical function which provides a means of comparing data from two different sources. The cost function gives a non-dimensional number which is indicative of the 'goodness of fit' between two sets of data. For model validation it can objectively quantify the difference between model results and measurement data. During the ASMO Eutrophication modelling workshop (Ospar et al., 1998) different cost-functions were defined for comparing the 'goodness of fit' of several North Sea ecological models, with respect to a common data set. In general, the cost function is the sum of the absolute deviations of the model values from the observations, normalized by the deviations of the observations for the chosen temporal and spatial range. Thus it is a standardized, relative mean error. The normalised deviation between model results and observations $(C_{x,t})$ for defined areas (x) and time intervals (t) is calculated as:

$$C_{x,t} = \frac{M_{x,t} - D_{x,t}}{sd_{x,t}}$$

where $C_{x,t}$ is the normalised (in sd units) deviation between model and data for area x and time interval t, $M_{x,t}$ is the mean value of the model results within

area x and time interval t, $D_{x,t}$ is the mean value of the in situ data within area x and time interval t, and $sd_{x,t}$ is the standard deviation of the in situ data within area x and time interval t.

In the ASMO workshop, different cost functions were defined to make the comparison between model and data on different spatial and temporal scales, including:

1. large scale, synoptic, long-term seasonal averaged,
2. large scale, specific regions (boxes), time resolving,
3. regional scale, specific stations, time resolving.

These cost functions were based on temporal and spatial means, calculated from both observations and model results in the same way. Radach and Moll (2006) also present results of this validation comparison and propose to make the cost function approach a standard method of model validation. This recommendation has been followed and cost function results for BLOOM/ GEM have been calculated for the regional scale (at specific stations and time resolving), calculated as:

$$C_x = \frac{\sum |M_{x,t} - D_{x,t}|/n}{sd_x} * ((1-c) + c(1-r_x))$$

where C_x is the normalised deviation per station, annual value, $M_{x,t}$ is mean value of the model results per station per month, $D_{x,t}$ is mean value of the in situ data per station per month, sd_x is standard deviation of the annual mean based on the monthly means of the in situ data (df=11), n is 12 months, c is 0.5 and r_x is the correlation between $M_{x,t}$ and $D_{x,t}$ over the time period considered (in this case one year).

Results are presented in Section 4. The sharpened ratings criteria proposed by of Radach and Moll for the cost functions (cf; below) have also been adopted.

Rating	Condition	
Very good	$0 < cf \leq 1$	Standard deviations
Good	$1 < cf \leq 2$	Standard deviations
Reasonable	$2 < cf \leq 3$	Standard deviations
Poor	$3 > cf$	Standard deviations

3.1.3. Objectives of modelling

Every modelling instrument and every model application is made for a specific purpose and certain modelling objective. It is important that this objective is clearly stated, as the interpretation of the model results and the evaluation of the model performance (validation) must be made with respect to the modelling objectives.

Considering BLOOM/GEM and its applications to the North Sea, the ability of the model to simulate the ecological status of the Dutch coastal waters and the southern North Sea and support relevant management issues for the Dutch Government are of importance. The model should first of all be able to describe the existing status of the system, with respect to the assumed key process of primary production, species composition, as well as nutrient and oxygen concentrations. Furthermore, it should also be able to predict the impacts of future situations, for instance, changes in anthropogenic nutrient loadings, climatic changes, sand mining, fisheries, windmill parks, harmful algae blooms, etc., on these key processes.

With these broad objectives, the adequate predictive management model needs to be based on a multi-disciplinary approach taking the key hydrodynamical, chemical and biological processes into account. The ideal model for these is 3-dimensional, it has historical forcing by weather, discharges and concentrationsat the open boundaries. The chemical and biological processes are represented in detail with respect to space, time and internal structure. Specific modelling goals are to successfully describe long-term changes to the system (trends) as well as short-term events (e.g. harmful algal blooms, calamities), the latter being of increasing importance in recent years. In principle, the same modelling instrument can be used for both objectives, though the specific application, model inputs, and evaluation of model results may differ.

3.2. *History of GEM/BLOOM applications for the North Sea – repeated validations*

Conceptual and mathematical ecological models have played a central role in marine biology and ecology as tools for synthesis, prediction, and understanding. It is clear that significant progress has been made over the last 20 years in the development of numerical process models of the marine environment (Moll and Radach, 2003; Prandle et al., 2005).

In the Netherlands, ecological modelling and model development occurred in several different groups, each with a different geographic and biological focus, e.g. primary production and macrobenthos in the Wadden Sea (ECOWASP, Brinkman, 1993) carbon flows and macrobenthos in the Eastern Scheldt Estuary (SMOES, Klepper et al., 1994; Scholten and Van der Tol, 1994) and benthic processes in the Western Scheldt (Soetaert et al., 1994; Soetaert and Herman, 1995a,b). There was also a considerable Dutch contribution to the development of the ERSEM ecological model (Baretta-Bekker, 1995; Baretta-Bekker and Baretta, 1997). WL | Delft Hydraulics together with the Netherlands Institute for Coastal and Marine Management/RIKZ have focused coupled

on hydrodynamic-water quality-ecology models and applications for the North Sea as a whole. This work began in the early 1980s when one of the main issues was to describe, understand and predict the development and abatement of eutrophication in the Dutch coastal zone (Van Pagee et al., 1988; Glas and Nauta, 1989; Nauta et al., 1992).

At that time, the transport modelling was based on 2D hydrodynamical calculations for average conditions (one representative day). As a compromise between computational performance and accuracy, a uniform grid size of 16×16 km was selected for the southern North Sea schematisation (up to 57 degrees N) totalling 1395 computational elements. The coupled water quality modelling instrument "DYNAMO" provided reasonable results on a seasonal basis for total algal biomass expressed as chlorophyll and nutrients in areas where the grid size was not too large to represent important horizontal gradients. DYNAMO considered only 2 algae types 'diatoms' and 'others', the first requiring silica as a nutrient.

The next big step in the Dutch North Sea ecological modelling was the implementation of the phytoplankton optimisation module BLOOM, first as a vertical water column application (De Groodt et al., 1991; Peeters et al., 1995), then as part of a coupled 2D hydrodynamic-water quality-ecology model (Los and Bokhorst, 1997; Ospar et al., 1998), followed by the model application to a new, curve-linear model grid of the Dutch North Sea coast. This coastline-following grid allowed transport to follow the coastal contours. Seasonal variations in flow conditions were mimicked by correcting the dispersive flow rates of the representative daily flow as a function of the historic wind direction and speed during a simulation of the water quality-ecology part of the model. The application, called North Sea Bloom, was used for several management evaluations (e.g. Boon and Bokhorst, 1995; Los and Bokhorst, 1997; Ospar et al., 1998). For the first applications, extensive comparison of model results and measurements was made to obtain the optimal set of model coefficients given the objective of the modelling and the application area. Subsequent advancements and improvements in this line of North Sea ecological modelling consisted of only minor changes in the ecological model formulations and process coefficients. More significant, were step-wise increases in the spatial resolution of the model. Developments in hydrodynamic modelling and increasing computational facilities also allowed more time-specific transport to be modelled, progressing from an 'average' daily residual flow (De Ruyter et al., 1987), to a representative spring-neap cycle (WL / Delft Hydraulics/MARE, 2001a,c), to 'historic' hydrodynamics based on spatially varying, daily wind and atmospheric pressure conditions (Ta b l e 1). These improvements were driven in part by the need for more accurate model results in regions where steep gradients for nutrients exists, as is the case in the Dutch coastal zone, the main region of interest.

Additional improvements in the model applications primarily concerned abiotic conditions that affect the ecology (suspended solids concentrations and light conditions), as well as the river nutrient loads. Since 1994, with cooperationof the leading Dutch ecological institutes,

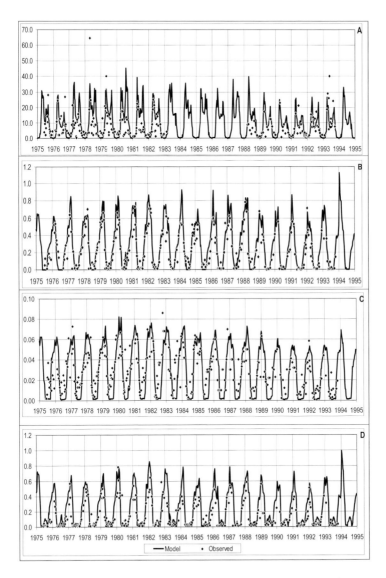

Fig. 2. Validation result for 2D BLOOM/GEM application: station Terschelling 4, 1975–1994 (A: chlorophyll-a in μg l^{-1}, B: nitrate in mg l^{-1} C: ortho-phosphate in mg l^{-1}, D: dissolved silicate in mg l^{-1}).

189

a number of generic ecological model formulations have been implemented, such as those for macrophytobenthos and other benthic processes. These specific formulations have been combined with BLOOM/GEM (WL | Delft Hydraulics, 2003a), but not all of these have yet been extensively tested. Further modifications in the ecological modules have been with respect to phytoplankton kinetics (WL / Delft Hydraulics/MARE, 2001a) and some coefficients for the extinction computation (Van Gils and Tatman, 2003). Ongoing developments include deriving forcing functions for suspended sediment concentrations from optical remote sensing data. The BLOOM/GEM model applications throughout this period have been primarily 2-dimensional (vertically mixed), due to generally well-mixed conditions in the area of interest, i.e. the Dutch coastal zone.

As a typical illustration of the performance of previous 2-D model simulations Fig. 2 shows the seasonal cycles for chlorophyll-a, nitrate, ortho-phosphate and dissolved silicate at the coastal station Terschelling 4 for a 20 year period, from 1975 till 1995 (Los and Bokhorst, 1997; De Vries et al., 1998). The model reproduces not only the seasonal concentration pattern of the four substances, but also the concentration variations from year to year. This simulation was based on a calibration of the model for the year 1990. The location of this station and the other sampling locations in the Dutch coastal monitoring programme are shown in Fig. 3.

3.2.1. Selected model options for North Sea application
As described in Section 2.2 on the modelling instrument, optional process descriptions are available for several processes, and each specific model application must make a decision as to which ecological processes to include. For reasons of robustness and predictive performance, as a general principle the simplest formulation that can still result in an overall model behaviour in accordance to the general objective of our model application is selected.

3.2.1.1. Benthic processes. The 'simple bottom' module for benthic mineralization has been adopted because in

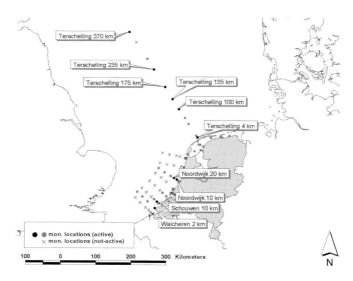

Fig. 3. Location of coastal water quality and ecology monitoring stations from the Dutch national (MWTL) monitoring programmes. All stations indicated by a circle were monitored in 1989, results are presented for stations mentioned by name. Stations marked by x were monitored in some but not all years between 1975 and 1989 and considered during the validation of earlier model versions.

the main areas of interest in the Dutch region of the North Sea, most organic matter mineralization occurs in the water column as opposed to the sediment. In the model, on average 77% of the mineralization of C occurs in the water and only 23% in the sediment. Numbers for nutrients are similar. The only areas of interest where these conditions are violated, is in shallow areas such as the Wadden Sea. For these areas, mineralization rates of water and sediment are equal and at the most shallow locations (less than 3 m) mineralization rates in the sediment exceed those of the water. Furthermore, microphytobenthos is not included in the North Sea applications. Consequently while the Wadden Sea is included in our model domain, its results in this area should be considered as indicative only.

3.2.1.2. Grazing. In the North Sea applications, the impact of grazing is included as an enhanced mortality term only. Because the mortality rate constant varies between the phytoplankton types included in the model, the effective mortality usually increases at the peak of the spring bloom when E-types with a low mortality are replaced by N- or P-types with high mortality and high sedimentation rates. This enhanced mortality therefore coincides with the period at which the grazing pressure is supposed to be large and to some extend mimics the impact of zooplankton grazing after the spring bloom.

3.2.2. Conclusion for this section

Since the early 1980s, the long-term process of repeated applications of the BLOOM/GEM instrument including repeated validations has resulted in a fixed (consistent and robust) set of ecological model process coefficients that produce good results for different years and for different areas within the North Sea, including good reproduction of spatial and temporal variations. This set of ecological equations and parameter values was already used for the Dutch Coastal zone North Sea BLOOM model (Los and Bokhorst, 1997) and have been kept nearly the same for more than five years (from 1996 through 2002). Through this development process, the main problems identified, (i.e. discrepancies between model results and data), have been shown to be caused primarily by inadequacies in the model forcing: hydrodynamics, suspended sediment concentrations and river loads. Improvements in these forcing factors have consistently led to improved results.

3.3. 3D application for the North Sea

To follow-up on the results provided by the 2D model applications (above), a 3D application of BLOOM/GEM

was prepared, as a final step in the validation of more local ecological model processes and a robust parameter set for this. The 3D application can locally give improvements compared to the 2D application, in terms of vertical gradients in the coastal zone due to the fresh-water Rhine plume as well as stratification in the north-central North Sea. The 3D application has been made without changes to the ecological process coefficients. The 3D model application to the southern North Sea is described below, and summarised in Table 2. In this example, the year 1989 was modelled (more precisely: November 1988-November 1989).

3.3.1. Schematisation

The hydrodynamic model application used the part of the North Sea included in the original "Zuidelijke NoordZee ('ZUNO') model" supplied by the Dutch Ministry of Public Works. The specific schematisation used is the so-called 'coarse ZUNO-model' (Fig. 4). Its horizontal resolution is relatively high in the coastal areas of interest, notably the Dutch coastal zone (approximately 1×1 km). The grid is much coarser in the northern part of the area included in the model (approximately 20 × 20 km). In the vertical direction, the model has 10 sigma layers. Near the bed and near the surface, the layer thickness is about 4% of the local water depth. At mid-depth, the layer thickness is approximately 20% of the local water depth (Table 3).

3.3.2. Hydrodynamics

Along the open boundaries of the hydrodynamic model water levels are prescribed based on astronomic tidal water level components. These boundary conditions originate from simulations with a large scale model covering the entire Continental Shelf (Gerritsen and Bijlsma, 1988). The eighteen main sources of fresh water are specified separately. Discharges from Dutch rivers are included on a decadal (10 day) basis for the actual simulation period, for the German rivers on a monthly basis for an average year and for all other rivers as an annual average. The initial conditions and boundary forcing of the salinity field are given in WL | Delft Hydraulics/MARE (2001c,d).

Earlier modelling studies, such as NOMADS and PROMISE (Boon et al., 1996, Gerritsen et al., 2000) revealed that varying wind conditions have a large impact on the long-shore and cross-shore currents, especially for those in the coastal areas. At the sea surface, the hydrodynamic model is forced by time and spatially varying wind and pressure fields originating from the NOMADS2 project (Proctor et al., 2002; Delhez et al., 2004) as well as the associated relative humidity, air

Table 2
Specifications of the BLOOM/GEM 3D model application

Characteristics	Model application: Delft3D-GEM
Spatial resolution	Variable (curve-linear grid) ranging from
Δh (km)	~1 km (smallest) to 20 km (largest)
Vertical resolution	Hydrodynamics D3D-FLOW is 3D (10 layers) Ecology BLOOM/GEM is also 3D
Longitude (degree)	-2.0, 10
Latitude (degree)	49.2, 57.0
Spatial extent (km)	950 from north to south
Temporal resolution	Transport time step (from D3D-FLOW) 1 h
Δt (s)	(3600 s) Ecological processes time step: 24 h Output written every 24 h or 7 days
Pelagic matter cycle	N, P, Si, O
No. of pelagic state variables	C: 4 major phytoplankton groups (comprising a total of 12 types), DetC N: NH4, NO3, DetN P: PO4, DetP Si: Si, DetSi O: Oxy Optional: Zooplankton or mussels Total: 21–22
Pelagic nutrients (bulk or explicit)	Explicit (NO3, NH4, DetN, PO4, DetP, Si, DetSi)
Phytoplankton	4 major functional groups: dinoflagellates, diatoms, flagellates, Phaeocystis (comprising a total of 12 types)
Zooplankton	Optional (Dynamically modeled or grazing function)
Benthic matter cycle	C, N, P, Si,
No. of benthic state variables	DetC, DetN, DetP, DetSi Total: 4
Benthic nutrients (bulk or explicit)	Explicit
Zoobenthos	Optional
DOM	Optional
Bacteria	No (modelled as mineralization)
Detritus/POM	Yes (DetC, DetN, DetP, DetSi)
Spin up time	1 year (The model is first run for 1 year and then these end results are used for the initial conditions of the real run)
Meteo: real data or climatological	Historic data for wind, atmospheric pressure, solar radiation, air temperature
Hydrodynamic Model	Delft3D-FLOW (fully 3D, 10 layers), Real forcings of wind and atmospheric pressure
SPM	Suspended particulate matter (SPM, sometimes also referred to as TSM) is either modeled or provided as a forcing function based on remote sensing data (For the OSPAR work-we used remote sensing)
Light	Light is a function of: inorganic suspended matter, yellow substances (freshwater), detritus, and phytoplankton SPM

temperature and cloudiness needed to compute the exchange of heat with the atmosphere. The exchange of heat with the atmosphere is included by means of formulations originating from Lane (1989). In the vertical direction, a k-epsilon turbulence closure scheme is applied to compute the exchange of momentum, salinity and temperature.

Prior to application of BLOOM/GEM, an extensive calibration of the hydrodynamic model was carried out. The calibration mainly focused on reproducing large and small scale transport patterns. Computational results were compared with measurements of, or general knowledge on: residual fluxes through the English Channel, the temperature distribution in the entire Southern North Sea, the horizontal and vertical salinity distribution within the Dutch Coastal River, residual flow velocities derived from measured velocities at a measurement location 10 km offshore the Dutch Coast ('Noordwijk 10'), and tidal and residual fluxes through the inlets to the Dutch Wadden Sea. An example of the model skill is given in Fig. 5, showing a comparison of computed and measured salinity values. Further details on the calibration and verification of the hydrodynamic model are given in Lesser et al. (2004); WL | Delft Hydraulics (2005); WL | Delft Hydraulics/MARE (2001a,b). Following a spin-up period of several months, the hydrodynamic results for the period November 1988 till November 1989 are stored in a database at an hourly interval, and form the basis for the BLOOM/GEM model simulations.

3.3.3. 3D GEM calculations

The primary production and nutrient model uses the same vertical and horizontal grid as the hydrodynamic model. For the substance transport, a time-step of 30 min is used, which means that the hourly hydrodynamic results for the advective transport component are divided into two equal steps of 30 min. Primary production and oxygen dynamics are also simulated at a 30 min time step, all other biological and chemical processes are simulated using a 24 h time step. A second order numerical integration scheme is used. Mass balance is checked as an initial validation of the simulations.

3.3.3.1. Meteorological forcing. The meteorological conditions are not constant over the area included in the model. In the case of the water temperature variations are considered to be significant. The input of BLOOM/GEM is therefore directly taken from the 3D temperature model, which is part of the Delft3D-Flow hydrodynamic model. Spatial variations in other meteorological forcings are not directly measured or otherwise readily available. Therefore the day length, solar irradiance and wind speed are adopted from historic measurements by the Royal Dutch Meteorological Institute at a single land-based station "de Kooy" near Den Helder in the north western part of the Netherlands. In the case of the total irradiance there are no

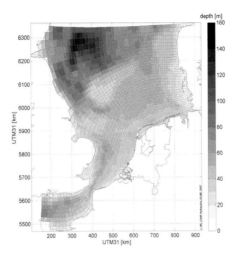

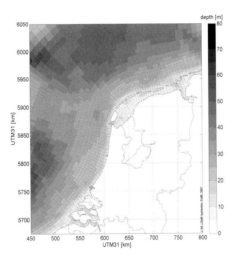

Fig. 4. Model grid and bathymetry of the Southern North Sea used in the hydrodynamic and BLOOM/GEM simulations (known as the 'coarse ZUNO grid'). Total number of cells in the horizontal schematisation is 8710 (m=65, n=134). Number of active cells in the horizontal schematisation=4350. Vertically, the grid has 10 sigma layers. The BLOOM/GEM calculations are made on the full model grid (no aggregation). Top: full model grid, Bottom: zoom of the Dutch coast.

Table 3
Specifications of the Zuno model grid

Horizontal dimensions	65 × 134 (hydrodynamics)
No. active horizontal grid cells	4350
Vertical distribution	layer 1 (top): 4%
	layer 2: 5.9%
	layer 3: 8.7%
	layer 4: 12.7%
	layer 5: 18.7%
	layer 6: 18.7%
	layer 7: 12.7%
	layer 8: 8.7%
	layer 9: 5.9%
	layer 10 (bottom): 4%

indications that spatial variations in different parts of the North Sea are significant on a seasonal basis. The day length varies as a function of the latitude, but in the model this is only used to distribute the measured, non-spatially variable daily irradiance on a 30 min basis. Using a slightly longer or shorter light period does not affect the simulated rate of primary production significantly. Moreover during the spring bloom, which usually occurs at the end of March or beginning of April, the day length is approximately 12 h irrespective of the latitude. From simultaneous measurements at sea and at the land it is known that the wind speed at sea is systematically higher than on the land. The wind speed in the model is scaled with respect to the land-based measurements to compensate for this effect.

3.3.3.2. Boundary conditions. The exchange of water masses through the English Channel and at the northern boundary (57 degrees N) are adopted from the hydrodynamic simulation. The concentrations of substances at these boundaries are based on the work of Laane et al. (1993). In the model, the northern boundary concentrations are constant in time. Data for the English Channel boundary concentrations are specified as a monthly time series. These same data have regularly been used previously (e.g. Los and Bokhorst, 1997).

3.3.3.3. Rivers and other nutrient sources. The model application contains the point sources of nutrients and fresh water, from the main Dutch, German, French and UK rivers. For the Dutch rivers, substance loads were derived from measured discharges and concentrations in rivers at decadal (10 day) intervals for the period November 1988 through November 1989 (http://www.waterbase.nl). Modelled substances that were not measured have been derived from measured data of other substances, using stoichiometric ratios and other knowledge rules that have been developed and proved

successful in previous modelling (e.g. Los and Wijsman, 2007). For other rivers, less frequent data were available, but their contribution to the total loading hence their significance is smaller. Fig. 6 shows the location of these river discharges as well as the less significant ones, which have not been taken into account at all.

3.3.3.4. Inorganic suspended matter. Light is an important limiting factor for phytoplankton in the North Sea particularly in the Dutch coastal zone. In contrast to variations in the surface irradiances, spatial and temporal variations in inorganic suspended matter, hence in the under water light climate, are considerable. In the model presented in this paper the steady state result from a 3D suspended matter simulation model is used as input for BLOOM/GEM. The overall seasonal variation is simulated by means of a cosine function with relative high values in winter and low values in summer. Its amplitude is based upon the level of variation in the measurements. This same method was applied previously during other modelling studies (Los and Bokhorst, 1997). Using this cosine function, the spring bloom in the model tends to be rather late because in reality short periods of quite

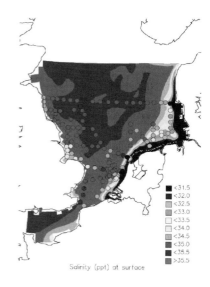

Fig. 5. Validation result for 3D hydrodynamic calculation (salinity) for the surface layer for March 1989. Coloured patterns are the model results, filled circles are measurements from the NERC cruise (NERC, 1992).

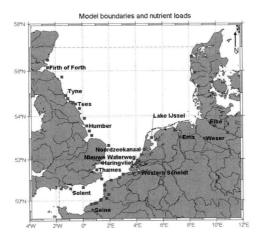

Fig. 6. Location of the northern (Atlantic) and southern (Channel) boundary of the model domain (dashed lines). Location of the major river discharges of the southern North Sea (squares). The ones specified by name are explicitly taken into account by the model. The others have been included in the latest model version (not presented here).

conditions with relatively low levels of suspended matter play an important role in triggering the onset of the spring bloom (unpublished results). To account for these short-term variations, we have assumed a relationship between the SPM concentration and the average wind speed to further adjust the cosine signal. The amplitude of this short term fluctuation is a multiplication factor varying between 0.3 and 1.7 depending on the difference between the actual and average wind speed (5.5 m s^{-1}). Again, these factors were determined empirically in such a way that the observed variability range could be reproduced sufficiently well. Sensitivity studies showed that in fact, any type of short-term fluctuation in the suspended matter concentration, even if it was purely random, was sufficient to trigger the reactions and improve the timing of the spring bloom in the model at those location where it is controlled by light (Fig. 7). In the autumn the growing season is extended in these simulated relative to the base case.

4. Results

Although an extensive data set for the North Sea and Dutch coastal zone is available for validation (locations shown in Fig. 3) the number of locations is much smaller than the number of computational elements of the model. Moreover most of this data is confined to the water surface. Samples at the middle of the water column and just above the bottom are, however, available for a limited number of stations for the years 1987 through 1989. During recent years, more non-surface data have been collected which will be used for future model validation. In this section, model results are presented for validation in several different formats:

- Tables of cost function results for 27 specific locations, for six ecological parameters;
- Concentrations trends in time, for specific (representative) locations; and
- Interpolated concentration fields at one time, shown either (i) spatially or (ii) as a transect.

The cost function values present an objective comparison of model results with the available monitoring data, while the graphs and spatial plots provide more qualitative insight into the functioning of the model (spatially and temporarily) and hence allow a more subjective assessment of model performance.

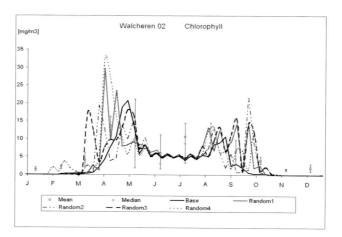

Fig. 7. Comparison of chlorophyll-a (mg m⁻³) model simulations with seasonal variation of suspended matter determined by a cosine function (base) or by four different random functions superimposed upon this function at the location Walcheren 2. Also shown are the mean, median and error bars for the 90 percentile of measurements for the years 1981–1994.

At most of the Dutch monitoring locations, samples are only collected near the surface, within the top meter. These monitoring data are compared with the model results from the top layer (top 4% of the water column).

The national Dutch monitoring program started in 1975. Some stations were temporarily or permanently added in 1987 along the Terschelling transact (i.e. Oystergrounds; Dogger bank), others were only sampled during a limited number of years (i.e. 1975–1981 or 1987–1989; see also Fig. 3). Analysis of the data (De Vries et al., 1998) shows that even in the coastal zone systematic trends over a period of 10 or 15 years are not obvious except for phosphate which shows some response in reaction to changes in management reflected by river loads. Hence the observations mainly reflect the natural variation which is why we have used all available data for the period of 1981 through 1994 in the cost function calculations and as an indication of the background variations for the error bars in the graphs showing model results and measurements.

4.1. Cost function results

The cost function values have been calculated for 27 specific locations, comparing the model results with the monitoring data for the period 1981–1994 (or a shorter period if less data were available). At each location, the cost function has been calculated for six substances: chlorophyll-a, NO_3, PO_4, SiO_2, salinity and suspended matter (SPM), for a total of 162 cost function results. For each substance, the result is classified as poor, reason-

able, good or very good based on the cost function value according to the criteria proposed by Radach and Moll (2006).

In Table 4, the number of results in each class, for each of the 27 locations is presented. In all, 80% of the calculated cost functions can be classified as good or very good. Only 7% are classified as poor. In Table 5, these same results are presented, but organised per substance. From these results it can be seen that for SPM and SiO_2, the cost function results at all 27 locations can be classified as good or very good. For salinity and chlorophyll-a, this is 24 and 21 locations, respectively.

4.2. Concentration trends in time, showing seasonal variation at different representative locations

Concentration trends in time are critical for many substances. In this paper the modelling results for a number of state variables are compared to the observations for four selected sampling stations (Figs. 8–11): Noordwijk 10, Schouwen 10, Terschelling 4, and Terschelling 235. These sampling stations are in four different areas of the Dutch coastal waters, each having specific characteristics that influence the ecological quality locally. Noordwijk 10 is situated 10 km offshore, in the near-coastal region influenced by the plume of the river Rhine, which has a low salinity, high turbidity and high nutrient concentrations. As regional concentration gradients for many substances tend to be steep from on-shore to off-shore. Schouwen 10 is located 10 km off-shore in the Delta region of the Netherlands and is

Table 4

Cost function results (goodness of fit) for 3D BLOOM/GEM model application at 27 stations for 6 substances chlorophyll-a, NO₃, PO₄, SiO₂, salinity and suspended matter (surface measurements and model results)

Station	Very good (0–1)	Good (1–2)	Reasonable (2–3)	Poor >3
Walcheren 2 km	3	1	2	0
Walcheren 30 km	3	3	0	0
Walcheren 50 km	3	2	1	0
Walcheren 70 km	4	2	0	0
Schouwen 10 km	3	3	0	0
Goeree 20 km	3	2	1	0
Noordwijk 2 km	3	2	1	0
Noordwijk 4 km	3	1	0	2
Noordwijk 10 km	3	2	1	0
Noordwijk 20 km	2	3	1	0
Noordwijk 30 km	3	1	2	0
Noordwijk 50 km	2	2	2	0
Noordwijk 70 km	3	2	1	0
Egmond 10 km	2	3	1	0
Calandsoog 2 km	4	2	0	0
Calandsoog 30 km	2	2	2	0
Calandsoog 50 km	4	0	0	2
Calandsoog 70 km	3	3	0	0
Marsdiep	2	3	1	0
Doove Balg West	3	2	0	1
Terschelling 4 km	4	2	0	0
Terschelling 10 km	6	0	0	0
Terschelling 100 km	1	3	0	2
Terschelling 135 km	3	1	2	0
Terschelling 175 km	3	1	2	0
Terschelling 235 km	2	1	1	2
Terschelling 370 km	2	2	0	2
Total	79 (49%)	51 (31%)	21 (13%)	11 (7%)

Model results and data are for the period November 1988–November 1989.

characterised by moderately high nutrient levels and very high concentrations of suspended matter. Terschelling 4 is 4 km offshore, in the coastal zone north of the Wadden Sea. With respect to salinity and nutrients this station is still clearly influenced by the river discharges, but not nearly to the same extend as Noordwijk 10. Suspended matter concentrations are lower compared to the other two coastal stations. Terschelling 235 is in the centre of the Dogger Bank, 235 km north of the island of Terschelling, This is located in deeper (33 m), but still well mixed waters. The salinity is nearly constant, nutrients are mainly determined by the background concentrations of the northern inflow. Suspended matter concentrations are very low. As such, the combination of these four stations can be considered as representative for a wide range of conditions in the Dutch part of the North Sea.

Each graph shows the model result November 1988–November 1989 (drawn line), the observations for

November 1988-November 1989 (solid circle), the monthly average (open circle), median (open square) and error bar (90% confidence interval) for all measurements collected between 1981 and 1994. Thus the figures give an impression of the ability of the model to specifically reproduce 1989 conditions, as well as the overall pattern of each state variable at each station. Not all stations were monitored during this entire period. Each figure first shows salinity and suspended matter, which are not simulated by BLOOM/GEM, but are important inputs/forcings. The salinity is calculated by the hydrodynamic model while suspended matter is derived from a steady state model result (see Section 3.3.3.4). If either of these parameters deviates consistently from the observations, errors in the simulated ecological state variables should also be expected. Figures further show the model results forNO3, PO4, SiO2 and chlorophyll-a. The terminology used in the description of the graphs is based on the cost function results for a particular substance at a particular location, i.e. 'very well' means a cost function results less than 1, 'well' a value between 1 and 2 etc. (see table in Section 3.1).

Fig. 8 shows the results for the Noordwijk 10 location, which has the highest sampling frequency of all stations (bi-weekly almost all year round). In general, the characteristic seasonal cycles between the model and the observations agree well or very well, particularly for the nutrients. The spring chlorophyll-a bloom is very well reproduced both in term of timing as well as peak value, but the summer chlorophyll-a levels exceed the measurements. The salinity simulation result is only reasonable which is mainly the result of a deviation between model and observations during the winter months hence it has little effect on the simulation of chlorophyll-a.

Table 5

Cost function results (goodness of fit) for the BLOOM/GEM application for 6 substances: chlorophyll-a, NO₃, PO₄, SiO₂, salinity and suspended matter

Substance	Very good (0–1)	Good (1–2)	Reasonable (2–3)	Poor >3
Salinity	4	17	5	1
SPM	20	5	1	1
NO₃	6	11	6	4
PO₄	2	15	5	5
SiO₂	25	1	1	1
Chlorophyll-a	22	2	3	0
Total	79 (49%)	51 (31%)	21 (13%)	11 (7%)

The number of cost function results in each class are given for 27 sampling stations in the Dutch region of North Sea (surface measurements and model results). See Ta b l e 4 and Fig. 3 for the names and locations of the 27 sampling stations.

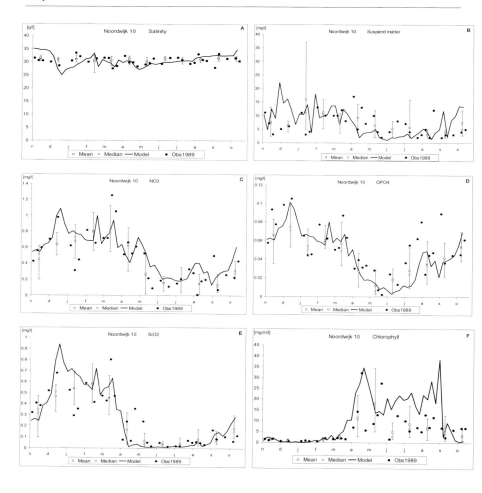

Fig. 8. Validation result for 3D BLOOM/GEM simulation at station Noordwijk 10: Salinity (A) (gl⁻¹), suspended matter (B), NO₃ (C), PO₄ (D), SiO₂ (E) (mg l⁻¹) and chlorophyll-a (F) (mg m⁻³). Circles are measurements for 1989, bars indicate 90 percentile of measurements for the years 1981–1994.

Fig. 9 shows the results for the location Schouwen 10. The calculated salinity and the suspended matter (forcing function) correspond well respectively very well with the measurements, which is essential for an accurate calculation of the ecologic parameters. As such, the simulated nutrient concentrations and chlorophyll-a agree well or very well with the measured values, and hence give a good representation of the characteristic seasonal cycles. Calculated winter and spring concentrations of NO₃, and SiO₂ somewhat lower than measurements for 1989, but are in almost all cases within the range of measurements of 1981–94.

At the location Terschelling 4 (Fig. 10) the salinity, and PO₄ agree well with the data while the results for suspended matter, NO₃, SiO₂ and chlorophyll-a all agree very well with the data. Forcing by suspended matter is below the measurements during summer, but since this deviation occurs in a period of nutrient limitation, the chlorophyll-a simulations are not affected. Calculated summer values of PO₄ are lower than observed, which is a common model characteristic in shallow areas due to the simple description of the sediment (see also Section 5).

Conditions are completely different at location Terschelling 235, the Dogger Bank (Fig. 11). Here the

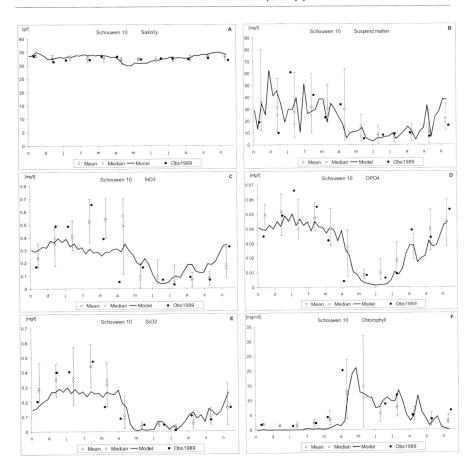

Fig. 9. Validation result for 3D BLOOM/GEM simulation at station at station Schouwen 10: Salinity (A) (g l⁻¹), suspended matter (B), NO₃ (C), PO₄ (D), SiO₂ (E) (mg l⁻¹) and chlorophyll-a (F) (mg m⁻³). Circles are measurements for 1989, bars indicate 90 percentile of measurements for the years 1981–1994.

salinity is nearly constant at about 35 indicating that there is hardly any influence from fresh water discharges by the rivers. Moreover, according to the model, the marine water here is of northern origin, whereas the salt water of the locations discussed so far originates from the Channel. Suspended matter and chlorophyll-a agree well or very well with the data, the result for NO₃ is reasonable. Notice in particular that the model reproduces the very early spring bloom, which in this part of the North Sea already starts in January. Levels of PO₄ again are too low during summer. The overall cost function score for this variable is poor. The same holds

for the simulation result of SiO₂: unlike at most other locations, summer SiO₂ levels at Terschelling 235 are under predicted by the model. In the model, some diatoms are present during summer and take up the available silicate. This may not have been the case in reality (there is no data).

4.3. Spatial distribution – transects and maps

To give some information of the spatial performance of the model, some results in the form of transects and spatial maps are given in Figs. 12–15. The two most

199

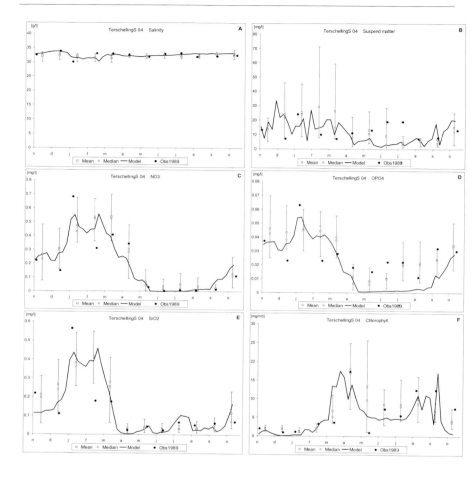

Fig. 10. Validation result for 3D BLOOM/GEM simulation at station Terschelling 4: Salinity (A) (gl^{-1}), suspended matter (B), NO$_3$ (C), PO$_4$ (D), SiO$_2$ (E) (mg l^{-1}) and chlorophyll-a (F) (mg m^{-3}). Circles are measurements for 1989, bars indicate 90 percentile of measurements for the years 1981–1994.

frequently sampled transects in the Dutch coast are at Noordwijk and Terschelling (Fig. 3). One of the best data sets showing spatial concentration distribution for many parameters for a large area of the North Sea was collected during the so-called NERC cruises (NERC, 1992). These data cover a vast area, but unfortunately the number of samples is relatively small in the Dutch coastal area, where the sharpest gradients exist.

Fig. 12 illustrates the ability of the model to reproduce spatial gradients, based on the data from the Noordwijk and Terschelling transects. The upper two graphs are for the Noordwijk transect with monitoring stations at 2, 4, 10, 20, 30, 50 and 70 km offshore, during May 1989.

Model results represent the mean value for the month. Results for NO$_3$ agree very well with the measurements, chlorophyll-a results are less precise but still capture the overall gradient in an adequate way. The lower 4 graphs represent results for the Terschelling transect, with monitoring stations at 4, 10, 50, 70, 100 135, 170 and 235 km offshore. Figures are included for February (middle panel) and June (lower panel). During winter, NO$_3$ levels near the coast (from 0 to 50 km) are relatively high due to river discharges. Further offshore, beyond the influence of the Rhine plume, the concentrations are consistently low, on the order of 0.1 µg l^{-1}. Chlorophyll-a in the winter shows an interesting distribution, with the highest

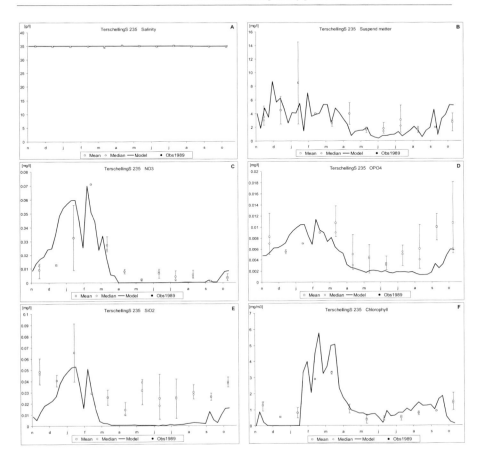

Fig. 11. Validation result for 3D BLOOM/GEM simulation station Terschelling 235 (Doggersbank): Salinity (A) (gl⁻¹), suspended matter (B), NO₃ (C), PO₄ (D), SiO₂ (E) (mg l⁻¹) and chlorophyll-a (F) (mg m⁻³). Circles are measurements for 1989, bars indicate 90 percentile of measurements for the years 1987–1994.

levels near the coast and at the Dogger Bank (approximately 235 km offshore). This area is relatively shallow, receives enough sunlight and has enough nutrients for winter phytoplankton growth. The model describes this gradient correctly. The summer concentrations at this transect are quite different. The summer NO₃ levels are almost exhausted everywhere due to uptake by phytoplankton, with the exception of Terschelling 100 (Oystergrounds). Modelled values are within the observed range at all but this station. Chlorophyll-a gradually declines from the shore to the central North Sea. Simulated values agree with the average observations, but are low relative to

the 1989 data, which are exceptionally high in comparison to the observations for other years at about half the stations. As for NO₃ modelled results deviate from the measurements at Terschelling 100. It seems that in the model there is too little transport of nutrients from the hypolimnion to the epilimnion which also causes an under prediction of chlorophyll-a due to nutrient limitation.

Fig. 13 shows model results at the Noordwijk transect for the total extinction coefficient. Simulations results at four different times of the year agree very well with the measurements. The offshore gradients are much stronger in winter than in summer, both in the model and in the

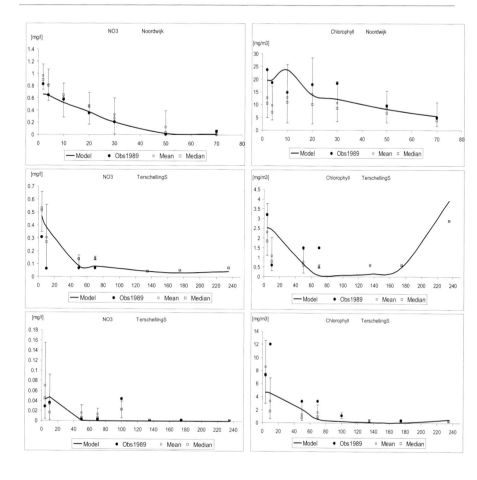

Fig. 12. Validation result for 3D BLOOM/GEM simulation for NO₃ (mg l⁻¹) (left panel) and chlorophyll-a (mg m⁻³) (right panel). Model results are from the top layer. Circles are measurements for 1989, bars indicate 90 percentile of measurements for the years 1981–1994. Top row: Noordwijk transect in spring (May); Middle row: Terschelling transect in winter (February); Bottom row: Terschelling transect in summer(June).

data. This is because in winter suspend matter and the amount of fresh water are relatively high near the coast. In summer these terms contribute less, but phytoplankton contributes more to the extinction.

The model also seems fit to reproduce spatial patterns on a larger scale. Fig. 14 shows that under-saturation of DO occurred in bottom waters on a fairly large scale in 1989 in model and measurements. The simulated oxygen concentrations for September 1989 (mean values for the month) are compared with the NERC data (NERC, 1992) near the surface and in the lower part of the water column. Near the surface, measured concen-

trations are all in the range of 7–9 mg l⁻¹, which the model is generally able to reproduce. Only in the Dutch coastal region, does the model predict slightly lower concentrations of 6–7 mg l⁻¹, which is due to a relatively low saturation values caused by relatively high temperatures and because in September local consumption exceeds production of. With respect to concentrations in the bottom waters, the model simulates concentration levels below saturation (4–7 mg l⁻¹) in many northern regions of the North Sea, agreeing fairly well with the observations. Simulated value are too low (by about 1 mg l⁻¹) in the German Bight and near the coast of

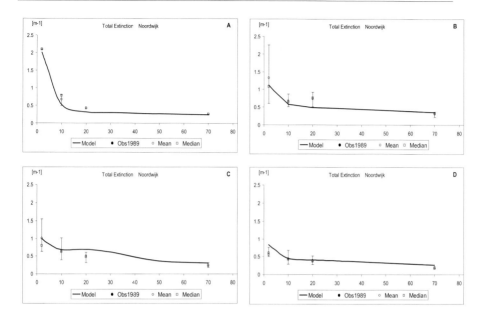

Fig. 13. Validation result for 3D BLOOM/GEM for total extinction coefficient (m⁻¹) at the Noordwijk transect in February (A), April (B), June (C) and August (D). Model results are from the top layer. Circles are measurements for 1989, bars indicate 90 percentile of measurements for the years 1987–1994.

Scotland. In the Northern German Bight, two measurements show a localized oxygen depletion with concentrations of 3–5 mg l⁻¹, which is not reproduced by the model. Earlier in the year in July both measured and simulated values were in the range of 7 to 7.5 mg l⁻¹. Most probably the model overestimates the vertical exchange in this area in September. The model does calculate a similar depletion further to the North in deeper waters.

Fig. 15 shows similar spatial maps of chlorophyll-a, NO₃, PO₄ and SiO₂, comparing simulated and measured concentrations near the surface of the water column for the period April–May 1989 (no bottom data were available). Model results represent the mean value over this period. Although the modelled results do not perfectly match the concentration range of every single monitoring, there is a good general agreement of the model results and data, considering the main spatial variations over the southern North Sea. For chlorophyll-a, a large region of the north and central North Sea has concentrations b2.5 µg l⁻¹. The model reproduces this to a large extent, except in the northwest, which is strongly influenced by the model boundary conditions. Monitoring results show that the concentrations near the Belgium and Dutch coasts and in the German Bight are

significantly higher, at some locations in the range of 25–50 µg l⁻¹. The model also calculates similar elevated concentrations in these coastal areas. For nitrate, the monitoring data shows a more complex spatial distribution of concentrations, with the highest values (0.25–0.5 mg N l⁻¹) being measured in the coastal waters of the UK, the Netherlands and Germany. The modelled results also represent this complex spatial distribution and the measured range of values. This illustrates the ability of the model application to simulate spatial variability and gradients of the key ecological parameters for the southern North Sea.

The validated 3D model application can give additional information on the vertical distribution of other model variables, such as chlorophyll-a and species composition. As an example, Fig. 16 shows the vertical profile of phytoplankton at four representative time periods at the location Terschelling 175, between the Oystergrounds and the Dogger Bank. These are model results only because no data were collected on the species composition of the phytoplankton community in 1989, however, results are in agreement with general expectations. During the spring bloom (April) there is no stratification, hence there is a more or less uniform distribution over the depth. Diatoms and *Phaeocystis*

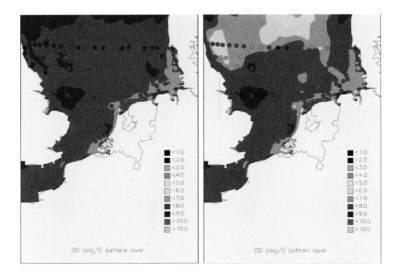

Fig. 14. Validation result for 3D BLOOM/GEM simulation. Simulated (coloured areas) and measured (circles) concentrations of Dissolved Oxygen (mg l^{-1}) in September 1989 are shown at the surface (left panel) and near the bottom (right panel). Data from the NERC cruises (NERC, 1992).

dominate the species composition at this time. The situation has completely changed by July, during a period of stratification. The model now simulates a peak of dinoflagellates just below the thermocline. It is interesting that this peak occurs simply because of the gradients of nutrients and light in the model; the dinoflagellates in the model are not positively buoyant. Due to the summer stratification, significant amounts of nutrients are trapped below the thermocline. These nutrients are derived from the decomposition of detritus which sinks to the lower part of the water column after the spring bloom. Since the water is very transparent in this area, a sufficient amount of light penetrates through the water column to allow growth of phytoplankton below the thermocline, which gives this interesting result. Above the thermocline light intensities are more favourable, but there are insufficient nutrients. Further below the thermocline even more nutrients are available, but light gets limiting. These phenomena are well known (Holligan et al., 1984; Peeters et al., 1995), but are sometimes attributed to the ability of dinoflagellates to actively regulate their vertical position. This simulation demonstrates that physical gradients are a sufficient condition to

generate these sub-surface phytoplankton peaks. In September the stratification is less pronounced and the irradiance levels are lower. Under these conditions, the model simulates a distribution pattern with highest concentrations in the upper part of the water column. Finally, in October, the stratification has disappeared and the phytoplankton distribution is uniform again, though at lower concentrations than in April. The species composition is dominated by dinoflagellates.

Until 1989 no data were collected on the species composition of the phytoplankton community so it is not possible to compare model simulations of individual species or function groups to measurements. *Phaeocystis* data are available for recent years, however. Fig. 17 shows the simulation results for the year 1998 (so not 1989) and compares them with the measurements for that particular year (1998) and the range for the years 1994–1998. For five out of six stations the model agrees sufficiently well with the measurements, specifically in terms of the seasonal dynamics. The only obvious difference occurs at station Noordwijk 2 where the observed levels are underestimated although it may also be noted that the observed level of variation as indicated by the 90

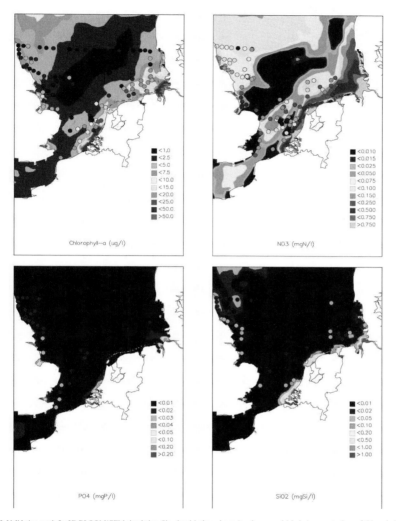

Fig. 15. Validation result for 3D BLOOM/GEM simulation. Simulated (coloured areas) and measured (circles) concentrations of chlorophyll-a, NO₃, PO₄ and SiO₂ for April–May 1989 at the surface (no bottom data available). Data from the NERC cruises (NERC, 1992).

percentile is exceptionally large indicating that the inter-annual variations are extremely large at this location.

5. Discussion

5.1. Validation results

The 2D applications of BLOOM/GEM for the North Sea have over many years provided a representation of the seasonal cycles and spatial gradients of interest, which were consistent with the available data, considering the objectives of the modelling. Although most of the comparisons were made on the basis of visual comparison between model results and data, these subjective analyses gave sufficient confidence that the model was validated. The purpose of the 3D model application was to (i) provide a further validation of the BLOOM/GEM modelling instrument and (ii) assess if there is additional value in making 3D applications, with respect to the modelling objectives.

While there are limited data available for validation of the vertical profiles simulated by the model, a number of different presentations of the 3D application results together with data have been made and have provided satisfactory results. The cost function has been applied for station specific, time varying comparison and has confirmed that a large majority of model results for key parameters can objectively be classified as 'good' and 'very good'. This can be considered the main 'proof' of the model validation. Additional, but more subjective confirmation of the model validation is given by plotted

results which are considered sufficiently accurate in terms of the predicted seasonal concentration patterns for the main ecological variables (Figs. 8–11) as well as the variations along transects (Figs. 12 and 13) and spatial distribution over the southern North Sea (Figs. 14 and 15). The model application also reproduces the measurements of low summer oxygen levels below the thermocline (Fig. 19H), for example in the Oystergrounds, a pelagic area in the central North Sea that is regularly stratified in the summer. In almost all cases, the model was able to reproduce observations of the state variables correctly, in that the simulated time-series fall into the range of observed variability. These combined results were achieved without any additional tuning of model process parameters and give further confirmation that the BLOOM/GEM model is well validated for simulating the seasonal cycles in the North Sea and notably the coastal zone, for the ecological processes of interest, in both 2D and 3D applications.

It is interesting but not very easy to compare the performance of the various North Sea models. Radach and Moll (2006) provide cost function results on the performance of several models, but these results were mainly obtained for relatively old model versions i.e. those presented during the Asmo modelling workshop in 1996. Probably models have been improved during the last decade. Lancelot et al. (2007) present low cost function numbers for a comparison between a recent version of MIRO and a single monitoring station off the Belgian coast. It should be noted though that these published cost function results are usually obtained for

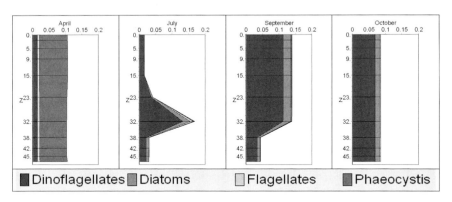

Fig. 16. Depth profiles of phytoplankton in mg C l^{-1} at Terschelling 175, between the Oystergrounds and the Dogger bank, in April (A), July (B), September (C) and October (D) of 1989. The sequence shows not only the total concentration but also the species composition in terms of dinoflagellates (red), diatoms (orange), flagellates (yellow) and *Phaeocystis* (grey). The concentration depth profiles and composition changes are in agreement with general expectations.

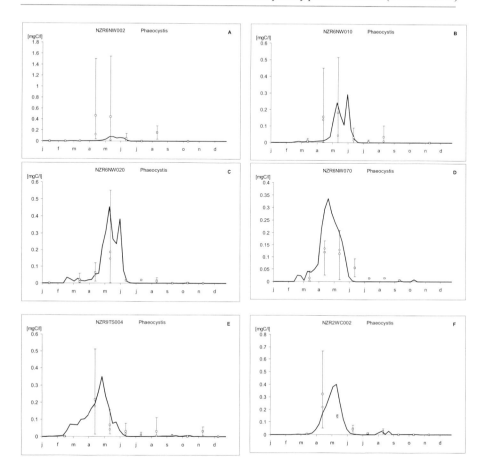

Fig. 17. Simulated and observed *Phaeocystis* (mg Cl) at the Noordwijk transect (A=2 km, B = 10 km, C = 20 km, D = 70 km) and at stations Terschelling 4 km (E) and Walcheren 2 km (F). Model results are for 1998 and monitoring data are for the period 1995-1998. This BLOOM/GEM model application was made with 'average' spring-neap hydrodynamic conditions, using average nutrient loads for the period 1995-1998.

large areas (boxes) rather than for individual stations. The cost function results presented in this paper were obtained for 27 individual stations, many of which are located in areas where gradients are steep. Cost function results of BLOOM/GEM computed for larger areas are consistently smaller indicating a better fit than those for individual stations. This is because in the larger boxes, existing spatial gradients which present a big challenge to the model, have largely been removed due to aver- aging. A score of 80% of results to be classified as either good or very good indicates that the application of

BLOOM/GEM to areas with steep gradients can be regarded as generally well validated.

5.2. Choice of ecological processes and spatial resolution

Because nature is complicated, it is tempting to aim for the construction and application of comprehensive 3D models with many vertical layers and a fine hori- zontal resolution that further include many biological and chemical processes. Although these kinds of models indisputably are a more detailed approximation of reality, they have several serious disadvantages as well: they require huge simulation times posing an upper limit on the amount of runs to be conducted within a certain time frame, it is increasingly difficult to analyse their results and there is a limited amount of (3D) data, hampering their validation (Beck, 1987). In the development of the BLOOM/GEM modelling instrument and its applications, explicit choices have been made with respect to the spatial and ecological resolution, given the modelling objectives.

5.2.1. Model domain (area) and spatial resolution
Choosing an appropriate spatial resolution involves (1) the area of the model, (2) the horizontal grid layout and (3) the number of vertical layers. If the area of the model is too small, its results will strongly depend on the boundary conditions, which may be a serious obstacle for the prediction of the future. Using a rectangular grid with elements of 10×10 km or even more may be adequate for a model application targeted towards the central North Sea, but is inappropriate for the simulation of gradients in the first 30 km of the Dutch coastal zone where the measurements show that steep gradients exist for depth, nutrients and suspended solids. To account for these gradients, grid cells should be in the order of 1×1 km. Moreover the grid layout should be able to follow the coast line, which is much easier with a curve-linear or finite element grid than with a rectangular grid. For most of the recent assessments using BLOOM/ GEM, the area and grid layout shown in Fig. 4 proved to be adequate as it satisfies the aforementioned requirements. Occasionally, a higher spatial resolution has been adopted for the Dutch coastal zone or the Wadden Sea in those cases where the representation of extremely localised ecological changes was anobjectiveofthe study. With respect to the vertical resolution, there is not a generic approach suitable for all purposes. The additional vertical variation in 3D hydrodynamics and suspend matter is essential for ecological modelling so these modules are always run in 3D mode for reasons of accuracy. The same is not true for the simulation of nutrients and primary production. When compared to the 14-day monitoring data, the 2D results of BLOOM/GEM are very similar to the surface layer results (upper 4% of the water column), as well as to the bottom layer results (bottom 4% of the water column) from the 3D model application at the coastal locations (at least up to approximately 30–50 km offshore), as shown in Fig. 18 for location Noordwijk 20. The measurement data at this location indicate that there is little or no difference in the salinity values for surface and bottom waters, and that

nitrate, ortho-phosphate and chlorophyll-a concentrations show only minor differences with depth. This gives justification for a 2D approach in the well-mixed coastal region. Furthermore, extensive sensitivity runs have shown that in the Dutch coastal waters up to 30 km offshore, a sufficiently refined horizontal resolution and a correct forcing of the light regime by suspended matter concentrations are more important to the overall model representation than the addition of the third dimension. Hence, for long-term historic 'trend' modelling, scenario simulations of the potential impacts of infrastructural works or for the assessment of measures undertaken for the Water Framework Directive, it is often justified to use the 2D version of BLOOM/GEM. The net gain is flexibility and efficiency during sensitivity analyses.

On the other hand, if the area of interest includes those parts of the North Sea that are regularly stratified in summer, such as the Oystergrounds, or the area north of the Dogger Bank, 3D simulations are essential for modelling key variables. This is illustrated in Fig. 19, showing 2D and 3D model results as well as data for the Oystergounds (Stations Terschelling 100 and Terschelling 135 km) where the depth is about 48 m. Due to the meteorological conditions, there was a relatively strong and long lasting stratification in this area in 1989 (Peeters et al., 1995). Hence the data show clear differences between surface and bottom layers for the parameters nitrate, ortho-phosphate, chlorophyll-a and oxygen. Although insufficient data were available to justify a cost function analyses, a visual comparison of by means of the graphs clearly demonstrate that the 3D model agrees better with the data than the 2D model (Fig. 19). This is true in particular for ortho-phosphate at the surface and for nitrate and oxygen in the bottom segment. For oxygen observed levels were lower than in the previous two years for which data had been collected. Measured and simulated concentrations in the bottom waters were less than 6 mg l^{-1} in late August (Fig. 19H).

The need for 3D ecological simulations is also apparent in the case of short term, operational model predictions of harmful algal blooms in coastal waters ('event' modelling). Particularly north of the Rhine River mouth, continuous monitoring indicates there is a band of frequently stratified water due to the salinity difference between North Sea water and the Rhine River plume. These temporary stratifications significantly affect short-lived bloom phenomena (primary production and transport) and hence should be explicitly modelled to assess bloom events even though these are quite localised in space and time. The ability of a 3D model application to resolve these very localised and possibly short-term stratification still needs to be further reviewed and

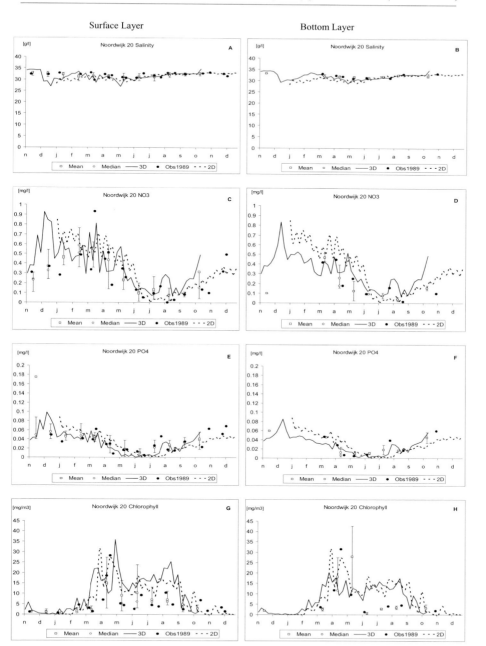

Fig. 18. Comparison of 3D (solid line) and 2D (hatched line) simulation results with measurements at station Noordwijk 20 km (Dutch coastal zone) for surface layer (left panel) and bottom layer (right panel). Salinity (A and B) (g l^{-1}), NO$_3$ (C and D), PO$_4$ (E and F), (mg l^{-1}) and chlorophyll-a (G and H) (mg m^{-3}). Circles are measurements for 1989, bars indicate 90 percentile of measurements for the years 1987–1989.

Surface Layer   Bottom Layer

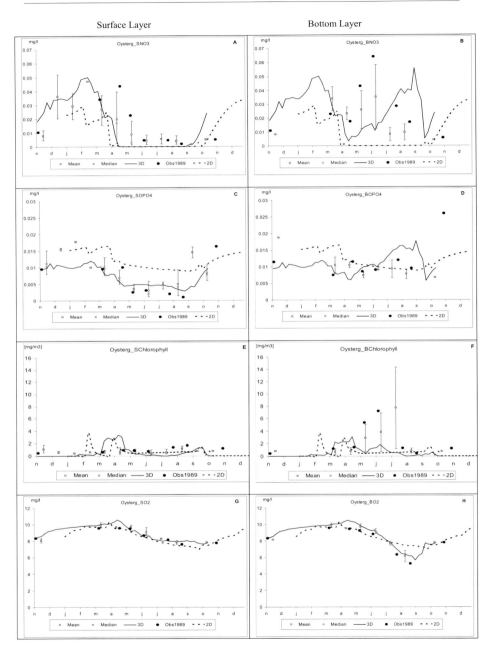

Fig. 19. Comparison of 3D (solid line) and 2D (hatched line) simulation results with measurements at the Oysterground (Central North Sea) for surface mixed layer (left panel) and bottom mixed layer (right panel). NO$_3$ (A and B), PO$_4$ (C and D), (mg l^{-1}), chlorophyll-a (E and F) (mg m^{-3}), dissolved oxygen (G and H) (mg l^{-1}). Circles are measurements for 1989, bars indicate 90 percentile of measurements for the years 1987–1989.

210

validated. Detailed monitoring data is necessary for such a comparison, for example from the "smartbuoy" results (Mills et al., 2003).

It is concluded that the choice of the domain area, the horizontal resolution and the application of a 2D versus a 3D model, are of key importance and should always be based on the specific objective of the modelling study.

5.2.2. Ecological resolution

BLOOM/GEM was developed with the purpose of providing a generic and robust predictive modelling instrument, applicable to many different water systems under a range of variable conditions without the necessity for 'fine-tuning' to local conditions for every single application. In terms of the number of ecological processes included, BLOOM/GEM is moderately complex (e.g. no microbial loop, no explicit grazing or higher trophic levels, simple benthic processes, but full nutrient cycles, advanced extinction module and complex phytoplankton kinetics). Given the objectives of the coastal water management issues and related modelling studies in the Netherlands, this choice of processes is appropriate. From an application point of view, it makes no sense to apply an over-dimensioned model system, as long as a simpler modelling approach meets the essential requirements.

Furthermore, identification of parameters in a model of high ecological complexity is problematic (Beck, 1987). Different combinations of parameter settings may seem to give acceptable calibration and validation results, but lack robustness. Dealing with such a model requires a thorough analysis of its predictive uncertainty as was done for instance for the prediction of the future filter feeder biomass in the Eastern Scheldt as a result of decreasing nutrient loads (Van der Tol and Scholten, 1998). They show that with a slightly more complex ecosystem model and a small computational grid, despite a relatively large dataset, the uncertainty in the predictions of their model, expressed by a set of 10 different but equally acceptable calibration results, can be very high. Because it is not feasible to perform such an analysis for a 2D let alone 3D computational grid for the North Sea, the choice was made for a more general ecological model rather than a detailed ecological model that would require more fine-tuning and increase the risk of generating unrealistic model results by using a specific, non-generic calibration result.

Nevertheless one should always be aware of the possible limitations of the model set-up. If the model is too simple ecologically because it ignores some processes which are essential to understand and describe the dynamics of the future ecosystem, the complexity of the model should be enhanced (Scholten and Van der Tol,

1993) even if these enhancements cannot be checked directly by comparison with the actual data. The ultimate test is the evaluation of the model results with respect to the available data.

5.3. Overall conclusion and continuing improvements

Many management questions can be addressed adequately by making justified simplifications in the model set-up, depending on the specific objective of the study. So, for each application, the appropriate spatial and ecological resolution has to be chosen carefully. There is an optimal balance between the required accuracy, predictability or robustness and operational aspects. The selection of processes presented in this paper is considered the minimum set with which a sufficient level of accuracy can be obtained both in the coastal areas as well in central parts of the North Sea such as the Dogger Bank and Oystergrounds. As such, this model application can be used as a tool to evaluate different policy scenarios, for example, those relevant to OSPAR and the EU Water Framework Directive.

With respect to further improvements of the model, a number of options will be explored in the near future. Recently the loads and boundaries of the model have been revised. Current and more accurate data have been gathered on German, French and UK rivers. Moreover the number of river loads has been tripled and now also includes many of the smaller point sources. Since the load of the major rivers for the Dutch part of the North Sea have not changed, the impact on the overall model behaviour will be limited. Simulation results for other areas may show moderate differences. Available literature data confirm the validity of the Channel boundary, but the North Atlantic boundary will be modified based on recent information (WL | Delft Hydraulics, 2006). Preliminary simulations suggest an impact in the results for the northern part of the model domain. This is also a region in which stratification is a regular phenomenon and for which additional 3D data have been collected since 1990. Hence in the near future there will be a stronger focus on validating the model for recent years for this part of the North Sea. New simulations will include both years with a strong stratification as well as years in which the stratification was less pronounced.

Since most BLOOM/GEM model applications in recent years were targetted towards the coastal zone, the model performance in this part of the North Sea is better than for other areas (i.e. Ta ble 4). Recently interest in those off shore regions is definitely increasing (Marine Strategy; Ospar). Hence future validation exercises will concentrate on improving the model performance in these areas to bring it up to the same level as in the coastal zone.

Since the beginning of 1991, biomass estimates of individual phytoplankton species have been collected at about a dozen stations. This data set has recently become available and will be used for a quantitative rather than qualitative validation of the predicted species composition of BLOOM. In order to improve the model's behaviour in very shallow areas such as the Wadden Sea, new model concepts have been formulated and implemented for benthic primary production, grazing and the sediment–water exchange. Validation of these new processes has recently started.

Acknowledgements

The authors wish to thank Drs. Hanneke Baretta-Bekker and Dr. Herman Gerritsen for critically reviewing the manuscript. We thank the two anonymous reviewers who madea numberof valuable commentstoimprove the paper. Development and application of the BLOOM/ GEM model and writing of this paper were partly financed by the National Institute for Coastal and Marine Management/RIKZ.

References

Allen, J.I., Blackford, J.C., Holt, J., Proctor, R., Ashworth, M., Siddorn, J., 2001. A highly spatially resolved ecosystem model for the North West European Continental Shelf. Sarsia 86, 423–440.

Atkins, R., Rose, T., Brown, R.S., Robb, M., 2001. The Microcystis cyanobacterial bloom in the Swan River – Feb. 2000. Water Science and Technology 43, 107–114.

Baretta-Bekker, J.G. (Ed.), 1995. European Regional Seas Ecosystem Model I. Neth. J. of Sea Research, vol. 33, p. 3/4.

Baretta-Bekker, J.G., Baretta, J.W., 1997. European Regional Seas Ecosystem Model II. Netherlands Journal of Sea Research 38 (3/4).

Baretta, J.W., Ebenhoh, W., Ruardij, P., 1995. The European Regional Seas Ecosystem model (ERSEM), a complex marine ecosystem model. Netherlands Journal of Sea Research 33 (3/4), 233–246.

Beck, M.B., 1987. Water Quality Modelling: a review of the analysis of uncertainty. Water Resources Research 23 (8), 1393–1442.

Berger, C., 1984. Consistent blooming of Oscillatoria agardhii gom. in shallow hypertrophic lakes. Verhandlungen – Internationale Vereinigung fuÉr Theoretische und Angewandte Limnologie 22, 910–916.

Berger, C., Bij de Vaate, A., 1983. Limnological studies on the eutrophication of Lake Wolderwijd A shallow hypertrophic Oscillatoria dominated lake in the Netherlands. Schweizerische Zeitschrift fuÉr Hydrologie, Hydrobiologie, Limnologie, Fischereiwissenschaft, Abwasserreinigung 45, 459–479.

Boon, J.G., Bokhorst, M., 1995. KSENOS, Adjustment and extension of the modelling suite for toxic substances and eutrophication in the North Sea and Dutch coastal waters. WL | Delft Hydraulics report T1236. (in Dutch).

Boon, J.G., De Kok, J.M., Salden, R., Gerritsen, H., 1996. NOMADS. North Sea Model Advection-Dispersion Study. The Dutch contribution to the simulation. RIKZ. 96.010. WL. Z854/Z995/ T1634.

Brinkman, A.G., 1993. Biological processes in the ECOWASP ecosystem model. IBN research report 93/6. Institute for Forestry and Nature Research. (IBN-DLO) 0928–6896.

Chorus, I., Falconer, I.R., Salas, H.J., Bartram, J., 2000. Health risks caused by freshwater cyanobacteria in recreational waters. Journal of Toxicology and Environmental Health, Part B 3, 323–347.

Crank, J., 1975. The Mathematics of Diffusion. Clarendon Press, Oxford.

Danzig, G.B., 1963. Linear programming and extensions. Princeton University Press, Princeton N.J.

Delhez, E.J.M., Damm, P., De Goede, E., De Kok, J.M., Dumas, F., Gerritsen, H., Jones, J.E., Ozer, J., Pohlmann, T., Rasch, P.S., Skogen, M., Proctor, R., 2004. Variability of shelf-seas hydrodynamic models: lessons from the NOMADS2 Project. Journal of Marine Systems 45, 39–53.

De Groodt, E.G., Los, F.J., Nauta, T.A., Markus, A.A., DeVries, I., 1991. Modelling cause-effect relationships in eutrophication of marine systems: an integral approach. Marine Eutrophication and Population Dynamics. In: Colombo, et al. (Eds.), Proceedings of the 25th EMBS. Olsen & Olsen, Fredensborg, Denmark. ISBN: 87-85215-19-8.

De Ruyter, W.P.M., Postma, L., De Kok, J.M., 1987. Transport Atlas of the Southern North Sea. Rijkswaterstaat and Delft Hydraulics, The Hague.

De Vries, I., Duin, R.N.M., Peeters, J.C.H., Los, F.J., Bokhorst, M., Laane, R.W.P.M., 1998. 'Patterns and trends in nutrients and phytoplankton in Dutch coastal waters: comparison of time-series analysis, ecological model simulation and mesocosm experiments. ICES Journal of Marine Science 55, 620–634.

EDF-DER, 1998. TELEMAC-3D Validation document version 2.1., Electicité de France. 74 pp.

EDF-DER, 2000. TELEMAC-2D Validation document version 5.0., Electicité deFrance. 124 pp.

Gerritsen, H., Bijlsma, A.C., 1988. Modelling of tidal and wind-driven flow; the Dutch continental shelf model. In: Schrefler, B.A. (Ed.), Proceedings ofthe International Conference on Computer Modelling in Ocean Engineering. A.A. Balkema, Rotterdam, pp. 331–338.

Gerritsen, H., Vos, R.J., Van der Kaaij, T., Lane, A., Boon, J.G., 2000. Suspended sediment modelling in a shelf sea (North Sea). Coastal Engineering 41 (1–3), 317–352.

Glas, P.C.G., Nauta, T.A., 1989. A North Sea computational framework for environmental and management studies: an application for eutrophication and nutrient cycles. Contribution to the international symposium on integrated approaches to water pollution problems (SISIPPA 89), Lisbon.

Harris, G.P., 1986. Phytoplankton ecology. Chapman and Hall, London. 1986.

Holligan, P.M., Williams, P.J., Le, B., Purdie, D., Harris, R.P., 1984. Photosynthesis, respiration and nitrogen supply of plankton populations in stratified, frontal and tidally mixed shelf waters. Marine Ecology. Progress Series 17, 201–213.

Jahnke, J., 1989. The light and temperature dependence of growth rate and elemental composition of Phaeocystis globosa Scherffel and P. Pouchetti (Har.) Lagerh. in batch cultures. Netherlands Journal of Sea Research 23 (1), 15–21.

Klepper, O., Van der Tol, W.M., Scholten, H., Herman, P.M.J., 1994. SMOES: a simulation model for the Oosterschelde ecosystem Part 1: Description and uncertainty analysis. Hydrobiology 282/283, 453–474.

Laane, R.W.P.M., Groeneveld, G., De Vries, A., Van Bennekom, A.J., Sydow, J.S., 1993. Nutrients (N, P, Si) in the Channel and the

Dover Strait: seasonal and year-to-year variation and fluxes to the North Sea. Oceanologica Acta 16, 607-616.

Lacroix, G., Ruddick, K., Park, Y, Gypens, N, Gypens, N, Lancelot, C, 2007. Validation of the 3D biogeochemical model MIRO&CO with field nutrient and phytoplankton data and MERIS-derived surface chlorophyll-a images. Journal of Marine Systems 64, 66-88.

Lancelot, C, Billen, G, Sournia, A., Weisse, T., Colijn, F., Veldhuis, M.J.W., Davies, A., Wassman, P., 1987. Phaeocystis blooms and nutrient enrichment in the continental coastal waters of the North Sea. Ambio 16, 38-47.

Lancelot, C, Rousseau, V, Billen, G, Van Eeckhout, B., 1995. Coastal Eutrophication of the Southern Bight of the North Sea: Assessment and modelling. NATO advanced Research work group on Sensitivity of North Sea, Baltic Sea and Black Sea to anthropogenic and climate changes, 14-18 November 1995. NATO-ASI series.

Lancelot, C, Hannon, E., Becqevort, S., Veth, C, De Baar, H.J.W., 2000. Modeling phytoplankton blooms and carbon export production in the Southern Ocean: dominant controls by light and iron in the Atlantic sector in Austral spring 1992. Deep-Sea Research. Part 1. Oceanographic Research Papers 47, 1621-1662.

Lancelot, C, Spitz, Y., Gypens, N, Ruddick, K., Becquevort, S., Rousseau, V, Lacroix, G, Billen, G, 2005. Modelling diatom and Phaeocystis blooms and nutrient cycles in the Southern Bight of the North Sea: the MIRO model. Marine Ecology Progress Series (289), 63-78.

Lancelot, C, Gypens, N, Billen, G, Garnier, J., Roubeix, V, 2007. Testing an integrated riverocean mathematical tool. Journal of Marine Systems 64, 216-228.

Lane, A., 1989. The heat balance of the North Sea. Proudman Oceanographic Laboratory, report no. 8.

Lesser, G.R., Roelvink, J.A., Van Kester, J.A.T.M., Stelling, G.S., 2004. Development and validation of a three-dimensional morphological model. Coastal Engineering (51), 883-915.

Los, F.J., 1991. Mathematical Simulation of Algae Blooms by the Model BLOOM II Version 2. WL | Delft Hydraulics Report T68.

Los, F.J., Bokhorst, M., 1997. Trend analysis Dutch coastal zone. New Challenges for North Sea Research. Zentrum for Meeres- und Klimaforschung. University of Hamburg, pp. 161-175.

Los, F.J., Brinkman, J.J., 1988. Phytoplankton modelling by means of optimization: a 10-year experience with BLOOM II. Verhandlungen - Internationale Vereinigung fuEr Theoretische und Angewandte Limnologie 23, 790-795.

Los, F.J., Gerritsen, H, 1995. Validation of water quality and ecological models. Presented at the 26th IAHR Conference, London, Delft Hydraulics, 11-15 September 1995. 8pp.

Los, F.J., Wijsman, J.W.M., 2007. Application of a validated primary production model (BLOOM) as a screening tool for marine, coastal and transitional waters. Journal of Marine Systems 64, 201-215.

Los, F.J., Smits, J.G.C., De Rooij, N.M., 1984. Application of an Algal Bloom Model (BLOOM II) to combat eutrophication. Verhandlungen - Internationale Vereinigung fuEr Theoretische und Angewandte Limnologie 22, 917-923.

Water resources systems planning and management - an introduction to methods, models and applications; Chapter 12 'Water quality modelling and prediction'. In: Loucks, D.P., Van Beek, E. (Eds.), Studies and Reports in Hydrology. UNESCO publishing. ISBN: 92-3-103998-9.

Luyten, P.J., Jones, J.E., Proctor, R., Tabor, A., Tett, P., Wild-Allen, K., 1999. COHERENS - a coupled hydrodynamical-ecological model for regional and shelf seas: user documentation. MUMM Report, Management Unit of the Mathematical Models of the North Sea, Brussels, Belgium.

Lynch, D.R., Davies, A.M. (Eds.), 1995. Quantitative Skill Assessment for Coastal Ocean Models. American Geophysical Union, Washington D.C.

Mankin, J.B., O'Neill, R.V., Sugart, H.H., Rust, B.W., 1975. The importance of validation in ecosystem analysis. In: Innis, G.S. (Ed.), New Directions in the Analysis of Ecological Systems, Part 1. Simulation Councils Proc. Ser., vol. 5. Calif.: Simulations Councils, Inc., pp. 309–317. No 1, Lajolla, 132 pp.

Ménesguen, A., Guillaud, F.F., Aminot, A., Hoch, T., 1995. Modelling the eutrophication process in a river plume; the Seine case study (France). Ophelia (42), 205–225.

Mills, D.K., Laane, R.W.P.M., Rees, J.M., Rutgers van der Loeff, M., Suylen, J.M., Pearce, D.J., Sivyer, D.B., Heins, C., Platt, K., Rawlinson, M., 2003. Smartbuoy: a marine environmental monitoring buoy with a difference. Building the European capacity in Operational Oceanography. Proceedings of the 3rd International Conference on EuroGOOS conference, 3–6th December 2002, Athens. Elsevier Oceanography Series, vol. 69, pp. 311–316.

Moll, A., 1997. Modelling primary production in the North Sea. Oceanography 10 (1), 24–26.

Moll, A., 1998. Regional distribution of primary production in the North Sea simulated by a three-dimensional model. Journal of Marine Systems 16 (1–2), 151–170.

Moll, A., Radach, G., 2003. Review of three-dimensional ecological modelling related to the North Sea shelf system. Part I: models and their results. Progress in Oceanography 57, 175–217.

Nauta, T.A., De Vries, I., Markus, A.A., De Groodt, E.G., 1992. An integral approach to assess cause–effect relationships in eutrophication of marine systems. Science of the Total Environment, Supplement. Elsevier Science Publishers B.V., Amsterdam, pp. 1133–1147.

NERC, 1992. North Sea Project data set, British Oceanographic Data Center, Proudman Oceanographic Laboratory, UK.

OSPAR, Villars, M., de Vries, I., Bokhorst, M., Ferreira, J., Gellers-Barkman, S., Kelly-Gerreyn, B., Lancelot, C., Menesguen, A., Moll, A., Patsch, J., Radach, G., Skogen, M., Soiland, H., Svendsen, E., Vested, H.J., 1998. Report of the ASMO modelling workshop on eutrophication issues, 5–8 November 1996, The Hague, The Netherlands, OSPAR Commission Report, Netherlands Institute for Coastal and Marine Management, RIKZ, The Hague, The Netherlands.

Peeters, J.C.H., Los, F.J., Jansen, R., Haas, H.A., Peperzak, L., De Vries, I., 1995. The oxygen dynamics of the Oyster Ground, North Sea. Impact of eutrophication and environmental conditions. Ophelia 42, 257–288.

Post, A.F., De Wit, R., Mur, L.R., 1985. Interactions between temperature and light intensity on growth and photosynthesis of the cyanobacterium Oscillatoria agardhii. Journal of Plankton Research 7, 487–495.

Prandle, D., Los, F.J., Pohlmann, T., De Roeck, Y.H., Stipa, T., 2005. Modelling in Coastal and Shelf Seas – European Challenges. European Science Foundation, European Marine Board Position Paper 7. www.esf.org.

Proctor, R., Damm, P., Delhez, E., Dumas, F., Gerritsen, H., DeGoede, E., Jones, J.E., De Kok, J., Ozer, J., Pohlmann, T., Rasch, P., Skogen, M., Sorensen, J.T., 2002. North Sea Model Advection Dispersion Study 2 (NOMADS-2): Assessments of model variability. Proudman Oceanographic Laboratory, Internal Document, No 144, 255 p. + CD-ROM in pocket at rear.

Radach, G., Moll, A., 2006. Review of three-dimensional ecological modelling related to the North Sea shelf system. Part II: model validation and data needs. Oceanography and Marine Biology: An Annual Review 44, 1–60.

213

Refsgaard, J.C., Henriksen, H.J., 2004. Modelling guidelines – terminology and guiding principles. Advances in Water Reserouces 27, 71–82.

Riegman, R., 1996. Species composition of harmful algal blooms in relation to macronutrient dynamics. Allan D. Cembella, Gustaaf M. Hallegraeff, Physiological Ecology of Harmful Algal Blooms, Donald M. Anderson. NATO ASI Series, vol. 41. Springer Verlag.

Riegman, R., Noordeloos, A.A.M., Cadee, G., 1992. *Phaeocystis* blooms and eutrophication of the continental coastal zones of the North Sea. Marine Biology 112, 479–484.

Riegman, R., De Boer, M., De Senerpont Domis, M., 1996. Growth of harmful marine algae in multispecies cultures. Journal of Plankton Research 18 (10), 1851–1866.

Rogers, S.I., Lockwood, S.J., 1990. Observations of coastal fish fauna during a spring bloom of *Phaeocystis* pouchetii in the Eastern Irish Sea. Journal of the Marine Biological Association of the United Kingdom 70, 249–253.

Ruardij, P., Van Raaphorst, W., 1995. Benthic nutrient regeneration in the ERSEM ecosystem model of the North Sea. Netherlands Journal of Sea Research 33 (3/4), 453–483.

Schlesinger, S., Crosbie, R.E., Gagné, R.E., Innis, G.S., Lalwani, C.S., Loch, J., Sylvester, J., Wright, R.D., Kheir, N., Bartos, D., 1979. Terminology for model credibility. SCS Technical Committee on Model Credibility. Simulation 32 (3), 103–104.

Scholten, H., Van der Tol, M.W.M., 1993. Towards a metrics for simulation model validation. In: Grasman, J., Van Straten, G. (Eds.), Predictability and Nonlinear Modelling in Natural Sciences and Economics. Kluwer Academic Publishers, Dordrecht, pp. 398–410. Proceedings of the 75th Anniversary Conf. Of WAU, April 5–7, 1993, Wageningen, The Netherlands.

Scholten, H., Van der Tol, M.W.M., 1994. SMOES: a simulation model for the Oostershelde ecosystem. Part 2: calibration and validation. Hydrobiology 282/283, 453–474.

Scholten, H., Van Waveren, R.H., Groot, S., Van Geer, F.C., Wösten, J.H.M., Koeze, R.D., 2000. Good modelling practice in water management. Paper presented at Hydroinformatics. Cedar Rapids, IA, USA.

Schrum, C., Alekseeva, I., St. John, M., 2006a. Development of a coupled physical–biological ecosystem model ECOSMO. Part 1: Model description and validation for the North Sea. Journal of Marine Systems 61, 79–99.

Schrum, C., St. John, M., Alekseeva, I., 2006b. ECOSMO, a coupled ecosystem model of the North Sea and Baltic Sea: Part 1. Spatial-seasonal characteristics in the North Sea as revealed by EOF analysis. Journal of Marine Systems 61, 100–113.

Skogen, M.D., Soiland, H., 1998. A user's guide to NORWECOM V2.0. The Norwegian ecological model system. Fisken og Havet 18, 42.

Smits, J.G.C., Van der Molen, D.T., 1993. Application of SWITCH, a model for sediment-water exchange of nutrients, to Lake Veluwe in the Netherlands. Hydrobiology 253, 281–300.

Soetaert, K., Herman, P.M.J., 1995a. Carbon flows in the Westerschelde estuary (the Netherlands) evaluated by means of an ecosystem model (MOSES). Hydrobiology 311, 247–266.

Soetaert, K., Herman, P.M.J., 1995b. Nitrogen dynamics in the Westerschelde estuary (SW Netherlands) estimated by means of the ecosystem model MOSES. Hydrobiology 311, 225–246.

Soetaert, K., Herman, P.M.J., Kromkamp, J., 1994. Living in the twilight – estimating net phytoplankton growth in the Westerschelde estuary (the Netherlands) by means of an ecosystem model (MOSES). Journal of Plankton Research 16, 1277–1301.

Van der Molen, D.T., Los, F.J., Van Ballegooijen, L., Van der Vat, M.P., 1994. Mathematical modelling as a tool for management in eutrophication control of shallow lakes. Hydrobiology 275/276, 479–492.

Van der Tol, M.W.M., Scholten, H., 1998. A model analysis on the effect of decreasing nutrient loads on the biomass of benthic suspension feeders in the Oosterschelde (SW Netherlands). Journal of Aquatic Ecology 31, 395–408.

Van Gils, J., Tatman, S., 2003. Light penetration in the water column. MARE Report, WL2003001 Z3379.

Van Pagee, J.A., Glas, P.C.G., Markus, A.A., Postma, L., 1988. Mathematical modelling as a tool for assessment of North Sea Pollution. In: Salomons, W., Bayne, B.L., Duursma, E.K., Forstner, U. (Eds.), Pollution of the North Sea, an Assessment. Springer-Verlag, London.

WL | Delft Hydraulics, 2002. GEM documentation and user manual. WL | Delft Hydraulics report Z3197, Delft, The Netherlands.

WL | Delft Hydraulics, 2003a. Delft3D-WAQ users manual. WL | Delft Hydraulics, Delft, The Netherlands.

WL | Delft Hydraulics, 2003b. DMI for modelling of total suspended matter in the North Sea. Delft Hydraulics Z3293.

WL | Delft Hydraulics, 2005. Delft3D-FLOW users manual, v 3.12. WL | Delft Hydraulics, Delft, The Netherlands.

WL | Delft Hydraulics/MARE, 2001a. Description and model representation of an artificial island and effects on transport and ecology. Delft Hydraulics Report WL2001103 Z3030.10.

WL | Delft Hydraulics/MARE, 2001b. Description and model representation T0 situation. Part 2: The transport of fine-grained sediments in the southern North Sea. Delft Hydraulics Report WL2001003 Z3030.10.

WL | Delft Hydraulics/MARE, 2001c. Large-scale hydrodynamic impacts of airport islands. Delft Hydraulics Z3029.13.

WL | Delft Hydraulics/MARE, 2001d. Reference scenarios and design alternatives, phase 1. Delft Hydraulics Z3029.20.

WL|Delft Hydraulics, 2006. Transboundary nutrient transports in the North Sea. Delft Hydraulics Z4188.00.

Zevenboom, W., Mur, L.R., 1981. Amonium-limited growth and uptake by Oscillatoria agardhii in chemostat cultures. Archives of Microbiology 129, 61–66.

Zevenboom, W., Mur, L.R., 1984. Growth and photosynthetic response of the cyanobacterium Microcystis aeruginosa in relation to photoperiodicity and irradiance. Archives of Microbiology 139, 232–239.

Zevenboom, W., Bij De Vaate, A., Mur, L.R., 1982. Assessment of factors limiting growth rate of Oscillatoria agarhii in hypertrophic Lake Wolderwijd, 1978, by use of physiological indicators. Limnology and Oceanography 27, 39–52.

Zevenboom, W., Post, A.F., Van Hes, U., Mur, L.R., 1983. A new incubator for measuring photosynthetic activity of phototrophic microorganisms, using the amperometric oxygen method. Limnology and Oceanography 28, 787–791.

Chapter 10

Complexity, accuracy and
practical applicability of different
biogeochemical model versions

F.J. Los, M.Blaas

Journal of Marine Systems, AMEMR 2008
Special Issue (Submitted)

Complexity, accuracy and practical applicability of different biogeochemical model versions

F.J. Los and M. Blaas

Deltares, P.O. Box 177, 2600 MH Delft, The Netherlands

Abstract

The construction of validated biogeochemical model applications to be used as prognostic tools for the marine environment involves a large number of choices particularly with respect to the level of details of the physical, chemical and biological aspects. In theory, enhanced complexity should promote enhanced realism, accuracy and credibility as well. Unfortunately, with growing complexity, simulation times increase drastically and may become prohibitive in practice. The amount of data necessary to force the model increases and the spatial and temporal coverage of monitoring data is usually poor relative to the outputs produced by even moderately complex models. In this paper we show the results of comparative modelling applications varying in spatial resolution (from coarse to fine), in vertical resolution (2D versus 3D), in forcing of transport, in turbidity forcing and in the number of phytoplankton species. Included models range from 15 years old relatively simple models to a relatively advanced 3D model. Results are compared to each other and to monitoring data for the Dutch part of the North Sea using different goodness of fit criteria as well as time series plots.
It is concluded that recent models are more consistent and have a smaller bias. From a graphical inspection the level of variability looks more realistic, but whether this is a real improvement cannot be judged by comparison to traditional monitoring data due to the low frequency of sampling. More specifically the overall results for chlorophyll-a are rather consistent throughout all models, but regionally recent models are better; resolution is crucial for the accuracy of transport and more important than the degree of realism in hydrodynamic forcing; SPM strongly affects the biomass simulation and species composition, but even the most recent SPM models do not yet obtain a good overall score; organic matter (CDOM) should be included in the calculation of the light regime; more complexity in the phytoplankton model improves the chlorophyll-a simulation, but the simulated species composition needs further improvement for some of the functional groups. From a technical point of view we propose using formal goodness of fit criteria which strongly discriminate between different simulations.

1. Introduction

The first eco-hydrodynamic models for aquatic systems were developed more than thirty years ago. Examples of these first generation models are found in DiToro et al. (1971; 1977). At present, many models exist, some with a relatively long history, while development of others has started more recently, but of course these also include many features from older models. Many papers describe the status i.e. the present version of a model application, demonstrating its strong points and discussing some of its weaker points. While these papers are certainly meaningful, it is often hard to determine which characteristics are of major importance and which characteristics actually do not contribute much to the quality of a particular model.

One could argue that the 'best' model by definition is the most advanced and complex model that can be constructed and operated at a certain moment of time. While more knowledge and computation power become available, the model could be extended further. However, in Los et al. (2008) we have pointed out that adding more complexity not necessarily improves the quality of the model's results in terms of realism or applicability. In stead we have argued that there should be a balance between ecological and physical resolution in relation to the specific question that should be addressed by a specific model application. For example, an appropriate model for assessing the impacts of sand mining in a coastal area is not necessarily adequate to assess the impacts of nutrient reduction or the probability of low oxygen conditions in an offshore area or the occurrence of undesirable blooms of *Phaeocystis* during the spring bloom.

In this paper, a comparison is made between several generations of the eco-hydrodynamic model applications developed for the North Sea at Deltares (formerly WL | Delft Hydraulics) during the last 15 years. There are many differences between these applications with respect to their forcing, resolution, complexity and parameterization. In order to find out how much each modification contributes to changes in model behaviour, we need to make a systematic comparison in terms of the spatial and ecological resolution of these models. To that purpose we have revitalized several distinguishable model versions, and run all of these alongside using forcings for a single, recent year (2003). Thus, the central question is: which factors matter most and which might look important, but actually contribute less to improvements in model behaviour? A secondary question is if, and if so how, we can quantify progress in model skill. Notice that progress does not necessarily occur in a linear fashion, so whereas the overall behaviour may improve relative to the measurements, results for some variables or at some locations or in parts of the year might actually deteriorate simultaneously. The following factors were considered during this study:

- the resolution of the grid,
- degree of realism in forcing of transport,
- forcing of the light climate by SPM and CDOM,
- the level of detail of the phytoplankton model.

Unlike the inter model comparison by Moll and Radach (2003) and Radach and Moll (2006), all of the applications presented in the present paper belong to the same model family and their set-up and forcing could be standardized to a large extend. Generalization of the results will be discussed at the end of this paper.

2. Main features of models

Our current model is called BLOOM/GEM. This is a generic model code with a long history, which in its current mode is applied to many different water systems such as the North Sea, a number of Dutch water bodies i.e. the saline lakes Grevelingen, and Veere, The Eastern Scheldt Estuary, and the future saline Lake Volkerak - Zoom. International applications include the Lagoon of Venice, the Sea of Marmara and the future saline Marina Reservoir in Singapore. An extensive description of the main features of the model is provided by Blauw et al. (2009); its application to the North Sea is described by Los et al. (2008). A more detailed description of the phytoplankton module (BLOOM) is presented by Los et al. (2007) and in Loucks and van Beek (2005).

2.1 General principle and similarities

Each of the models discussed here was originally applied to explain observed phenomena and to predict some future conditions. Later model versions were usually run with some new processes, parameters settings and forcings for a more recent period of time. Simply comparing the existing output of previous model simulations therefore leaves many questions open on how to explain the differences between them. For this study many differences were eliminated in order to be able to concentrate on those modifications that matter most. Occasionally different combinations of forcings were run to check their impacts one at a time. For instance the latest models were also run using the suspended matter (SPM) field of the oldest models to force the under water light climate. We did not try to do a full recalibration of all of these models by harmonizing their parameter values because we assume these models had been validated during their previous applications.

Year specific forcings i.e. nutrient loads from rivers and climatology were adopted from data for a single, recent year, 2003, for all simulations by all models. At the time this study was performed, (almost) complete datasets for forcing and monitoring were available for the entire period 1996 - 2003. This last year was chosen not just because it is the most recent one, but also because it is an a-typical year with a wet spring and a dry, warm summer and autumn. We expected that such a year would be more suitable for finding differences between models than a more average year. Notice that consequently all of the pre-2003 models were actually run for a set of conditions for which they had never been validated in the past. An overview of the differences between the models as they were applied here is given in Table 1. There also a key to the codes that indicate the different application variants is given. An extensive overview of historic model versions is given by Los et al. (2008). The grids are shown in Fig. (1).

Table 1   Description of model runs and explanation of codes

ID	Application	Grid	Hydrodyn. transport & dispersion	Boundary condition	imposed turbidity field	Sea water temperature	Eco model
GDGSD	GENO-DYNAMO	16 km southern North Sea rectilinear; 1395 elements	repeated single tidal cycle + wind-driven residual (SW wind 4.5 m/s)	climatology 1980s, Channel annual cycle, Atlantic constant in time, constant in space	Based on SPM from observations, space dependent, stationary annual mean	Spatially uniform time series (2003) from station Noordwijk 10	Michaelis-Menten kinetics, 2 functional groups (diatoms and 'green'), 12 state variables
GDGSB	GENO-NZB				As above, but modulated with seasonal cycle		BLOOM96: 4 species, 12 types, 26 state variables
CDGSB	CSM	8-16 km entire North Sea rectilinear; irregular; 3915 elements	repeated single tidal cycle + wind-driven residual (SW wind 9 m/s)	Channel as above; Atlantic: recomputed from GENO-NZB, constant in time and space.	as GDGSB	Spatially uniform time series (2003) from station Noordwijk 10	BLOOM96: 4 species, 12 types, 26 state variables
KSKCB	COAST	1-10 km Dutch coast <70 km, curvilinear; 2153 elements	as CSM but with additional, seasonal upwelling parameterization in horizontal dispersion coefficient based on 30-year average wind data	derived from measurements at near-boundary monitoring stations	spatial pattern from separate SPM model; harmonic annual cycle; also CDOM parameterised as function of fresh water fraction		
KSGCB					as GDGSB		
ZNZCB	ZUNO-2D	2-20 km southern NS, curvilinear; 4350 elements	repeated spring-neap cycle	Channel and Atlantic: As GENO-NZB and GENO-DYNAMO	WL | Delft Hydraulics/MARE., 2001. Seasonal cycle, wind dependent noise	as GDGSB	BLOOM with Further refined algal parameters (lab studies, see text)
ZRZCB			Full year, 2DH			Spatially varying, calculated from Delft3D hydrodynamic model	
ZRZCD							BLOOM with 2 species: diatom and flagellates
ZRGCB					as GDGSB		As ZNZCB
Z3ZCB	ZUNO-3D	as ZUNO 2D 10 layers; 43500 elements	Full year, 3D		as ZNZCB		
Z3VCB					From new SPM model (Van Kessel et al., 2008)		

Loads are all consistent since they are all based on the same data set Blauw et al. (2006) with data for the major rivers in the domains for 2003.

Description of runids

First letter:	Grid.	G = GENO, C = CSM, K = COAST, Z = ZUNO
Second letter:	Forcing.	D = average day, S = average day, seasonal correction dispersion, N = characteristic spring – neap, R = realistic (actual), 3 = realistic (actual) 3D mode.
Third letter:	SPM.	G = GENO, K = COAST, Z = ZUNO steady state + seasonal harmonic + noise, V = ZUNO dynamic
Fourth letter:	Extinction	S = only SPM, no CDOM, C = SPM + Salinity as proxy for CDOM
Fifth letter:	Algal model	D = Dynamo, B = BLOOM

220

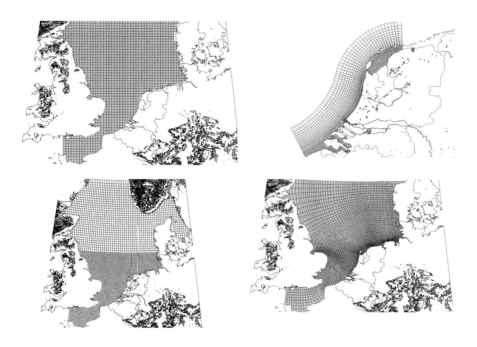

Figure 1   Grid of models: GENO (upper left), Coastal zone (upper right), CSM (lower left) and ZUNO (lower right).

Meteorological forcing: The day length and solar irradiance for 2003 are adopted from historic measurements by the Royal Dutch Meteorological Institute at a single land-based station (De Kooy) near Den Helder in the north western part of the Netherlands. In the GENO, CSM(CDGSB) and Coastal zone models a spatially uniform, seasonally varying sea water temperature was adopted based on measurements at station Noordwijk 10 km (see Fig. 2) for 2003. In both ZUNO models temperature is specified by a spatially varying temperature field taken from the simulations by the hydrodynamic model Delft3D-Flow (Lesser et al., 2004; WL | Delft Hydraulics, 2005).

Rivers and other nutrient sources: The nutrient loads of all models are basically the same. The model input contains the point sources of nutrients and fresh water from the main Belgian, Dutch, German, French and UK rivers in as far as they are part of the model domain. In the Coastal zone model only the Dutch rivers are explicitly included. For the Dutch rivers, substance loads were derived from measured discharges per day and concentrations in rivers at 10-day intervals for the year 2003. Data for the other main rivers is usually also available per decade (Blauw wt al., 2006). Modelled substances that were not measured have been derived from measured data of other substances, using stoichiometric ratios and other knowledge rules that have been developed and proven successful in previous studies (e.g. Los and Wijsman, 2007).

Boundary conditions: There is considerable overlap between the domains included in most of the models presented here. The position of the Southern boundary is the same for GENO, CSM and ZUNO. The concentrations of substances at this boundary are based on the work of Laane et al. (1993) and are the same for each of these models. Concentrations are specified as a monthly time series, uniformly distributed over the cross section. The north boundary of the GENO and ZUNO grids is at 57 degrees N and the substance concentrations are assumed to be constant during the year. The north boundary of CSM and all boundaries of COAST(KSKCB) were constructed in a different way and will be described below.

2.2 SPM forcing

Light is the main limiting factor of phytoplankton for most of the time during the winter half year in the southern North Sea. Also the onset of the spring bloom is controlled by the availability of light at many locations. Light availability depends on irradiance, the mixing depth and the turbidity (the extinction coefficient). The irradiance and depth are the same in all of the models discussed here and so is the vertical mixing in all but the ZUNO-3D(Z3ZCB) model. In all models the attenuation of light is computed according to the well known Lambert - Beer equation relating the overall absorption to substance concentrations, ignoring scattering, but the substance fractions taken into account and their contribution to the vertical extinction are not the same in all models. In particular the contribution of yellow substance (CDOM) is only taken into account in later models (see Table 1). In offshore North Sea waters, SPM usually does not contribute much to the light attenuation, but in the continental coastal waters, typically 25 to 75 percent of the light extinction is caused by SPM. The distribution of SPM in the North Sea is determined by coastal erosion, sea-bed resuspension by waves and tides and by the residual transport due to tidal, wind-driven, and density-driven currents (see e.g., Eisma and Kalf 1987, Eleveld et al, 2004). Because the spatial and temporal distribution of SPM varies, the SPM fields of all models discussed are adopted from SPM submodels, which are different for different versions of the primary production models. To assess the sensitivity of the modelling results to differences in SPM fields, all models have also been run using the oldest (although less accurate) SPM field. The later ones are forced by more realistic SPM values, which should be reflected in their ability to correctly model phytoplankton.

2.3 Initial conditions

For the sake of intercomparison of the model results, all runs have been spun up to statistical equilibrium by repeated recycling of the forcing conditions of 2003. Since 2003 is not an average year, conditions at the end of a the simulation do not exactly match those at the start. This means that some discrepancies between model and observation could occur at the beginning of the time series of output, but considering the relationship between the models, there is no reason to assume that this will cause systematic differences between various models.

2.4 Set up of individual models

GENO-DYNAMO(GDGSD): The development of GENO-DYNAMO(GDGSD) began in the middle of the 1980s when one of the main issues was to describe, understand and predict the development and abatement of eutrophication in the Dutch coastal zone (Van Pagee et al., 1988; Glas and Nauta, 1989; Nauta et al., 1992; Los et al., 1994; Peeters et al., 1995). Transport modelling is based on 2D hydrodynamic calculations for average conditions (one representative day). The GENO (Generic North Sea) grid has a uniform 16x16 km mesh size covering the southern North Sea (up to 57 degrees N), totaling 1395 computational elements (see Fig. 1). A constant SW wind of 4.5m.s^{-1} is applied to force the transport in addition to the semidiurnal tide.

In this model a linear light model is assumed including terms for back ground extinction, suspended matter (SPM), detritus and phytoplankton. The background extinction is constant in time and space, but SPM varies spatially. The SPM field was simulated by a steady state model (Los et al., 1994).

GENO-DYNAMO(GDGSD) considers closed nutrient cycles for N, P and Si and two functional groups of phytoplankton: 'diatoms' and 'others', the first requiring silica as a nutrient. The model includes only twelve state variables and includes a simple sediment module and processes such as mineralization, nitrification and denitrification.

GENO-NZB(GDGSB): This may be considered as an extended version of GENO-DYNAMO(GDGSD). There are two differences between these models. First: in GENO-NZB(GDGSB) the constant SPM field is transformed into a time variable field. The overall seasonal variation is simulated by means of a harmonic function with relative high values in winter and low values in summer. Its amplitude is based upon the observed level of variation for several years. This method was first described by Los and Bokhorst (1997). Second, the simple algal module was replaced by the much more advanced BLOOM module. This module considers 12 phytoplankton types grouped into 4 functional groups: diatoms, (micro) flagellates, dinoflagellates and *Phaeocystis*. BLOOM had already been applied to many fresh water systems since the end of the 1970s and as a 1Dv model to the North Sea (Peeters et al. 1995), but GENO-NZB(GDGSB) was the first extensive marine application of this model. For each species and species type, there is a different factor for converting biomass to chlorophyll concentration. This factor is depending on the limiting condition (light; nutrients) and ranges from 0.0067 to 0.0533 mg chlorophyll-*a* per mg C. A complete overview is given in Los and Wijsman (2007).

The original set of phytoplankton model coefficients was determined using preliminary results from laboratory experiments by Roel Riegman at the Netherlands Institute of Sea Research (NIOZ), which were later published in revised form. These are not used for the simulations presented here, however. During later model applications these coefficients were adapted based on work by Riegman et al., 1992; Riegman et al., 1996; Riegman, 1996; Jahnke, 1989) and simulation results of the model. This set of coefficients, which was originally used for the CSM(CDGSB) and COAST(KSKCB) model applications (see below), has also been used for GENO-NZB(GDGSB) in the present study in order to allow for a direct comparison between these models. Regarding all other aspects (hydrodynamics, forcings, non-algal model coefficients etc.), GENO-DYNAMO(GDGSD) and GENO-NZB(GDGSB) are identical.

CSM(CDGSB): During the mid 1990s two new model applications were developed: the large area Continental Shelf Model (CSM(CDGSB)) and a near coastal application COAST(KSKCB), see Fig. 1. The domain of CSM(CDGSB) extends much further to the north in comparison to GENO-NZB(GDGSB). These models were used for the OSPAR international model intercomparison (Villars and De Vries et al., 1998). In CSM two grid resolutions are used. In the northern part of the domain the grid is similar to GENO-NZB(GDGSB), but in the southern part, the elements are four times smaller (i.e. 8x8 km). Several combinations of transport fields, dispersion coefficients and wind speeds have been tested, but the most realistic results in transport were obtained by simply forcing the model with a constant SW wind of 9ms^{-1} without an additional dispersion coefficient. The total amount of computational elements of CSM(CDGSB) is 3915.

The Channel boundary of CSM(CDGSB) is at the same position as for GENO and the same concentrations are used. Notice that differences in residual flows could still lead to different fluxes of for instance nutrients though. The (new) north boundary was constructed by an iterative procedure in such a way that concentrations of total nutrients along the transect at 57 degrees north are similar to those specified as north boundary for GENO-NZB(GDGSB). In the CSM model the exchange with the Baltic through the Kattegat is specified as an additional discharge with a constant flow rate of about 14000 m^3 s^{-1} and concentrations similar to those of the Channel boundary.

The light model is the same as in GENO-NZB(GDGSB), the same SPM field is used but the specific extinction coefficient of SPM was increased. During the construction of the GENO models only a small number of light extinction measurements had been available. In later years, when these numbers started to be collected more regularly, it became obvious that the contribution by SPM to the extinction had been underestimated in the earlier models. Consequently: in CSM(CDGSB) a larger portion of the incident radiation is absorbed by the SPM compared to GENO-NZB(GDGSB) even though the same SPM forcing is imposed as forcing.

With respect to the other model equations and parameters, differences with GENO-NZB(GDGSB) are minute. The set-up of the phytoplankton model of CSM(CDGSB) is the same as in GENO-NZB(GDGSB).

COAST(KSKCB): COAST(KSKCB) is the first model application using a curve-linear, fine resolution grid along the Dutch North Sea coast. This coastline-following grid allows transport to follow the coastal contours (De Kok et al., 2001). The original hydrodynamic model grid was aggregated resulting in a total number of 2153 computational elements used for the ecological simulations (Fig. 1). Seasonal variations in flow conditions are not explicitly modelled, but mimicked by correcting the dispersive flow rates of a single representative daily flow as a function of the 30 year averaged historic wind direction and speed. The horizontal dispersion is relatively small when the average wind is from the south west and largest for winds from the north west (March and April). This application, originally called 'North Sea Bloom', was extensively calibrated to obtain the optimal set of model coefficients given the objective of the modelling and the application area (Los & Bokhorst, 1997; De Vries et al, 1998; Blauw et al.; 2009). It was also used for many management evaluations (e.g. Boon and Bokhorst, 1995; Los and Bokhorst, 1997; Villars and de Vries et al., 1998).

In comparison to all other applications presented here, the boundaries of COAST(KSKCB) are situated much closer to the continental coast. The model domain has open boundaries to the south, west and north (See Fig. 1). Since the prevailing transport direction along the Dutch coast is from south west to north east, the southern boundary the most important. For the original model application concentration values for all substances were adopted from the long year averaged measurements at the nearby *Appelzak* transect (Fig. 2) between 1975 and 1985. In the simulation results presented here, the PO_4 concentrations at this boundary were reduced by 35 percent to account for the about 50 percent PO_4 reduction of river discharges, which has been achieved since the mid 1980s. Using this correction factor simulated winter PO_4 values along the Dutch coast agreed well with the observations for 2003 (Rijkswaterstaat, 2003).

Values for computational boundary elements in which no monitoring stations are located, were obtained by fitting an exponential curve through the observations. Little variations are observed along the western boundary. Therefore concentration values here were obtained by simply taking the average of all observations at all stations 70 kilometres off the coast. Concentrations at the northern boundary were obtained from measurements at the Rottum transact. Notice that this is by far the least important boundary due to the prevailing direction of the currents. Substances which are not measured directly (i.e. detritus) were computed according to same procedure that was used for the boundaries of the other models (see Los et al., 2007 for more details).

For the application of the COAST(KSKCB) model, a new SPM field was generated (De Kok et al., 1995). In general the agreement with observed SPM values has improved, although this is not the case at all stations. To distinguish between differences due to changes in hydrodynamics and those due to changes in SPM an additional simulation was performed with the COAST(KSKCB) model using the SPM field from GENO-NZB(GDGSB) (see Table 1). Time series plots of these additional runs are not shown in this paper, but the overall results are included in the Goodness of Fit scores (GOFs) presented below.

In the coastal zone coloured dissolved organic matter (CDOM) from riverrine sources contributes significantly to the attenuation of light. To first order, the CDOM concentration can be approximated as a linear function of the fresh water content (Peters et al., 1991; Los et al., 1997; Van Gils and Tatman, 2003; Los et al., 2007). In COAST(KSKCB) therefore a salinity dependent term was added to the extinction model, resulting in improved simulation results of the extinction coefficient (see Los et al. 1997 for more details). With respect to all other model equations and coefficient values, COAST(KSKCB) and CSM(CDGSB) are identical.

ZUNO-2D(ZNZCB): The ZUNO (*Zuidelijke Noordzee*, i.e. Southern North Sea) grid was developed as a follow-up of both the CSM and COAST grid. This is a curve-linear grid with a moderately high resolution in the Dutch coastal zone of ca. 2x2km and a lower resolution of up to 20x20km in the most north westerly part of the domain. The version of ZUNO presented here consists of 4350 active elements horizontally and 10 vertical sigma layers. The transport fields applied in this paper all stem from 3D hydrodynamic simulations using the Delft3D flow code ((Lesser et al., 2004; WL | Delft Hydraulics, 2005). Depending on the specific question to be addressed, the BLOOM/GEM transport model is either run in 2D or in 3D mode. Unless stated otherwise ZUNO-2D(ZNZCB) simulation results presented in this paper are based on a single spring - neap cycle forcing. Besides, additional results are shown in which the climatological forcing of the 2D model is the same as for the 3D results shown here. This model will be denoted by ZUNO-2DR(ZRZCB).

The Channel boundary of ZUNO-2D(ZNZCB) is the same as for GENO and CSM. Although concentration values at the north Atlantic boundary have recently been updated based on an extensive literature study (Blauw et al., 2006), these new boundary conditions have not been used during the simulations reported here for the sake of comparison to the older models. Hence, the same northern boundary was used for all simulations by all models (except COAST(KSKCB)).

The SPM forcing of ZUNO is based on simulation results of an improved steady state model (WL|Delft Hydraulics, 2001). Simulations were performed on the 4x4 refined version of the ZUNO grid in 3D mode and projected on the coarser grid used here. In comparison to the previously used SPM fields, locally the agreement with the measurements has improved. This is particularly true in the Dutch Coastal zone due to the refined resolution. As in the other models the overall seasonal variation is simulated by means of a harmonic function with relative high values in winter and low values in summer. However, using this function, the spring bloom in the model tends to be rather late because in reality short periods of quiet conditions with relatively low levels of suspended matter play an important role in triggering the onset of the spring bloom. To account for these short-term variations, we have assumed a relationship between the SPM concentration and the average wind speed to further adjust the harmonic signal. The amplitude of this short term fluctuation is a multiplication factor varying between 0.3 and 1.7 depending on the difference between the actual and average wind speed (5.5 m s^{-1}). Again, these factors were determined empirically in such a way that the observed interannual variability could be reproduced sufficiently well (see Los et al. 2008 for more details). To distinguish between effects of changes in resolution and in SPM field, an additional simulation was also performed by which the ZUNO 2D model was forced by the same SPM field as the two GENO models and CSM(CDGSB). This simulation is denoted by ZUNO-2D(ZRGCB).

As for the previously discussed models, sea water temperature in the default ZUNO-2D(ZNZCB) applications is derived from measurements from station Noordwijk 10. In the simulations using actual transport fields (i.e. ZUNO-2DR(ZRZCB)), however, the temperatures are adopted from the Delft3D-Flow hydrodynamic model (spatially and temporarily varying).

The model set-up is basically the same as in COAST(KSKCB), but some model parameters were modified according to recent insights. Some coefficients of the light model were adjusted based on an extensive data analysis (van Gils & Tatman 2003). Several phytoplankton related parameter of the BLOOM module were updated to accommodate new experimental results on the functional groups in the model. Particularly for nutrient-stressed species the stoichiometric ratios of the model were modified (less nutrients per unit of biomass) (Riegman unpublished results). The optimum light intensity of several species was reduced (Jahnke, 1989; Ferris et al. 1991; Garcia et al., 1992) meaning enhanced growth rates at low light intensities.

ZUNO-3D(Z3ZCB): The main difference between ZUNO-3D(Z3ZCB) and ZUNO-2D(ZNZCB) is the vertical resolution. Furthermore historic atmospheric data force the transport in ZUNO-3D(Z3ZCB), while ZUNO-2D(ZNZCB) is forced by an average spring - neap cycle in the simulations reported here. Because the 3D model takes stratification into account, temperature adopted from the Delft3D Flow hydrodynamic model is specified as a 3D, time variable forcing to BLOOM/GEM. Other forcings and model parameters are the same as in ZUNO-2D(ZNZCB). To investigate the impact of advances in SPM modelling, an additional simulation (ZUNO-3D(Z3VCB)) was performed using results from a dynamic, fine resolution sediment transport model (Van Kessel et al, 2008).

2.5 Monitoring program

Half way the 1970s the Dutch national government initiated an extensive monitoring network covering all national waters including the North Sea which was rather unique at that time. The network covers sampling stations in river branches, estuaries and marine waters. For the present study only the marine stations are considered. These stations are visited by survey vessels every 2 to 4 weeks that collect water samples from the surface layer (nominally 1 m below sea surface level). Visits to the stations are randomly timed with respect to the tidal phase but are always carried out under relatively calm weather conditions (wind strength less than 7 Bft).

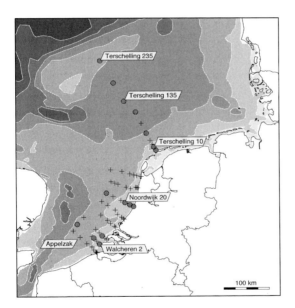

Figure 2.   Location of coastal water quality and ecology monitoring stations from the Dutch national (*MWTL*) monitoring programmes in the Southern North Sea. All stations indicated by a circle were monitored in 2003, results are presented for stations mentioned by name. Stations marked by + were monitored in some but not all years between 1975 and 2003 and considered during the calibration and validation of the original model applications. Bathymetry of ZUNO models is included for indication, contour lines start a 10 m depth, interval 15 m

Unfortunately, the original network was stepwise minimized rather than maintained or extended in later years. Hence, for 2003, 12 variables at 17 stations have been considered for direct comparison with model results (Rijkswaterstaat, 2003). Besides, all relatively recent (less than 10 years old) data that were available for about a dozen additional stations have been used as supporting evidence to evaluate the performance of the various model versions. An overview of the locations is shown in Fig. 2. The relevant substances for the present paper are listed in Table (2).

Table 2   Monitoring information on substances used in this paper.

	Substance	Unit	Remark
Chlorophyll-*a*	Chlorophyll-*a*	mg m^{-3}	
NO$_3$	nitrate NO$_3^-$	gN m^{-3}	NO$_2$ is measured seperately but ignored
PO$_4$	ortho phosphate PO$_4^-$	gP m^{-3}	
totN	total dissolved Nitrogen	gN m^{-3}	NO$_2$+NO$_3$+NH$_4$+organic N No refractory N in model
totP	total dissolved Phosphorus	gP m^{-3}	PO4 + particulate organic and inorganic. No refractory P in model
SiO$_2$	dissolved silica SiO$_2$	gSi m^{-3}	after filtration
SPM	suspended particulate matter	g m^{-3}	filter residue, anorganic
Ext	extinction coefficient of visible light	m^{-1}	
Phaeocystis	biomass of *Phaeocystis globosa*	gC m^{-3}	Cell counts converted to biovolume converted to gC
Diatoms	biomass of all diatom species	gC m^{-3}	Cell counts converted to biovolume converted to gC
micro flagellates	biomass of all pico phytoplankton and micro flagellate species	gC m^{-3}	Cell counts converted to biovolume converted to gC
Dinoflagellates	biomass of all dino flagellate species	gC m^{-3}	Cell counts converted to biovolume converted to gC

2.6 Calibration procedure original models

For the setup and calibration of the original models, the following procedure has been followed in general. First, salinity and tracer simulations were performed to check the main transport characteristics. Next calculated concentrations of chlorophyll-*a* and nutrients have been compared with measurements in the following manner: (1) The most important measure for phytoplankton biomass chlorophyll-*a* was compared graphically with measurements. Usually this was done using observations for several years plotted as a single set of data points. (2) The analysis of limiting factors and phytoplankton species and types was made. For phytoplankton only a qualitative analysis was possible. (3) The calculated dissolved nutrients were compared graphically with measurements. In this comparison it is important to know if a nutrient is (sometimes) limiting or not (see step 2). (4) The calculated total nutrients were compared graphically with measurements. In case of discrepancies with measurements, the comparison of individual terms (phytoplankton and dissolved species) also had to be reconsidered. (5) The calculated light extinction was compared graphically with measurements (only for later model versions as little data were available when the oldest models were calibrated).

2.7 Validation and model intercomparison: goodness of fit criteria

Apart from calibration, which, as outlined above, tended to focus on reproducing the multi-year mean, spatial and seasonal patterns, past validations have been carried out for particular years. Typical validation years were 1985 (GENO models), 1990 (CSM(CDGSB) and COAST(KSKCB)) and 1988 - 1989 (ZUNO-2D(ZNZCB) and ZUNO-3D(Z3ZCB)). In this paper, we will discuss the various model applications by comparing their results to the *in situ* monitoring data in the Dutch coastal zone by Rijkswaterstaat for 2003 (Rijkswaterstaat 2003).

Traditionally, the validation consisted of visual inspection of time series output and spatial maps of model results. Over time, however, more formal, quantitative validation methods gained attention. This became more common practice in particular due to the 1996 OSPAR ASMO Eutrophication modelling workshop (Villars and De Vries et al., 1998). Since then, the so-called OSPAR cost function has been applied to quantify the performance or skill of coastal biogeochemical models, see e.g. Los et al., 2008 and Blauw et al, 2009 for a discussion on the BLOOM-GEM model, but also Radach and Moll (2006).

The OSPAR cost function *CF* is one of the options. It is in fact the normalized mean absolute error (MAE) between model and observation, defined here as

$$CF = \frac{1}{N} \sum_{n=1}^{N} \frac{|M_n - D_n|}{\sigma_D} \qquad (1)$$

where the average over discrete time n spans an annual interval is determined for observations D_n at each individual station and model output M_n at its matching model grid cell; σ_D is the annual standard deviation of the observational data.

The normalization has been chosen to express the goodness of fit in terms of multiples of the standard deviation, with $CF<1$ being classified as 'very good', $1<CF<2$ as 'good', $2<CF<3$ as 'reasonable' and any CF beyond this upper limit as 'poor'. Note that the upper limit of 3 is chosen according to Radach and Moll, 2006, whereas Villars and the Vries adopted a limit of 5 to separate 'reasonable' from 'poor'.

Selection of GOF score criterion: As a first step in the analysis, cost function results have been computed for a number of substances and stations. Fig. 3 shows a typical example for chlorophyll-*a* at a number of representative stations. Based on the criteria proposed by Radach and Moll (2006), almost all model results at almost all stations could be qualified as 'very good' or 'good'. Moreover, scores for different models per station are often rather similar. In contrast, if we plot the model results against the measurements, there are sometimes clear and consistent differences between individual models which we think should be reflected by the GOF scores. Clearly, the OSPAR Cost Function has little distinctive power. For scrutinous model performance evaluation the OSPAR Cost Function seems hardly suitable.

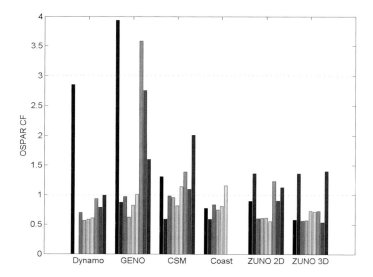

Figure 3   OSPAR Costfunction (CF, see eq. (1)) results all 6 base 2003 model cases at 9 representative
monitoring stations for Chlorophyll-*a* using the observational data of 2003. Bars correspond to the
scores for the individual stations (from left to right: Walcheren 2; Noordwijk 2, 10, 20, 70;
Terschelling 10, 135, 175, 235). Not all stations are represented in all applications. Dashed lines
indicate the class limits (see text).

The issue of defining appropriate measures to quantify model skill and aid model
intercomparison has received increasing attention in recent publications. Following the
atmospheric modelling community (e.g. Taylor, 2001), the biogeochemical modelling
community is entering a phase of growing need and possibility of quantifying model skill due
to the increasing interest in operational model applications and the growing amount of
observational data. Allen et al (2007), pointed out the issue of matching the traditional visual
inspection of time series to skill measures and explores a range of measures. In subsequent
studies by Jolliff et al. (2008), Friedrichs et al. (2008), Stow et al. (2008), various measures
have been applied more extensively. From these, we adopted the target diagram as presented
by Jolliff et al. (2008) as it turned out to best convey the message otherwise extracted from
the comparison of model and observation time series.

The target diagram displays the difference between data sets in terms of both the bias B and
the unbiased root-mean-squared difference $RMSD$':

$$B = \overline{M} - \overline{D} \tag{2}$$

$$RMSD' = \left[\overline{(M' - D')^2} \right]^{1/2} \tag{3}$$

where the overbar denotes averaging in the same sense as in (1) and primes denote the residue, e.g. for M':

$$M' = \frac{1}{N} \sum_{n=1}^{N} \left(M_n - \overline{M} \right) \tag{4}$$

The bias B is indicative of the match between model and observations in the annual mean sense, the unbiased root-mean-squared difference $RMSD'$ is a measure for the match between the residues of both time series after removal of the bias. As discussed by Jolliff et al. (2008) in more detail, $RMSD'$ is a measure of the overall agreement in both phase and amplitude of the variability of the compared time series.

Upon plotting both measures in a single graph, the overall root-mean-square difference ($RMSD$) is indicated by the radial distance from the origin. Since, by definition, the following holds for $RMSD$:

$$RMSD^2 = B^2 + RMSD'^2 \tag{5}$$

Following Jolliff et al. 2008, we adopted the convention to normalize B and $RMSD'$ by the standard deviation of the observations σ_D and to multiply $RMSD'$ by the sign of the standard deviation difference sgn($\sigma_M - \sigma_D$). Due to the normalization, both B and $RMSD'$ are non-dimensional and readily interpreted with respect to the variability within a given signal. The use of the signum (sgn) function adds information on whether the model over- or underpredicts the observed variability to $RMSD'$ which is positive definite.

In summary, the normalized bias will be shown on the ordinate of the target diagram:

$$B^* = \frac{\overline{M} - \overline{D}}{\sigma_D} \tag{6}$$

And the normalized, signed, unbiased root-mean-square difference on the abscissa:

$$RMSD'^* = \frac{\mathrm{sgn}(\sigma_M - \sigma_D)}{\sigma_D} \left[\overline{(M' - D')^2} \right]^{1/2} \tag{7}$$

As discussed by Taylor (2001), and reiterated by Jolliff et al, (2008), there is a relation between the linear correlation coefficient R and the unbiased $RMSD'$ which is helpful in defining additional goodness of fit criteria. Following equation (4), a circle with radius 1 on the target diagram corresponds to a total normalized $RMSD$ equal to the standard deviation of the observations. A model that would merely reproduce the annual mean of the observations would score $RMSD^*=1$ and zero bias. Hence any model result outside this circle with $RMSD^*=1$

can be considered as poor. Moreover, any result on the target diagram with total $RMSD^*<1$, cannot be negatively correlated to the observations (see Jolliff et al, 2008 for more details).

Because $RMSD^{*'}$ is related to the correlation coefficient R by

$$RMSD'^* = \sqrt{1 + \sigma^{*2} - 2\sigma^* R} \tag{8}$$

where $\sigma^* = \sigma_M / \sigma_D$ and $R = \overline{M'D'} / \sigma_M \sigma_D$, $RMSD'^*$ attains a minimum when $\sigma^* = R$. Since this is also the minimum total $RMSD^*$, a circle with radius $M_{Ro} = \sqrt{1 - R_0^2}$ denotes the minimum total $RMSD^*$ possible for a given R_0 and all points between this circle and the origin correspond to $R > R_0$. This leads to the definition of a second, relative rather than absolute, criterion. Since $R_0 = 0.67$ (i.e. $M_{Ro} = 0.74$) matches the 15% percentile of all model chlorophyll-a scores at all stations, we identify model results with $RMSD^* < M_{Ro}$ as 'good'. Any score with $M_{Ro} < RMSD^* < 1$ we refer to as 'reasonable' The choice for chlorophyll-a is motivated by the fact that this is the main objective variable for which the models have been developed. For the other variables the same classification has been adopted.

2.8 Illustration

Fig. 4 below illustrates the use of the Target Diagram for a particular station (Walcheren 02) for chlorophyll-a in comparison to the time series plot. Evaluating the time series in Fig. 4a leads to the impression that for example the COAST model applications do not capture the temporal pattern which leads to overprediction in summer and autumn. On the other hand, GENO-DYNAMO(GDGSD) and GENO-NZB(GDGSB) show an undeprediction of the signal and also a mismatch in timing. Relatively speaking, the 3D ZUNO models appear to perform best, as they exhibit a distinct spring autumn blooms, albeit that the timing could still be improved.

These general statements are reflected in Fig. 4b. Clearly COAST(KSGCB) (with SPM from GENO) is an outlier and performs poorest in relative sense. It is the only result with $RMSD'^* > 0$, i.e. where the model standard deviation exceeds the standard deviation of observations. Both the normalized bias and unbiased $RMSD$ are larger than one. The bias of the base COAST(KSKCB) model is remarkably smaller, but it is clear that the capture of variations is still poor. COAST(KSKCB) in this case is more or less comparable to a model merely describing the annual mean. The DYNAMO and GENO-NZB models also perform poorly and exhibit a relatively large negative bias (underprediction). The 3D ZUNO model (Z3VGB) performs best, whereas ZUNO (Z3ZCB) suffers somewhat from a larger underprediction of the mean and of the variability.

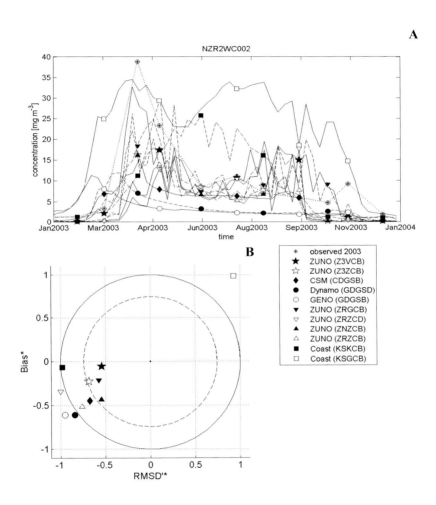

Figure 4   (A) Time series of Chlorophyll-a at station Walcheren 02 for all model runs discussed in this paper. Lines are shown to aid distinguishing the time series; they connect data at various intervals (models weekly, observations at least 2-weekly), but ignore actual shorter-term variability. (B) Corresponding target diagram, showing the normalised bias and signed, normalised, unbiased root-mean square difference of the model results with respect to the observations. Results within the drawn circle with overall $RMSD^*=1$ score at least 'reasonable', results within the dashed circle $RMSD^* = M_{Ro} = 0.74$, score 'good' (see text for further details).

3. Results

This series of model applications is intended to compare the model versions with each other and with the observations. First the overall model performance is assessed based on the Target Diagrams as a formal goodness of fit criterion using clustered results for all monitoring stations sampled in 2003 (see also Fig. 2). Notice that by considering all stations simultaneously, discrepancies in model behaviour might be obscured for instance due to compensating errors at different locations. Findings will be illustrated by some graphs for one representative monitoring location. Next, time series simulation results are presented for some typical stations in order to demonstrate the importance of various factors i.e. resolution, historic forcing, SPM forcing etc.

3.1 Overall model performance

Chlorophyll-a and species composition: As explained in 'Calibration procedure original models' during the calibration of all models an attempt was made to optimize the overall result for chlorophyll-*a* at a number of representative stations. Fig. 5 shows the results for the present application of the models. Indeed the overall score for all models shows little variation. Almost all models have a reasonable *RMSD* score and the differences within are rather small. The typical bias B^* of the models is between -0.2 and 0.2, which is much smaller than the unbiased *RMSD* scores which are typically between -1 and +1. In other words: in all models the annual average is close to the measurement, but the level of variation shows stronger discrepancies and is typically smaller than observed. This result is in agreement with the method of calibration, which focused on reproducing seasonally averaged results. Notice that even the *RMSD* score of the oldest, least advanced DYNAMO(GDGSD) (black circle) model does not differ much from the score of the latest ZUNO-3D(Z3ZCB) (white star). It should be pointed out, however, that for many individual stations, the overall RMSD score of chlorophyll-*a* did improve in later model versions. The singular position of the ZUNO-3D(Z3VCB) model in the target diagram indicates that the results are sensitive to the forcing by SPM: its overall *RMSD* is comparable its the companion ZUNO-3D(Z3ZCB), but its variability exceeds the observed.

For diatoms (Fig. 6a) the bias between model results and biomass estimates from cell counts is typically between 0 and 1. The ability of the models to reproduce the variability of the data is considerably smaller. The absolute value of all *RMSD* scores are larger than 1.0. There are two clusters with models. The *RMSD'** score is negative for all models which are forced with the SPM field from GENO and positive for all other models. So the SPM field, hence the light climate, is a crucial factor for the simulation of diatoms. Resolution and forcing of transport are less important considering the clustering of models.

The biases for *Phaeocystis* (Fig. 6b) are usually less than 0.5, which is considerably smaller than for diatoms. The *RMSD'** scores are also smaller. The differences within the model results are rather small; interestingly enough the oldest and coarsest GENO-NZB(GDGSB) model has the best overall *RMSD'* score.

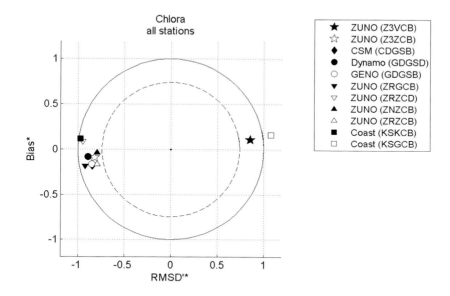

Figure 5 Target diagram for Chlorophyll-*a*, all stations in all applications.

Biases for flagellates (Fig. 6c) are usually positive and range from 0 to about 2, moreover all *RMSD'** scores are larger than 2. So there is a tendency to overpredict the overall biomasses and the level of variation produced by the models by far exceeds the level of variation of the data.

In the case of dinoflagellates (Fig. 6d) all biases are positive and always greater than 0.5. The *RMSD'** scores are positive and greater than 1.5, indicating that the models strongly overpredict the level of variation for this group.

So while the overall scores for diatoms and *Phaeocystis* could still be called reasonable or good, the score for flagellates and dinoflagellates are poor. This means that for all models the overall scores for chlorophyll-*a* are much better than for individual species groups. This issue will later be addressed in the discussion of this paper.

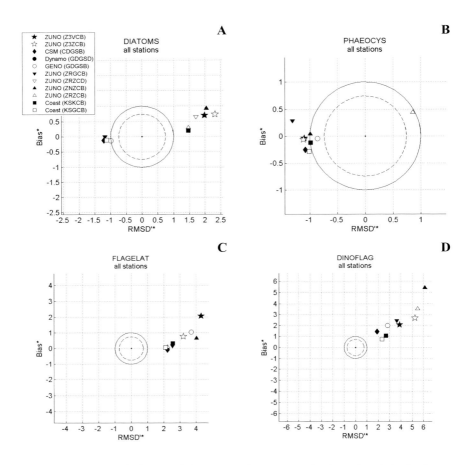

Figure 6   Target diagram for biomass of Diatoms (A), *Phaeocystis* (B), Flagellates (C) and Dinoflagellates (D), all stations in all applications.

Total nutrients: Often a comparison between the simulated and observed salinity is adopted to demonstrate the level of accuracy of the transport modelling. However, because in some of the older models salinity is not prognostically computed, we choose to show total nitrogen (TotN) and total phosphorus (TotP), which are included in all models, as proxies.

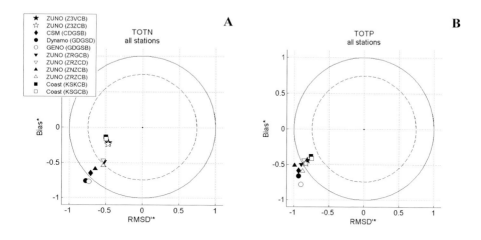

Figure 7   Target diagram for total Nitrogen (A) and total Phosphorus (B) concentration, all stations in all applications.

Fig. 7b shows the target plot for TotP. In comparison to TotN, the scores of individual models vary less. There is a negative bias for all models ranging from -0.8 for GENO-NZB(GDGSB) to less than -0.5 for COAST(KSKCB) and most of the ZUNO based models. This negative bias might be caused by an underestimation of the external P-inputs (loads and boundaries), or by an underestimation of the P release from the sediment. All models were rerun until they had achieved equilibrium for all state variables, but in reality sediments in the coastal regions might still contain significant amounts of reactive P stored during the 1970s and 1980s, when the external loads were considerably higher than in 2003. In comparison to TotN, the *RMSD'* scores are larger and all of them are negative, indicating that there is less variation in all of these models in comparison to the data. The overall score for the two 3D ZUNO models, the COAST models, and one of the 2D ZUNO(ZRZCD) models are reasonable

The ranking of the models clearly reflects the level of detail in resolution of the grids so this seems to be a critical factor for the accuracy of the overall transport model. Whether the model is 2D or 3D or what forcing is imposed does affect the results, but seems less critical on an annual basis when all stations are taken into account. For instance the score of COAST(KSKCB) with a daily forcing is on a par with the score of ZUNO-3D(Z3ZCB). Furthermore the scores of the ZUNO 2D model with historic forcing (ZUNO-2DR(ZRZCB)) and with a representative spring neap cycle (ZUNO-2D(ZNZCB)) are also about the same.

Notice that the *RMSD* score of the total nutrients is better for later models, but this has not resulted in a similar improvement of the overall score for chlorophyll-*a*.

SPM and total extinction: Fig. 8a shows the target diagram for the SPM forcing of the models. On average all models overestimate SPM and the average bias was hardly reduced during development of the SPM models up till the ones used here. Only the latest ZUNO-3D(Z3VCB) model using results from a new, dynamic SPM model (Van Kessel et al, 2008), shows an improvement in the overall bias. With the exception of the COAST(KSKCB) model the residues are larger than observed for all models. Two additional remarks need to be made. First, high-frequency OBS measurements indicate that the level a variation and the average value in SPM is underestimated by traditional monitoring (Blaas et al., 2007). Second, the SPM scores of some models, which are forced by the same SPM field, are not exactly identical as they should be in theory. This is caused by small imperfections in the procedures to project the SPM field from one grid onto another.

The total extinction is one of the most complex outputs from the models because it is influenced by both forcings (i.e. SPM) as well as by output variables (i.e. the phytoplankton biomass). Moreover, the individual terms are negatively correlated i.e. an overestimation of the contribution of SPM may be compensated by an underestimation of the amount of phytoplankton in case of light limitation.

The bias of the total extinction (Fig. 8b) has been reduced from 0.8 in the two GENO models to less than 0.1 in the latest ZUNO models. Two factors contribute to this improvement: (1) more recent models take the salinity as an approximation for dissolved organic matter (CDOM) into account and (2) new values for parameters of the light module were obtained from a statistical analysis of all available data (v. Gils et al., 2003). The *RMSD'* score has also been reduced, but not to the same extent as the bias. In the latest models the level of variation exceeds the measurements. This is caused by an exceedance of the level of variation of SPM, which has already been mentioned in the previous section. Nevertheless, the overall *RMSD* score has improved from poor to good over time.

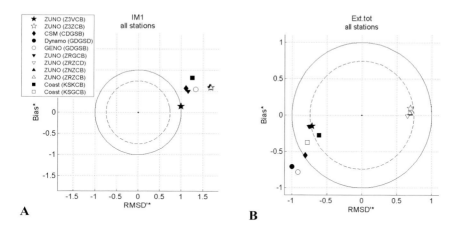

Figure 8 Target diagram for SPM (A) and vertical extinction (B), all stations in all applications.

Dissolved nutrients: The location of the models in the target diagrams for NO_3 and SiO_2 (Fig. 9a and 9b) are quite similar in terms of clustering of results, but both the biases and unbiased RMS-differences for most models are slightly smaller for NO_3. For NO_3 the score of all models except GENO-NZB is good, for SiO_2, most scores are reasonable, or marginally good. For both variables the sign of *RMSD'** is positive for the two ZUNO-3D and the two COAST models. The 3D model and the high resolution 2D model generate more variability by nature. There is no significant difference between the 2D ZUNO model with spring neap forcing (ZUNO-2D(ZNZCB)) and with actual forcing (ZUNO-2DR(ZRZCB)) so we may conclude that indeed the 3D transport phenomena cause the increase in variation of ZUNO-3D(Z3ZCB). It is less obvious why the unbiased *RMSD* for COAST(KSKCB) is similar to the one for ZUNO-3D(Z3ZCB). Most probably this is due to the seasonal correction of the horizontal dispersion, which was intended to mimic 3D phenomena. For SiO_2 one of the ZUNO 2D models has a positive RMS difference as well. Here, the simulated result for diatoms is different because a different SPM field was used as forcing which affects the development of the spring bloom.

For PO_4 (Fig. 9c) the overall *RMSD* scores are higher and only the GENO-NZB and DYNAMO models have a negative score. The high scores are not caused by a large bias, but by a relatively large value of the unbiased RMS difference. It is known from all these models that they tend to underpredict the PO_4 release from the sediment during summer which causes seasonal deviations relative to the observations (Los et al., 2008).

In spite of the obvious differences between the models, all scores for NO_3 are good, for SiO_2 one half ranks as good, the other half as reasonable and for PO_4 only GENO-DYNAMO(GDGSD) marginally passes the good criterion, while all others are reasonable or marginally poor.

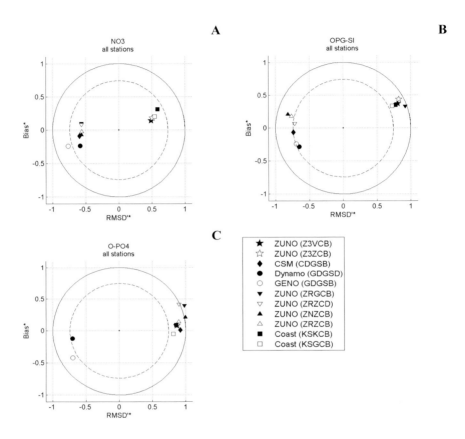

Figure 9 Target diagram for total NO₃(A), SiO₂ (B) and PO₄ (C) concentration, all stations in all applications.

3.2 Simulation results characteristic stations

As an illustration of the performances of the individual models at a typical location, results are presented here for three typical stations: Terschelling 10km, Walcheren 2km and Terschelling 235km (Dogger bank) (See Fig. 2). In all graphs the simulation results of some models are plotted against the measurements for 2003. Also included is the monthly mean, median and 90 percentile of the measurements for the years 1996 - 2002. These give an impression of the long term typical values for a particular station. Given the advances in the models, one might expect a gradual shift in model behaviour with an improved fit of later models with respect to the actual 2003 measurements.

Terschelling 10km: Terschelling 10km was selected for two reasons. First it is situated 10km offshore hence also the coarser models could still produce reasonable results here. Second it is more than 100 km north of the river Rhine outflow so observed nutrient levels are still clearly elevated in comparison to offshore locations, but the water is nearly always vertically mixed and horizontal gradients are not as strong as along the Noordwijk transect.

Fig. 10 shows the results of chlorophyll-*a* for all models. Although simulated chlorophyll-*a* levels of the two GENO based models are mostly below the 2003 observations, all models results are usually within the range of the long term observations. So if the purpose of the model application is to produce a multi-year mean result, chlorophyll-*a* simulations from all models should be regarded as reasonable or good. A closer inspection of the results does show some differences in the ability of the models to reproduce seasonality. The variability of the results tends to increase in later model versions. The spring peak gets more pronounced and while in the GENO based models chlorophyll-*a* declines almost monotonously following the spring bloom, elevated chlorophyll-*a* levels later in the year are simulated by some of the other models in particular by COAST(KSKCB) and the ZUNO models. This improves their visual appearance relative to the measurements.

Fig. 11 shows the results for total nitrogen and total phosphorus for all models at this station. Obviously the January observation differs considerably from the one in December. Remembering that all models were run with initial conditions taken from the end of the simulation, so the January measurements can hardly be reproduced by any of the models and should be ignored. In the case of total nitrogen, simulation results for the two GENO models and for CSM(CDGSB) are consistently below the measurements (Fig. 11a). Results of ZUNO-2D(ZNZCB) are also below the measurements, but the difference is smaller. ZUNO-3D(Z3ZCB) and COAST(KSKCB) clearly show the best performance (Fig. 11c). Notice that there is an exceptionally large river outflow at the beginning of the year, which is accounted for in the load of all models, but only in the residual transport due to density distributions and corresponding atmospheric conditions of ZUNO-3D(Z3ZCB). This explains why ZUNO-2D(ZNZCB) with its average spring - neap forcing doest not match the high observations in March and April, which ZUNO-3D(Z3ZCB) does. In spite of its daily forcing, COAST(KSKCB) does reproduce these peaks values and as a matter of fact even seems to do better than the ZUNO-3D(Z3ZCB) model.

For total phosphorus differences between models are smaller than for total nitrogen. The most plausible explanation for this is a difference in the relative importance of the nutrient sources. In the case of nitrogen, there is still a dominant contribution by the rivers. In contrast, phosphorus reductions in the river basin of the Rhine since the end of the 1980s have diminished the concentration differences with the Channel boundary. Hence TotN, with its larger spatial gradients, will be more susceptible to imperfections in the transport modelling along the Dutch coast than TotP. The underestimation by all models during the summer is caused by imperfections of the relatively simple sediment-bed model adopted in all these simulations.

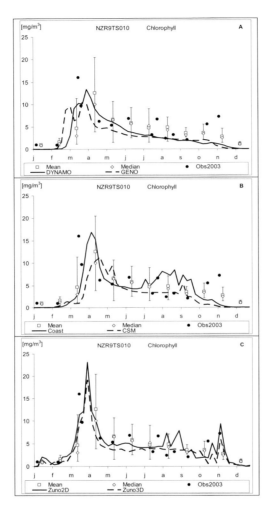

Figure 10   Comparison of model results for chlorophyll (mg.m^{-3}) for different model versions at station Terschelling 10km. Circles are measurements for 2003, bars indicate 90 percentile of measurements for the years 1996 – 2002.

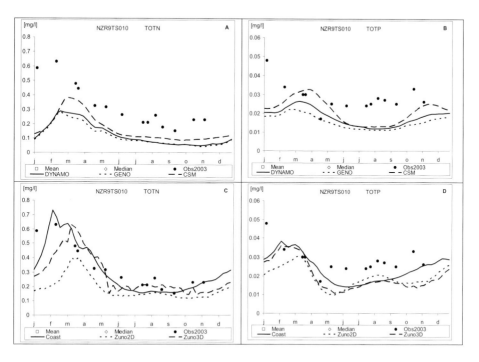

Figure 11   Comparison of model results for TOTN (mg.l^{-1}) and TOTP (mg.l^{-1}) for different model versions at station Terschelling 10km. Dots are measurements for 2003. Notice that the first measurement should be disregarded due to the spin-up procedure.

Fig. 12 shows the SPM forcing and the simulated vertical extinction coefficient. In the non-ZUNO based models, the SPM forcing at this location systematically exceeds the observations. Nevertheless the simulated extinction coefficients of these models are typically lower than observed. The main reason is that the contribution by CDOM (approximated by salinity) was not taken into account in these models. Average results for the ZUNO based models agree much better with the observations. The wind-based seasonal variation (See 'Main features of models') is similar in amplitude to the long term observed variation at this station. The simulated vertical extinction coefficients by COAST(KSKCB) and the two ZUNO models agree better with the observations compared to those of the previous models. The enhanced

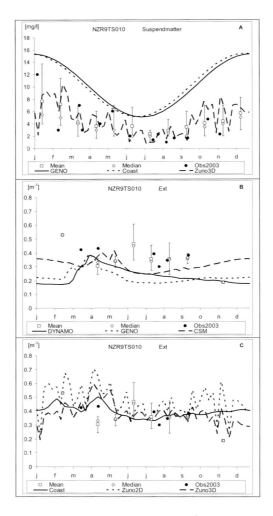

Figure 12   Comparison of model forcing by suspended matter (mg.l^{-1}) (A) and the resulting vertical extinction
coefficient (m^{-1}) for the two GENO based models and CSM (B) and for the Coastal zone model and
the two ZUNO based models (C) at station Terschelling 10km. Forcings for DYNAMO and CSM
(not shown) are the same as for GENO. Circles are measurements for 2003, bars indicate 90
percentile of measurements for the years 1996 – 2002.

variability of SPM is also reflected in an enhanced variability of the extinction relative to
COAST(KSKCB). Notice that the relatively high value of SPM of the COAST(KSKCB) model does
not result in an overprediction of the extinction coefficient. This is because during the
calibration of this model a lower specific extinction was adopted which compensates for high

SPM values. The specific extinction coefficient of the ZUNO models is based on a statistical analysis (Van Gils et al., 2003) and should be considered as more realistic.

From a visual inspection of the results for this station it may be concluded that they are in line with the general results of the target plots presented previously. Resolution affects the transport hence the total nutrients, but even the coarsest model does not appear to be far off. The same holds for chlorophyll-*a*. There have been clear advances in the simulation of the vertical extinction coefficient due to improvements in the light module.

Walcheren 2km: In an area with horizontal gradients, one might expect that the grid resolution has a clear impact on the results of the models. To investigate the importance of resolution at a coastal station, results are shown for the location Walcheren 2. This station is situated 2km offshore, just north of the Western Scheldt estuary (Fig. 2). It is relatively shallow with an average depth of about 12m and is characterized by high SPM levels. Because the residual current are towards the north east, the typical salinity is about 31 ppt, which is high in comparison to the coastal stations on the Noordwijk transect north of the Rhine-Meuse river mouth. Occasionally high fresh water discharges from the Rhine-Meuse system do protrude southward, resulting in a reduction of the salinity by about 1-3 ppt.

Fig. 13 shows the results for the coarse models (DYNAMO(GDGSD), GENO (GDGSB), CSM(CDGSB)), Fig. 14 or the fine resolution models (COAST(KSKCB), ZUNO-2D(ZNZCB) and ZUNO-3D(Z3ZCB)) for NO_3, SiO_2, chlorophyll-*a* and the extinction coefficient. In the coarse models, winter levels of the two nutrients are far below the measurements and should be qualified as reasonable or poor. Summer levels of both NO_3 and SiO_2 are limiting for a long period in all models. Consequently chlorophyll-*a* levels are strongly underpredicted by the two GENO models and so is the extinction coefficient. This is not just because chlorophyll-*a* (and detritus) are too low, but also because the forcing generated by the SPM sub-model on the same coarse grid is also far below the measurements. The chlorophyll-*a* result for CSM(CDGSB) is clearly better in terms of phasing, but still systematically below the measurements.

In all three fine resolution models shown here, results for NO_3, SiO_2 agree much better to the observations (Fig. 14). In combination with a better forcing of the light climate by SPM, this results in a better simulation result of chlorophyll-*a* in comparison to the coarse models as well. The overall result of the ZUNO based models is better than for COAST(KSKCB) because in the latter summer levels are overpredicted. So while for all stations together, results of chlorophyll-*a* do not vary much between the different models, the differences at this coastal station are considerable. Results for other coastal stations consistently show the same pattern. Hence we conclude that a fine resolution is a necessary condition to adequately describe conditions at coastal stations.

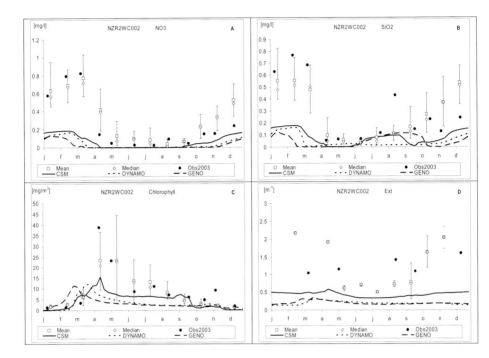

Figure 13   Comparison of NO$_3$(A), SiO$_2$ (B) (mg.l^{-1}), Chlorophyll-a (C) (mg.m^{-3}) and vertical extinction
coefficient (D) (m^{-1}) for coarse resolution models at station Walcheren 2km. Circles are
measurements for 2003, bars indicate 90 percentile of measurements for the years 1996 – 2002.

But how important is the forcing of the transport? The three fine resolution models differ strongly with respect to their transport forcing. The COAST(KSKCB) model is basically a residual-current-driven model with a seasonal correction term of the dispersion for the climatological wind speed and direction, ZUNO-2D(ZNZCB) is forced by a representative spring - neap cycle and ZUNO-3D(Z3ZCB) is driven by time and space varying historic forcing for 2003 (see the section 'Main features of models' for more details). The overall $RMSD$ scores (not shown) for NO$_3$ and SiO$_2$ for these three models are almost the same and qualify as good. On a seasonal basis ZUNO-3D(Z3ZCB) and COAST(KSKCB) give the best results for dissolved nutrients in spring during the period when the Rhine - Meuse discharges are extremely high, but during the rest of the year, nutrient levels exceed the measurement and the best fit is obtained by ZUNO-2D(ZNZCB). Due to non-linearities in the processes described by the model, the differences in chlorophyll-a and extinction between the three fine models are even smaller. So results are rather insensitive to the way of climatological forcing.

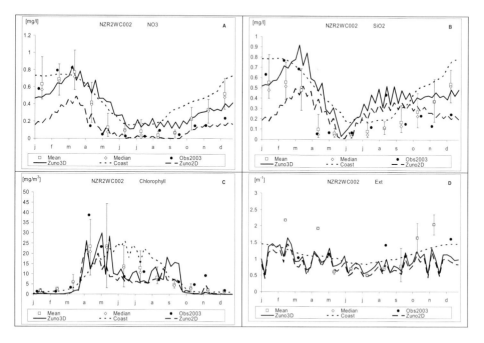

Figure 14   Comparison of NO₃(A), SiO₂ (B) (mg.l⁻¹), Chlorophyll-*a* (C) (mg.m⁻³) and vertical extinction coefficient (D) (m⁻¹) for fine resolution models at station Walcheren 2km. Circles are measurements for 2003, bars indicate 90 percentile of measurements for the years 1996 – 2002.

Dogger Bank (Terschelling 235 km): In the central North Sea, the residence times are relatively long. In particular in the vicinity of the Dogger Bank, which is approximately 235 km north west along the Terschelling transect (Fig. 2) tracer simulations indicate that after a year, between 50 and 80 percent of the water in this area was already within the southern and central North Sea domain when the simulation started. Hence external sources are less important here in comparison to the coastal zone. The Dogger Bank area is relatively shallow (about 23 m deep) and usually well mixed. Turbidity is low and, hence, the spring bloom at this location may occur quite early in the season and even in winter it is not uncommon to find chlorophyll-*a* levels of about 1 μg l⁻¹, which in this area is a typical summer value.

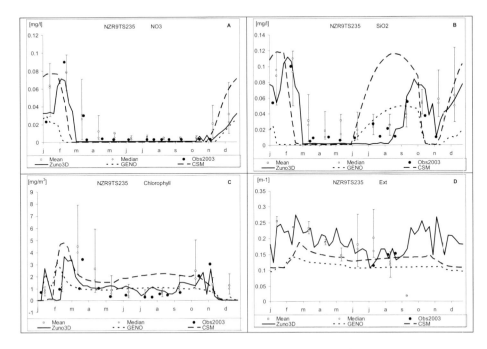

Figure 15   Comparison of three model generations for NO₃ (A), PO₄ (B), (mg.l⁻¹), Chlorophyll-a (C)
(mg.m⁻³) and vertical extinction coefficient (D) (m⁻¹) (right pannel) at the Dogger Bank
(Terschelling 235km). Circles are measurements for 2003, bars indicate 90 percentile of
measurements for the years 1996 – 2002.

Given the characteristics of this location, one might expect that differences in simulation
results between the models will be mainly due to the choice, formulation and parameterization
of the ecological processes in particular with respect to the stoichiometric ratios. In Fig. 15
results are shown for three different models, one from each model generation: GENO-
NZB(GDGSB), CSM(CDGSB) and ZUNO-3D(Z3ZCB). In GENO-NZB(GDGSB) winter levels of the
two most important nutrients, NO₃ and SiO₂ are low in comparison the measurements. Winter
levels in CSM(CDGSB) are highest, ZUNO-3D(Z3ZCB) is in between. All three models correctly
indicate that NO₃ is limiting for a very large part of the year. All models indicate correctly
that SiO₂ is also an important limiting nutrient. The recovery and autumn decline is well
reproduced by ZUNO-3D(Z3ZCB). Summer levels are much more smoothed in the other models
although they correctly describe that SiO₂ declines again in autumn. Summer SiO₂ levels of
CSM(CDGSB) are much higher compared to the other two models and clearly exceed the
measurements. ZUNO-3D(Z3ZCB) gives the most accurate result for chlorophyll-*a* and although
its curve is much smoother, levels simulated by GENO-NZB(GDGSB) are usually close to the
measurements as well. Results by CSM(CDGSB) systematically exceed the measurements.
ZUNO-3D(Z3ZCB) gives the best simulation result for the vertical extinction coefficient;
simulated levels by the other two models are usually too low.

In conclusion with respect to TS235, indeed improvements in the ecological model, and more specifically in the parameterization of BLOOM and the extinction module, lead to a better description of chlorophyll-*a* and the extinction coefficient. Nevertheless, some results cannot be explained by these improvements. In particular, the seasonal patterns of NO₃ and SiO₂ simulated by ZUNO-3D(Z3ZCB) are better described by ZUNO-3D(Z3ZCB) than by the two other models. Since there are no major differences with respect to resolution in the Dogger Bank area, there must be another cause for this difference. To investigate the impact of climatological forcing, two additional simulations were conducted with the ZUNO 2D model, one with spring - neap cycle forcing (ZUNO-2D(ZNZCB)), one with actual climatological forcing (ZUNO-2DR(ZRZCB)). In Fig. 16 a typical example is shown for SiO₂. Obviously, results of ZUNO-2DR(ZRZCB) closely resemble the 3D model, while results of ZUNO-2D(ZNZCB) resemble those by CSM(CDGSB) and GENO(GDGSB). In particular the pattern of SiO₂ increase towards the end of summer is replaced by a better matching of phase and amplitude once more realistic atmospheric forcing is applied. So in addition to improved parameterization of the model, taking more realistic forcing into account does improve the model performance as well at this location.

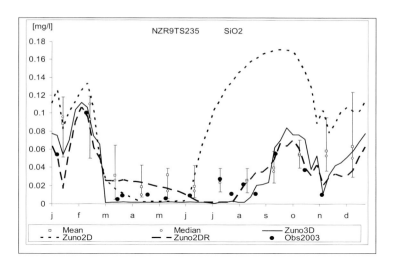

Figure 16   Comparison of SiO₂ (mg.l⁻¹) simulated by ZUNO-2D(ZNZCB) (spring – neap forcing), ZUNO-2DR(ZRZCB) (historic forcing vertically averaged) and ZUNO-3D(Z3ZCB) (historic forcing in 3D mode) at the Dogger Bank (Terschelling 235km). Circles are measurements for 2003, bars indicate 90 percentile of measurements for the years 1996 – 2002.

4. Discussion

4.1 GOF criteria

At the beginning of our analysis, we have applied the OSPAR cost function ((Villars and De Vries et al., 1998; Radach and Moll, 2006) to compare the scores of individual models. Previously, we had used this function to assess the performance of the ZUNO-3D(Z3ZCB) model as applied for the year 1989 (Los et al., 2008). However, during this model intercomparison we noticed that the discrimination of different model results by the cost function was rather poor. On several occasions models obtained a similar score for a particular output, although a visual inspection of the results revealed rather large differences which we felt should be reflected in their score. In particular for chlorophyll-a, almost all scores for all models at all considered monitoring locations pass the 'good' criterion (Fig. 4b), in spite of obvious differences noticed in the time series graphs. In this paper therefore, we have adopted the target diagram by Jolliff et al. (2008) which provided a concise but more contrasted picture of model performance on the entire model domain and for the entire period of the simulations. As with any method that aggregates results, local nuances in time and space are lost as flaws may obscure each other or overshadow good local performances. The evaluation of the bias in the target diagram focused on the annual mean, whereas the variability expressed by the unbiased RMS difference is mostly determined by the seasonal cycle. Given the limited number of observations in the present validation dataset, we judged it was not sensible to further refine the evaluation of the results on subseason or regional scale, although in principle this would be advisable. For the present paper, the focus is on the overall performance given the present data supported by a discussion of a selection of time series of local stations in the traditional sense. A follow-up of this study might include additional mooring and remote sensing data such that a spatial and temporal breakdown of the results is feasible.

4.2 General observation

During model development every modeller attempts to improve the overall performance in such a way that obvious shortcomings are corrected while maintaining the quality of those results that already qualify as good according to the criteria adopted by the modeller. For relatively complex models with many interactions, parameters and forcings this proofs to be difficult, if not impossible. The most recent and most advanced model that we have used during this comparison (ZUNO-3D(Z3ZCB)) indeed obtains the best overall scores, but it is easy to find exceptions, locally or temporarily, where some other model showed a better match with the observations. Several of those examples are included in this paper.

More specifically, we have noticed that the overall bias in modelling results has been reduced more than the ability to reproduce observed levels of variation. For instance an improper resolution at a particular region such as the Dutch coastal zone, causes biases in many variables, which cannot be completely 'corrected' by the parameterization of the models. Hence the bias in results of the least advanced models is relatively large and often also in one particular direction (i.e. model results are too low throughout a large area). In general improvements in forcings and physical representation have resulted in a decrease in bias of the models. With respect to the level of variability assessed here by the RMSD score, we noticed that while early generation models usually have a negative score (too little variation), recent models tend to have a positive score (too much variation). This increase in the level of variation of the models is due to refinements in resolution and enhanced seasonal variations in some of the forcings (i.e. SPM). The absolute values of these scores are remarkably similar so the contribution to the overall GOF score is of the same order of magnitude. While a visual inspection of the results of models with historic forcing often gives a more realistic impression, this cannot be substantiated with data obtained by traditional low-frequency monitoring programs. In contrast, data obtained from alternative measuring devices such as smartmoorings and satellites suggest that the level of variability is underestimated by standard monitoring programs. So use of these kind of data is essential to determine if the variability generated by models with actual forcing is realistic or rather some kind of noise.

4.3 Overall model performance

The overall model performance was assessed in this paper by a comparison of the *RMSD* scores of a number of variables simulated by the models at a large number of locations simultaneously. With respect to the main output variable of the models, chlorophyll-*a*, the overall scores are rather similar. This basically means that all models were rather well calibrated with respect chlorophyll-*a*. From a detailed analysis of the results of different models it also appears that this variable is well buffered due to internal compensation mechanisms. So chlorophyll-*a* is also a robust model output. Scores for other substances such as total and dissolved nutrients show more variations. More specifically, the bias clearly declines with enhanced resolution of the grid. Thus resolution not only affects local results as one might expect, but also the global transport throughout the model domain. From a comparison of different combinations of resolution and forcing we also concluded that on an annual basis an appropriate resolution is more important than the level of realism of the transport forcing. To put it another way: in this part of the North Sea having the appropriate residual flows is more important than a detailed simulation of historic events.

Improvements in modelling of the underwater light regime have resulted in a better *RMSD* score for the total extinction coefficient in more recent models. In particular, the bias was rather strongly reduced. This improvement would not have been possible if the salinity dependent approximation of CDOM had not been taken into account. The importance of this factor was also clearly demonstrated by Van Gils et al. (2003), who performed a statistical analyses of all regular observations in the Dutch part of the North Sea. So, although the overall score for chlorophyll-*a* did not improve much from one model generation to the next, the scores for nutrients and underwater light climate improved (rather) strongly. Basically this means that model development was particularly successful in increasing the consistency of its various components.

4.4 Local model performance

On a local scale, differences between models are more, sometimes much more apparent. Particularly in regions with steep gradients in external conditions such as in the Dutch coastal zone, a relatively fine resolution, preferably well adapted to the orientation of these gradients, is essential. On a local scale, not only the scores for nutrients and the extinction vary with resolution, but also the scores for chlorophyll-*a*. Again overall consistency increases. The importance of this becomes apparent when the models are applied for scenario simulations (nutrient reductions; dredging activities etc.). For instance, in the models with a coarse resolution tested here, nutrient limitation appears to control the phytoplankton biomass in the Dutch coastal zone too strongly. Fig. 17 shows a typical example of this type of result for the models COAST(KSKCB) and CSM(CDGSB) at station Walcheren 2km. Remember that both models are similar with respect to kinetic processes and parameterization. The cost function scores of both models for chlorophyll-*a* is 'very good'. Using the more restrictive *RMSD* scores as proposed in this paper, the result by COAST(KSKCB) is good and by CSM(CDGSB) is reasonable. However, in CSM(CDGSB) nutrient limitation is too severe. To assess the importance of this difference, a nutrient reduction scenario was run with both models in the fashion of the reduction scenario studies by the OSPAR Intercessional Correspondence Group on Eutrophication Modelling (ICG-EMO) reported by Lenhart et al. (*this issue*). In this example river loads of N and P (both inorganic and organic) were reduced by 50 respectively 20 percent. Both models were restarted several times until they were sufficiently well adapted to the new conditions. The scenario results are shown in Table 3. In both models winter levels of NO_3 are reduced by almost the same percentage (between 28 and 29 percent). In the case of CSM(CDGSB), nitrogen controls the biomass to such extent that the decline of the summer average chlorophyll-*a* in the scenario is almost the same as the reduction of inorganic nitrogen: 27 percent. So this model's response is nearly linear. In contrast in the COAST(KSKCB) model, light rather than nitrogen is the main limiting factor and the summer average of chlorophyll-*a* decreases by less than 5 percent for the reduction scenario. Given the results for NO_3, PO_4 and SPM in the base case, this last result seems more realistic. This difference in response would probably be considered as 'significant' during an actual study on nutrient reductions.

In order to assess the adequacy of a particular model for a particular task it is therefore insufficient to consider the GOF score(s) for only a single key output such as chlorophyll-*a*. To check the consistency also the scores for controlling factors such as SPM, CDOM and nutrients should be taken into account. A low score for a factor, which is actually controlling the model behaviour to a large extent, means that scenarios affecting that factor will not be simulated correctly.

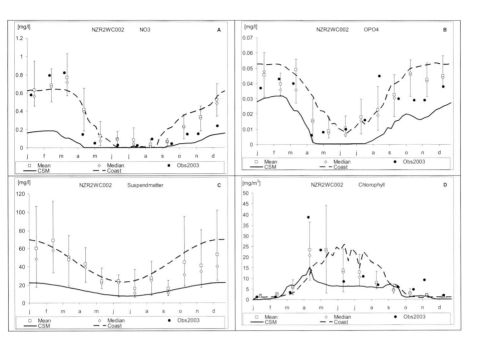

Figure 17   Comparison of NO₃ and PO₄, suspended matter (mg.l⁻¹) and Chlorophyll-*a* (mg.m⁻³) for COAST - KSKCB and CSM - CDGSB model at station Walcheren 2km. Circles are measurements for 2003, bars indicate 90 percentile of measurements for the years 1996 – 2002.

Table 3   Impact of nutrient reduction scenario on NO₃ and Chlorophyll-*a* at station Walcheren 2km for Coastal zone and for CSM model.

Case	Base			Nutrient Reduction Scenario			Percent Effect Scenario		
Model	NO₃ Winter	NO₃ Summer	Chl-*a* Summer	NO3 Winter	NO₃ Summer	Chl-*a* Summer	NO₃ Winter	NO₃ Summer	Chl-*a* Summer
COAST	0.46	0.3	16.7	0.33	0.19	15.9	-28.26	-36.67	-4.79
CSM	0.07	0.01	7.3	0.05	0	5.3	-28.57	-100	-27.39

4.5 Total inorganic matter forcing of light

As explained previously, the SPM forcing is very important, particularly in determining the onset of the spring bloom. Locally, particularly in the parts of the Dutch coastal zone, light remains limiting to phytoplankton all year round due to high values of SPM, but the area where this is the case has decreased recently. Due to riverrine reductions in phosphorus since the 1980s, PO_4 is now often limiting during summer.

So how sensitive are the modelling results to the forcing by SPM? This question cannot easily be answered by comparing the results of the base simulations considered here because changes in SPM forcings coincide with other modifications notably in resolution and transport. For the 3D model two results are available that differ only with respect to the forcing by SPM. From Fig. 8a and Fig. 5 it may be concluded that the better *RMSD* score of the ZUNO-3D(Z3VCB) simulation for SPM also results in a better score of chlorophyll-*a*.

The effect of SPM variations on the timing of the spring phytoplankton bloom can also be demonstrated differently as was done in Los et al. (2008). Sensitivity studies presented in that paper showed that any type of short-term fluctuation in the suspended matter concentration, even if it was purely random, was sufficient to trigger an improvement in the timing of the spring bloom in the model at those location where it is controlled by light.

We conclude that improvements in SPM modelling clearly contribute to an improvement of the chlorophyll-*a* simulation.

4.6 Complex vs. simple phytoplankton modelling

Development and application of the freshwater implementation of BLOOM had already started in the 1970s and 1980s (Los, 1982; Los et al., 1984; Los et al., 1988). Modelling several functional groups, species and types is considered interesting for two reasons: (1) species dominance is an important ecosystem characteristic, not in the least since a number of species have been denoted objectionable for one reason or another and (2) it was demonstrated by for instance Zevenboom et al. (1982) that even if one and the same species is dominant in a particular waterbody for a prolonged period of time, its characteristics in terms of stoichiometry, chlorophyll-*a* contents and maximum growth rate vary considerably in response to changes in external conditions. For management of the North Sea, *Phaeocystis* is considered to be an important species, for which separate target values have been defined e.g., for the EU Water Framework Directive.

With respect to the species composition it was previously concluded in this paper that the overall, seasonally averaged performance of the models is reasonable or good for diatoms and *Phaeocystis*, but poor for the other two groups. Also there is little difference between model generations: there has not been much progress in this domain during development of the models.

In contrast, Los (1991) demonstrated that the freshwater implementation of BLOOM reproduced the observed phytoplankton species groups adequately in most of 30 lakes that were modelled. Van der Molen et al. (1994) showed that the model is capable of reproducing a major shift in species dominance in the well investigated Lake Veluwe in the Netherlands after a series of management measures.

A seasonal example of the species composition simulated by the models GENO-NZB(GDGSB), COAST(KSKCB), CSM(CDGSB) and ZUNO-3D(Z3ZCB) and of the biomass estimates from cell counts at the coastal station Noordwijk 20km is shown in Fig. 18. As for the *RMSD* scores of all station, the agreement between model results and observations look reasonable or good for diatoms and *Phaeocystis*, results are mixed for flagellates and timing and size of the simulated dinoflagellates biomass is rather poor. As was previously remarked, results of different model versions show some obvious differences, but there is not much progression.

There are several reasons why the scores for the simulated species biomasses in the North Sea models do not yet equalize those by the freshwater model. Firstly, monitoring data were lacking during the development of the marine models. In fact quantitative data have only become available *after* the calibration of the models presented here. Secondly, for the freshwater species more information is available on species growth and nutrient characteristics, both from chemostats and from field data. This is particularly true for the cyanobacteria species, which dominate in many Dutch lakes. Thirdly, biomass estimates for different years and locations show a lot of scatter. This in combination with a low (nominally monthly) sampling frequency, makes it difficult for any model to faithfully reproduce these numbers. In the example presented here, the 2003 observations for flagellates are highest in April and May, but these values are far out of the range of observations during all other years. In contrast the low values observed in July and August are at the very low end of the range obtained from all other years. Fourthly, Fig. 18e shows that there is no systematic difference between the measured and simulated total algal biomass during winter and spring, but there is a clear difference in summer. Instead, a corresponding discrepancy is not found for chlorophyll-*a* (Fig. 18f). This means that either the stoichiometric ratios adopted in all model versions for the typical summer species are incorrect, or that biomass estimates from cell counts and the chlorophyll-*a* measurements are inconsistent during the summer. This could be due to errors in the conversions from cell numbers to biomass, or because a significant part of the actual biomass consists of (small) species which are not observed under the microscope. Clearly this issue needs to be further investigated to improve the BLOOM model.

Notice that with respect to the usefulness of the model, the best species scores are obtained for what is from a management point of view considered to be the most important species: *Phaeocystis*.

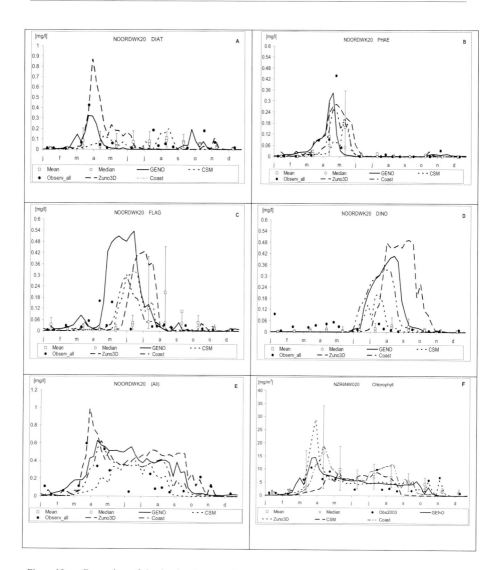

Figure 18   Comparison of simulated and measured biomasses of diatoms, *Phaeocystis*, microflagellates and dinoflagellates (mgC.l⁻¹) and for total biomass (mgC.l⁻¹) and chlrophyll-*a* (mg.m⁻³) of four different models at station Noordwijk 20km. Circles are measurements for 2003, bars indicate 90 percentile of measurements for the years 1996 – 2002.

The complexity of the phytoplankton module also affects the chlorophyll-*a* simulation of the model. The 12-type BLOOM phytoplankton module has the ability to display a wider range of variation both spatially as well as temporarily compared to a 2 species model because it selects its composition from a wider range of possible stoichiometric ratios. A comparison of the results of GENO-DYNAMO(GDGSD) and GENO-NZB(GDGSB) is, however, hampered by the dominant impact of the grid resolution on the modelling results. Therefore a comparison was made between the original ZUNO-2D(ZRZCB) and the ZUNO-2D(ZRZCB) application where the 12-type plankton model has been replaced by a 2-type model. Since this model is not valid for the sometimes stratified Oysterground region, results for only 9 well-mixed stations could be compared. At those stations, the overall *RMSD* score for chlorophyll-*a* of the 12 type simulation was reasonable whereas the 2 type model version scored poor. We conclude that the overall ability of the model to simulate chlorophyll-*a* improves if the more complex BLOOM module is adopted.

4.7 One model or several related models?

Modellers usually concentrate their activities on their latest, most advanced and best performing model version. In our case ZUNO-3D(Z3ZCB) and ZUNO-3D(Z3VCB) with improved SPM forcing have the best overall score and hence will be the basis for further model development. But this model intercomparison has also demonstrated that the GOF scores of other model versions are sometimes equally good, sometimes even better and at least accurate enough to answer specific questions. So there are a number of reasons why we intend to keep more than one model version operational. (1) Ensemble modelling generates insight even if the models are as closely related as those used during this study. (2) For long forecast simulations, realistic forcings are not available so using a schematic, repetitive forcing of the transport is often necessary. (3) Quick scans on for instance nutrient reductions in decision support systems in the Netherlands are now usually based on expert rules. The fastest performing deterministic model application shown here, simulates an entire year in a few minutes and one might argue that they are more accurate than the knowledge rules. (4) Related to the previous point, in many management studies the total number of simulations requested by the clients makes it more practical to adopt a somewhat simplified approach (i.e. 2D in stead of 3D; limited model domain etc). (5) A formal parameter sensitivity analysis of the complete 3D model is not very practical due to the necessity to perform a very large number of simulations. Instead, recently an extensive analysis was completed for ZUNO-2D(ZNZCB) (Salacinska, 2008), the results of which are also meaningful for ZUNO-3D(Z3ZCB).

5 Overall Conclusions

During this study the following factors were considered:

- the resolution of the grid,
- degree of realism in forcing of transport,
- influence on the light climate by SPM and CDOM,
- the level of detail of the phytoplankton model.

With respect to resolution we concluded that refinement has a clear and positive effect on the transport of substances, not just locally as might be expected but also on a global scale, taking all measurements into account simultaneously. Results for the degree of realism in the transports are less conclusive. In the Dutch coastal zone it appears that if the residual flow is correct, 2D models with daily tidal forcing, or with spring–neap forcing and a steady wind forcing do not differ much in performance compared to the 3D model with actual wind and density forcing. It should be noted though, that validation of models with real time forcing is hampered by the lack of high-frequency observation data. In contrast to the coastal zone, at the Doggerbank the 2D and 3D models with actual forcing outperform the models with schematic forcing. With respect to the underwater light climate, it is obvious that the results of the primary production models are sensitive to this forcing in particular with respect to the timing, size and species composition of the spring bloom. Unfortunately neither of the SPM models considered here performs very well. The SPM modelling therefore should be improved. The contribution of CDOM to the attenuation of light is significant. The accuracy with which models simulate the extinction coefficient, clearly improves when CDOM (parameterized in terms of salinity) is taken into account. Modelling 12 rather than 2 types of phytoplankton improves the ability of the model to correctly simulate chlorophyll-*a* because a wider range in nutrient and biomass to chlorophyll-*a* ratios is considered. This range is less well described by a 2 species model with fixed coefficients. Using the BLOOM module, results for *Phaeocystis* are good, but the match between simulation results for micro-flagellates and dinoflagellates should be improved.

The use of objective Goodness of Fit criteria depends on the nature of the available validation data. If only low frequency data are available, obviously a certain degree of spatial and temporal aggregation is required in order to arrive at statistically sound results. The use of target diagrams facilitates to discern improvements in the mean and in the variability and is preferred over a cost function of the mean-absolute error type. Classification still remains a partially subjective issue depending on the nature of the observations (resolution, representativity, uncertainty) and the purpose for which a model application is developed (i.e., time and spatial scale, bulk quantities or species composition etc.). In that context, we conclude that clearly the model applications discussed here have been developing from representing multi-annual mean seasonal signals to seasonal patterns within a particular year, but validation of further refinement is only possible with further refined observations.

Acknowledgements

The research activities presented here have been partly funded by the Deltares Applied Research Programme BTO43 (contract Z4575) funded by Rijkswaterstaat and partly by the Deltares Strategic Research Programme SO9.3 (contract Z4753) and SO12.1 (contract Z4672).

References

Allen, J.I., Holt J.T., Blackford, J, and Proctor, R., 2007, Error quantification of a high-resolution coupled hydrodynamic ecosystem coastal-ocean model: Part 2. Chlorophyll-a, nutrients and SPM, *J. Mar. Sys. (in press),* doi:10.1016/j.jmarsys.2007.01.005

Baretta-Bekker, J.G., J.W. Baretta, M.J. Latuhihin, X. Desmit and T.C. Prins, *in press*, Description of the long-term temporal and spatial distribution of phytoplankton carbon biomass in the Dutch North Sea, Journal of Sea Research (*accepted*).

Blaas, M., El Serafy, G.Y.H., van Kessel, T, Eleveld, M.A. 2007. Data Model Integration of SPM transport in the Dutch coastal zone. *Proceedings of the Joint 2007 EUMETSAT / AMS Conference*, Sep. 2007, Amsterdam, The Netherlands.

Blauw, A.N, Van de Wolfhaar, K., Meuwese, H., 2006. Transboundary nutrient transports in the North Sea: model study. WL | Delft Hydraulics Report, Z4188, December 2006, WL | Delft Hydraulics, Delft.

Blauw, Anouk N. , Hans F. J. Los, Marinus Bokhorst, and Paul L. A. Erftemeijer., 2009. GEM: a generic ecological model for estuaries and coastal waters. *Hydrobiologia* DOI 10.1007/s10750-008-9575-x

Boon, J.G. and Bokhorst, M., 1995. KSENOS, Adjustment and extension of the modelling suite for toxic substances and eutrophication in the North Sea and Dutch coastal waters. WL | Delft Hydraulics report T1236 (in Dutch).

De Kok, J.M., C. de Valk, J.H.Th.M. van Kester, E. de Goede and R.E. Uittenbogaard , 2001. Salinity and temperature stratification in the Rhine plume. A model study. *Estuarine, Coastal and Shelf Science*, 53: 467-475.

De Kok, J.M., R.Salden, I.Rozendaal, P.Blokland, J.Lander, 1995. Transport paths of suspended matter along the Dutch coast. In:*Computer Modelling of seas and Coastal Regions*,75-86. Computational Mechanics Publications, Southampton.

De Vries, I., Duin, R.N.M., Peeters, J.C.H., Los, F.J., Bokhorst, M. and Laane. R.W.P.M., 1998. 'Patterns and trends in nutrients and phytoplankton in Dutch coastal waters: comparison of time-series analysis, ecological model simulation and mesocosm experiments.' In *ICES Journal of Marine Science,* 55: 620-634.

Di Toro, D.M., Fitzpatrick, J.J. and Thomann, R.V., 1971. A Dynamic Model of the Phytoplankton Population in the Sacramento San Joaquim Delta. *Adv. Chem. Ser.*, 106:131-180.

Di Toro, D.M., Thomann, R.V., O'Connor, D.J., and Mancini, J.L., 1977. Estuarine phytoplankton models-verification analysis and preliminary applications. In: Goldberg, E.D. (Ed.), *The sea. Marine modeling*, New York.

Eisma, D. and J. Kalf, 1987, Dispersal, concentration and deposition of suspended matter in the North Sea, *Journal of the Geological Society*; 144: 161-178

Eleveld, M.A., R. Pasterkamp, H.J. van der Woerd, 2004, A Survey of Total Suspended Matter in the Southern North Sea Based on the 2001 Seawifs Data, *EARSeL eProceedings 3*, 2/2004 166

Ferris, J.M., Christian, R. (1991). Aquatic primary production in relation to microalgal responses to changing light: a review. *Aquatic Sciences* 53, 2/3, 1991.

Friedrichs, M.A.M., M.-E. Carr, R.T. Barber, M. Scardi, D. Antoine, R.A. Armstrong, I. Asanuma, M.J. Behrenfeld, E.. Buitenhuis, F. Chai, J.R. Christian, A.M. Ciotti, S.C. Doney, M. Dowell, J. Dunne, B. Gentili, W. Gregg, N. Hoepffner, J. Ishizaka, T. Kameda, I. Lima, J. Marra, F. Mélin, J.K. Moore, A. Morel, R.T. O'Malley, J. O'Reilly, V.S. Saba, M. Schmeltz, T.J. Smyth, J. Tjiputra K. Waters, T.K. Westberry, A. Winguth, 2008, Assessing the uncertainties of model estimates of primary productivity in the tropical Pacific Ocean, *Journal of Marine Systems* (2008), doi:10.1016/j.jmarsys.2008.05.010

Garcia, V.M.T., Purdie, D.A. (1992). The influence of irradiance on growth, photosynthesis and respiration of Gyrodinium cf. aureolum. Marine Microbiology Group, Department of Oceanography, The University, Southampton. *Journal of Plankton Research* Vol. 14 no. 9 pp 1251-1265.

Glas, P.C.G. and Nauta, T.A., 1989. A North Sea computational framework for environmental and management studies: an application for eutrophication and nutrient cycles. Contribution to the international symposium on integrated approaches to water pollution problems (SISIPPA 89), Lisbon.

Jahnke, J., 1989, The light and temperature dependence of growth rate and elemental composition of *Phaeocystis globosa Scherffel* and *P. Pouchetti (Har.) Lagerh.* in batch cultures. *Neth. J. of Sea Research*, 23(1): 15-21.

Jolliff, J.K, J.C. Kindle, I. Shulman, B. Penta, M.A.M. Friedrichs, R. Helber, R.A. Arnone, 2008, Summary diagrams for coupled hydrodynamic-ecosystem model skill assessment. *J. Mar. Sys.* (2008) doi: 10.1016/j.jmarsys.2008.05.014

Laane, R.W.P.M., Groeneveld, G., De Vries, A., Van Bennekom, A.J. and Sydow, J.S., 1993. Nutrients (N, P, Si) in the Channel and the Dover Strait: seasonal and year-to-year variation and fluxes to the North Sea. *Oceanologia Acta*, 16: 607-616.

Lesser, G.R., Roelvink. J.A., Van Kester, J.A.T.M., and Stelling, G.S., 2004. Development and validation of a three-dimensional morphological model. *Coastal Engi*neering, (51): 883-915.

Los, F.J, An Algal Bloom model as a tool to simulate management measures. In: Barica, J. and Mur, L.R. (eds.), *Hypertrophic Ecosystems*, Junk, Dr. W. BV Publishers, The Hague-Boston-London, 1982.

Los, F.J., 1991. Mathematical Simulation of Algae Blooms by the Model BLOOM II
Version 2. WL | Delft Hydraulics Report T68.

Los, F.J. and Bokhorst, M., 1997. Trend analysis Dutch coastal zone. In: New Challenges for North Sea Research. Zentrum for Meeres- und Klimaforschung, University of Hamburg, pp .161-175.

Los, F.J. and Brinkman, J.J., 1988. Phytoplankton modelling by means of optimization:
A 10-year experience with BLOOM II. *Verh. Internat. Verein. Limnol.*, 23: 790-795.

Los, Hans, Rene Jansen and Sandra Cramer. 1994. MANS Eutrophication Modelling System. National Institute for Coastal and Marine Management (RIKZ).

Los, F.J., Smits, J.G.C. and De Rooij, N.M., 1984. Application of an Algal Bloom Model (BLOOM II) to combat eutrophication. Verh. Internat. Verein. *Limnol.*, 22: 917-923.

Los, F. J., M. T. Villars, & M. W. M. Van der Tol, 2008. A 3- dimensional primary production model (BLOOM/GEM) and its applications to the (southern) North Sea (coupled physical–chemical–ecological model). *Journal of Marine Systems*, 74: 259-294

Los, F.J. and Wijsman, J.W.M., 2007. Application of a validated primary production model (BLOOM) as a screening tool for marine, coastal and transitional waters. *Journal of Marine Systems*, 64: 201-215.

Loucks, D.P. and Van Beek, E. (Eds.), 2005. *Water Resources Systems Planning and Management - an introduction to methods, models and applications*; Chapter 12 'Water quality modelling and prediction'. Studies and Reports in Hydrology, UNESCO publishing, ISBN 92-3-103998-9.

Menden-Deuer, S. & E. Lessard, 2000. Carbon to volume relationships for dinoflagellates, diatoms and other protist plankton. *Limnol. Oceanogr.* 45: 569-579.

Moll, A. and Radach, G., 2003. Review of three-dimensional ecological modelling related to the North Sea shelf system. Part I: Models and their results. *Progress in Oceanography*, 57: 175-217.

Nauta T.A., De Vries, I., Markus, A.A. and De Groodt, E.G., 1992. An integral approach to assess cause-effect relationships in eutrophication of marine systems. *Science of the Total Environment, Supplement*, 1992, Elsevier Science Publishers B.V., Amsterdam, pp. 1133-1147.

Villars, M., de Vries, I. (eds.),1998 , Villars, M., de Vries, I. Bokhorst, M., Ferreira, J., Gellers-Barkman, S., Kelly-Gerreyn, B., Lancelot, C., Menesguen, A., Moll, A., Patsch, J., Radach, G., Skogen, M., Soiland, H., Svendsen, E. and Vested, H.J., 1998. Report of the ASMO modelling workshop on eutrophication issues, 5-8 November 1996, The Hague, The Netherlands, OSPAR Commission Report, Netherlands Institute for Coastal and Marine Management, RIKZ, The Hague, The Netherlands.

Peeters, J.C.H., H.A. Haas, and L. Peperzak (1991). Eutrofiering, primaire produktie en zuurstofhuishouding in de Noordzee. National Institute of Coastal and Marine Management. Report GWAO-91.083 (in Dutch)

Peeters, J.C.H., Los, F.J., Jansen, R., Haas, H.A., Peperzak, L. and De Vries, I., 1995. The oxygen dynamics of the Oyster Ground, North Sea. Impact of eutrophication and environmental conditions. *Ophelia*, 42: 257-288.

Radach, G. and Moll, A., 2006. Review of three-dimensional ecological modelling related to the North Sea shelf system. Part II: Model validation and data needs. *Oceanography and Marine Biology: an Annual Review*, 44, 1-60.

Riegman, R., Noordeloos, A.A.M., and Cadee, G., 1992. *Phaeocystis* blooms and eutrophication of the continental coastal zones of the North Sea. *Marine Biology*, 112: 479-484.

Riegman, R., De Boer, M. and De Senerpont Domis, M., 1996. Growth of harmful marine algae in multispecies cultures. *Journal of Plankton Research*, 18 (10): 1851-1866.

Riegman, R., 1996. Species Composition of Harmful Algal Blooms in Relation to Macronutrient Dynamics, In: Allan D. Cembella and Gustaaf M. Hallegraeff, Physiological Ecology of Harmful Algal Blooms, Donald M. Anderson, NATO ASI Series, Vol. 41., Springer Verlag, 1996.

Rijkswaterstaat, 2003, Waterbase, Database of the Dutch National Surface Water Monitoring Network, http://www.waterbase.nl

Salacinska, K., 2008. Sensitivity analysis of the 2D application of the Generic Ecological Model to the North Sea. MSc thesis, Delft University of Technology, Delft, The Netherlands, 84 pp.

Stow, C.A., J. Jolliff, D.J. McGillicuddy Jr, S.C. Doney, J.I. Allen, M.A.M. Friedrichs, K.A. Rose, Ph. Wallhead, 2008, Skill assessment for coupled biological/physical models of marine systems, *Journal of Marine Systems*, doi:10.1016/j.jmarsys.2008.03.011

Taylor, K.E. (2001), Summarizing multiple aspects of model performance in a single diagram, *J. Geophys.Res.* 106(D7), 7183--7192

Van der Molen, D.T., Los, F.J., Van Ballegooijen, L. and Van der Vat, M.P., 1994. Mathematical modelling as a tool for management in eutrophication control of shallow lakes. *Hydrobiol.*, 275/276: 479-492.

Van Gils, J. and Tatman, S., 2003. Light penetration in the water column. MARE Report, WL2003001 Z3379 WL | Delft Hydraulics, Delft, The Netherlands

Van Pagee, J.A., Glas, P.C.G., Markus, A.A. and Postma, L., 1988. Mathematical modelling as a tool for assessment of North Sea Pollution, in *"Pollution of the North Sea, an Assessment"*, W. Salomons, B.L. Bayne, E.K. Duursma and U. Forstner (Eds.). Springer-Verlag, London.

Van Kessel, T., J.C. Winterwerp, B. van Prooijen, M. van Ledden, W.G. Borst, 2008, modelling the seasonal dynamics of SPM with a simple algorithm for the buffering of fines in a sandy seabed, Proc. INTERCOH 2007, Brest

WL | Delft Hydraulics, 2005. Delft3D-FLOW users manual, v 3.12. WL | Delft Hydraulics, Delft, The Netherlands.

WL | Delft Hydraulics/MARE., 2001. Description and model representation T0 situation. Part 2 : The transport of fine-grained sediments in the southern North Sea. Delft Hydraulics Report WL2001003 Z3030.10.

Zevenboom, W., A. Bij De Vaate, and L.R. Mur, Assesment of factors limiting growth rate of Oscillatoria agarhii in hypertrophic Lake Wolderwijd, 1978, by use of physiological indicators, *Limnol. and Oceanogr.*, Vol. 27, No. 1, 1982, pp. 39-52.

Summary

Hans Los

Summary

General introduction

There is a long tradition in the application of models as tools to understand the behaviour of ecosystems and to predict their future developments based on assumptions about environmental conditions. Results of these models must be 'credible' and 'acceptable'. Validating the scientific quality of models is an essential activity to promote credibility. It should be noted though that ecological models cannot be calibrated and validated in the tradition of physical models: too many interactions exist, too many relationships cannot be described formally. A 'calibrated' ecological model i.e. one that fits the data, is not necessarily 'valid' under a different set of conditions. Validation as described here means: proving that the ecological model is good enough to address a specific, well defined question. In other words: ecological models should be 'fit for purpose'. This can be achieved by applying fully documented models taking strict rules into account regarding the way modellers deal with the parameters.

Many management questions can be addressed adequately by making justified simplifications in the model set-up, depending on the specific objective of the study. So, for each application, the appropriate 'spatial' and 'ecological resolution' has to be chosen carefully. There is an optimal balance between the required accuracy, predictability or robustness and operational aspects. Models with relatively complex spatial resolution and (overly) simple ecology or complex ecological models without an appropriate description of transport are bound to produce biased results. Essentially any model application should be backed up by a thorough description of its results demonstrating that the model does indeed capture the main features of the natural system. This is called the 'system approach' in this thesis.

The work described in this thesis is concentrated on the interface between the abiotic components (the physical and chemical compartments) and the lower part of the food chain. In particular it is focused on algal biomass and algal species in relation to hydrodynamics, suspended sediments and light, nutrients and bottom sediments. The family of models, which has been developed for that purpose at Deltares (formerly Delft Hydraulics), is described here. Although for historic reasons different names have been used for these models i.e. DBS, GEM, BLOOM, DELFT3D-ECO, it should be stressed there is actually only one model code, only the input and the mode of operation varies between different applications. Biological interactions, such as competition with macrophytes and grazing are an integral part of many studies and have been addressed with DELFT3D-ECO, but only one example is included here.

The aim of this thesis is the development and application of mathematical models for primary production by phytoplankton and macro algae that are 'credible' and 'fit for purpose'. Credibility is enhanced by using generic equations and fixed parameters as much as possible, by adopting well defined methodologies for validation, and by demonstrating that the results of the model fit in with other knowledge and contributes to obtaining a consistent, comprehensive view of the natural systems for which the models are applied.

The BLOOM module

At the heart of the eco-hydrodynamic model is the BLOOM module. The main principles and equations of this module are described in Chapter 1. It is a model with a long history, a very long list of both freshwater and marine applications and some unusual features.

In BLOOM each phytoplankton species is represented by various (usually three) types with distinct characteristics definened in the input. The model can mix the mass of these types and thus vary the average characteristics of the functional species group. This proves to be a simple, yet robust way to deal with adaptation and intraspecific variation of characteristics. The ratio between types is determined by the external conditions such as surplus or shortage of nutrients, light etc. The introduction of types has definitely enhanced the model's ability to reproduce empirical results for nutrients and chlorophyll-*a*.

The overall purpose of the BLOOM model may be described as: *Selecting the best adapted combination of phytoplankton types at a certain moment and at a certain location consistent with the available resources, the existing biomass levels at the beginning of a time interval and the potential rates of change of each type.*

BLOOM differs from many existing models in that it uses a rather strict Liebig's Law approach: each ecotype is limited only by the resource with the lowest availability. Also, instead of using a Monod or Droop equation to growth rates as a function of nutrient levels, BLOOM does not let nutrient shortage reduce growth rates, other than setting growth to zero when the concentration of a dissolved nutrient has become zero. With respect to shading effects, all species for which light is below a fixed specific tolerance limit are set to zero. The higher the total biomass permitted by the availability of nutrients, the higher the potential turbidity and hence the smaller the number of species that can still maintain a positive energy balance. Light is limiting to at most one of the species in the model.

Another difference to many other ecological models is the use of linear programming to predict the state of the system at the next time-step. The algorithm first defines the different possible states at which one of the nutrients or light halts growth of one of the ecotypes. Subsequently in accordance to the general LP methodology, from those states, the one is selected at which the the potential growth rate of all ecotypes is maximal and the requirement for the resources is minimal. It can be shown analytically that effectively a high potential growth capacity as well as a low requirements for nutrients and light are equally weighted in determining the algal composition of the predicted steady state of the system.

To prevent unrealistically fast jumps towards such steady state solutions, BLOOM also computes the biomass increment each species can potentially realize in the given time-step from the temperature and light dependent 'potential net growth'. If the result is smaller than the steady state solution, the growth-limited new state is assigned. Similarly, the model imposes a limit on mortality, to prevent unrealistically rapid declines.

All computations are made at each time-step for all phytoplankton types, and the resulting assemblage is typically a mix of various types, some of which are resource limited, and some of which are controlled by growth and mortality rates. Since a trade-off between resource requirement and potential growth rate is assumed, r-selected species are predicted to dominate

266

under dynamic conditions where potential growth rates are high, while K-selected species are predicted to dominate under more stable conditions. Overall, the relative weight given to growth rates produces some bias towards r-select species compared to traditional equilibrium solutions. On the other hand, such equilibrium solutions may be argued to be biased towards K-select species given that a model will typically not capture all loss processes and environmental fluctuations well. The fact that predictions by BLOOM match observed plankton communities so well for a wide range of systems, suggests that indeed the weighting of growth rates in the algorithm for selecting amongst the set of potential solutions is useful for capturing some aspects of complex reality that tend to be missed in other ways.

Although the approach in BLOOM may seem radically different from that taken in most other competition models, comparisons between BLOOM predictions and predictions from traditionally solved sets of equations in simple as well as complex systems illustrate that the particular approach taken in BLOOM involving the linear programming procedure and the strict Liebig's approach do not result in markedly different outcomes with respect to the total biomass. The main merits of the approach are its computational speed and stability and its straightforward way in dealing with limiting factors and variable characteristics of members of the plankton community, shown to produce realistic predictions of total biomass as well as species composition over a wide range of conditions and systems.

Different modes of complexity are possible with BLOOM ranging from a straightforward 0-D screening tool to a much more detailed 3-D eco-hydrodynamic model. Depending on the specific questions to be addressed the kind of model application must be carefully selected. In Chapters 2 and 3 some applications of BLOOM as a screening tool are described for both freshwater and marine situations. For these applications the environmental conditions (nutrients, background turbidity, irradiance and temperature) are all based upon measurements. Applied in this way the model proves to be capable of reproducing observed levels of chlorophyll-*a* very accurately. It also indicates the limiting resources and species composition and hence can be applied to solve typical 'what if questions'. What will happen if the concentration in the water of a particular nutrient can be reduced by 50 percent? Or a related question: by how much should a nutrient be reduced in order to meet a specific standard such as imposed by the EU Water Framework Directive? It appears there are large differences in response between individual waters, some reacting more strongly to phosphorus reductions, others to a decrease in nitrogen. Sometimes the response is nearly linear, but more often the response with respect to simulated chlorophyll-*a* levels is less than proportional.

As a result of this type of application, a more detailed and focused management strategy can be developed for a particular water body. Occasionally it might be concluded that none of the easily controllable management options will be successful. In Chapter 3 an example is shown for the Tagus estuary in Portugal, where phytoplankton production is so heavily controlled by the turbidity, that even very strong nutrient reductions will have no effect at all. In such a case more advanced modelling is not necessary from a management point of view.

Integrated modelling

In the next chapters applications are discussed in which BLOOM is integrated with modules for transport and chemical processes. In this type of application individual sources, tributaries and the sediment are all explicitly included. Different segments are usually distinguished with some predefined exchange between them. For hydrologically simple, well mixed systems, the exchanges are often simply computed with a water balance program. Usually, however, the complexity of the natural system requires the usage of hydrodynamic models which are run in advance of the primary production model. In Chapter 4 a long term hindcast simulation of DBS (DELWAQ - BLOOM - SWITCH) to Lake Veluwe (the Netherlands) is described. The simulation covers a period during which conditions in the lake changed drastically due to a revised management strategy. Model results for nutrients, chlorophyll-*a*, species composition and transparency all agree convincingly with the measurements. Unlike observed in many other systems, Lake Veluwe responded very quickly to the reduction of external loadings. The model provides a mechanistic explanation for this: an increase of the oxidized layer of the sediment resulted in a decrease of its phosphorus release and hence in an extra reduction of the phosphorus concentrations in the water. This invoked a chain of effects, resulting in lower phytoplankton biomasses, a higher transparency and a strong reduction in the relative importance of cyanobacteria. Chapter 5 deals with the methodology for validating DBS. A general validation methodology is proposed, which is applied to the complex Rijnland water network in the Netherlands. Basically model coefficients are divided into three categories:

1. System independent, fixed coefficients, which should not be changed. Most model coefficients fall into this category.
2. System dependent, fixed coefficients, which are known to be different in different types of waters but should not be varied i.e. their values are taken from a table indicating the appropriate value for a particular type of water.
3. System dependent, variable coefficients.

This methodology is used throughout the model applications presented in this thesis.

In Chapter 6 results are presented of a DBS application to a shallow lake, Botshol during a period of more than ten years. In about half of the years macrophytes (*Chara*) dominate, in the other half of the years phytoplankton is the dominant primary producer. This alternating pattern could be reproduced by the model, which also provides a plausible explanation for this behaviour: variations in nutrient loads due to interannual differences in weather. Wet winters promote high levels of P run-off leading to an early development of phytoplankton in the lake and an unfavourable light climate for the macrophytes. The opposite is true following a dry winter.

Another long term hindcast simulation for the Dutch coastal zone is presented in Chapter 7. It starts in 1975, when phosphate loads are still increasing, includes the mid 1980s, when phosphate loads to the North Sea peaked and continues all the way up to 1994, when phosphorus loads by rivers clearly started declining. While in the DBS applications for Lake Veluwe and Botshol the 'ecological resolution' was crucial to the model result, in this case the physical schematization proved to be very important. To obtain an appropriate 'spatial resolution' a relatively refined grid was developed with computational elements near the coast of less than 1 square kilometer. Using previous, much coarser models, the observed gradients in the Dutch coastal zone could not be reproduced with sufficient accuracy.

Though it had been expected by some of the managers that chlorophyll-*a* levels in the coastal zone would react strongly to the long term trends in phosphate loads by the main rivers, in reality this was not the case. This observation was reproduced by the model and can be explained as (1) other factors (nitrogen and light) were limiting phytoplankton most of the time and (2) about one half of the phosphorus load originates from sources that were not affected by the load reductions i.e. the Channel boundary. These results demonstrate that targets formulated in terms of a concentration reduction relative to some base condition, might be unrealistic because they overlook the contribution of uncontrollable sources. Notice that from the results presented for the year 2003 in Chapter 10, it seems that more recently phosphorus loads were reduced to such an extend that they do start affecting chlorophyll-a in the coastal zone.

The successor of the model version presented in the previous Chapter is called GEM (Generic Ecological Model for estuaries). It was created in response to the desire of the Dutch Rijkswaterstaat to harmonize the prognostic tools for the Dutch marine waters. Indeed operational applications of GEM now exist or will be created in the near future for all of these, the Ems – Dollard estuary being the last. Its basic principles, characteristics, equations and parameterization are extensively described in Chapter 8. As its name indicates, GEM is intended to be a generic modelling framework, which can be applied to a wide range of water systems without the need for major reformulation or recalibration. This claim is substantiated by presenting results for four systems: the North Sea, Venice Lagoon, Lake Veere and the Sea of Marmara, which all challenge the abilities of the model in a very distinctive way. The North Sea application covers a long time span, Venice Lagoon is very shallow but large and horizontally very complex, in Lake Veere the vertical gradients are crucial and finally the Sea of Marmara is extremely deep and hydrodynamically complex.

In Chapter 9 the first full 3D validation of GEM to the North Sea is presented. Also included in this chapter is a historic overview of previous model applications. The concepts of the validation procedure are discussed at length. Accuracy for a large number of variables are statistically validated for many monitoring stations using the internationally accepted Ospar cost function. According to this criterion, results for 80 percent of all model variables at all monitoring locations were classified as 'good' or 'very good'. Graphically model results are compared station wise, for transects perpendicular to the coast, for area maps and for vertical profiles. These findings are a strong indication that results by GEM provide the Dutch North Sea managers with a solid background for the development and implementation of international standards and management strategies for the North Sea i.e. within OSPAR (the Oslo Paris convention), the EU Water Framework Directive or the EU Marine Strategy or as a tool for environmental studies on infrastructural works in the coastal zone.

Lessons learned

The construction of validated biogeochemical model applications to be used as prognostic tools involves a large number of choices particularly with respect to the level of details of the physical, chemical and biological aspects. In theory enhanced complexity should promote enhanced realism, accuracy and credibility as well. Unfortunately with growing complexity, simulation times increase drastically and may become prohibitive in practice. The amount of data necessary to force the model increases and the spatial and temporal coverage of monitoring data is usually poor relative to the outputs produced by even moderately complex models, which makes it increasingly difficult to demonstrate that the model is valid. In Chapter 10 the results of comparative modelling applications to the North Sea are presented varying in spatial resolution (from coarse to fine), in vertical resolution (2D versus 3D), in climatological forcing of transport, in turbidity forcing and in the number of phytoplankton species. Included models range from 15 years old relatively simple models to the relatively advanced 3D BLOOM/GEM model described in Chapter 9. Results are compared to each other and to monitoring data using different goodness of fit criteria as well as seasonal plots.

This intercomparison clearly demonstrates that progression has not been linear, nor does a single model outcompetes all others everywhere and at all times. An overall improvement in performance or an improvement in a certain area, often comes at a price: some results become worse. Usually this can be explained and often the reason is rather trivial such as a new background turbidity field, which is better in most areas, but at some places is not as good as the previous one. Without a proper explanation, both the credibility and acceptability of the results will be aggravated by these kind of results. In general it was observed that the overall bias (the difference between measured and simulated values) has been reduced in the latest, more advanced models. So they perform better than older models on an annual bases. There is also more consistency in these models. This is typically the result of an overall improvement in obtaining a proper balance between spatial and ecological resolution. For instance old models were too coarse to resolve the transport and consequently phytoplankton biomasses were frequently limited by the wrong environmental condition (i.e. a nutrient in stead of turbidity). Improved consistency also implies improved predictive capabilities. It was demonstrated that models with statistically acceptable base simulations could produce quite different scenario results if one of them was originally 'right for the wrong reason'. So the base case results of models should not be judged only by looking at a limited number of statistics at a limited number of locations, but by their overall ability to capture the correct driving forces as well.

While the bias has been reduced, there has been less verifiable progression in the simulation of the temporal gradients: typically the level of variation produced by recent models exceeds the observed level of variation. This could be a problem of the models, but also of the monitoring data: taking water samples every 2 or 4 weeks is insufficient to correctly determine the variability within the actual water system. High frequency measurements using buoys or automatic sensors consistently indicate that actual variations are underestimated by conventional monitoring. So it cannot be concluded that the level of variations in recent models is unrealistically large.

It is interesting that chlorophyll-a is one of the most robust outputs of the models. Usually its goodness of fit score is among the highest of any simulated parameter; often it exceeds the score for (semi) conservative substances such as total nitrogen or salinity, which are a measure for the accuracy of the transport simulation. These results contradict the assertion that in general hydrodynamic models are more accurate than ecological models. From these and many similar results it appears that the accuracy in predicting some hydrodynamic features such as the water levels is very good indeed and clearly exceeds the accuracy of existing ecological models for nutrients and primary production. But as a driver for ecological and water quality models, transport is much more important than water levels. The direction and velocity of transport depends on highly variable factors such as the vertical gradients, waves, short term variations in climatology (wind direction and speed) etc. Given the sensitivity of hydrodynamic models to these factors, their accuracy in predicting transport is not necessarily better (and sometimes worse) than the accuracy with which primary production models simulate chlorophyll-a. There is another issue, however: Why is the score for chlorophyll-*a* often still (reasonably) good in models with obvious flaws in resolution or forcing? This could be because primary production models are always judged on their ability to reproduce this variable. Hence models are probably optimized to perform well for this output variable. By comparing differently forced models with identical phytoplankton kinetics, it appeared that the accuracy of the chlorophyll-*a* simulation remained relatively high in all models, including those with relatively poor physical forcing. So the high goodness of fit score for chlorophyll-*a* is also an intrinsic feature of the models applied here due to their non-linearity and build in compensations.

Some remarks about the future

The modelling work presented here has contributed to understanding many aquatic systems and has often been of help to managers in deciding what to do. The DELFT3D-ECO family of models is particularly strong in the way it interacts with the physical system, it is robust and generally produces reasonable or good results for primary producers, nutrients and under water light climate whatever criteria are used to measure its performance and last but not least: the list of worldwide applications is very long. But as indicated in the introduction: ecosystems are complex and often an integrated judgment is required which includes secondary producers, fish, mammals and birds. There is no unanimity among ecologist in general and modellers in particular how this judgment can be obtained. In models such as ERSEM for marine systems or PCLAKE for fresh water systems, equations for at least some of these ecosystem components are explicitly included. This is not the case in DELFT3D-ECO. This model strictly complies with the rule of thumb that for theoretical and operational reasons models should not simultaneously try to solve processes whose characteristic time scales are several orders of magnitude apart. In general models that try to do so are sensitive to the initial conditions and parameters and take a long time (many simulation years) to achieve equilibrium after some forcing or parameter was changed. As an alternative impacts on the upper part of the food web in our practice are usually assessed by different techniques (i.e. habitat evaluation procedures: HEPs) that operate upon the results of the primary production model among other. Needless to say this subject is suitable for further research among ecological modellers.

Other aspects which should be addressed in the near future, include criteria for measuring the goodness of fit of ecological models (Chapter 10), the species composition of the North Sea model (Chapter 10), real time forecasting of algal blooms both for lakes (cyanobacteria scums; toxin production) and coastal systems (*Phaeocystis* blooms; blooms by harmful dinoflagellates causing shell fish poisoning) and assessing the impacts of climate change on primary production (not just direct temperature effects but also indirect effects on transport, loads, synchronization with grazers etc.).

Fortunately in addition to a number of management applications, some interesting R&D oriented projects are about to start which may help in finding or improving integrated methods for dealing with the entire ecosystem. Three of them are particularly attractive. One project will focus on the ecosystems of Lake Marken and Lake IJssel; it includes several research groups and half a dozen PhD students. Similar in size is a project on deep lakes in cooperation with the National University of Singapore (NUS). The marine model will be further advanced by the obligation to develop an ecosystem approach in support of the EU marine strategy by the end of 2012. These major studies all build upon the work presented here and will extend the capabilities of the modelling framework in producing credible and acceptable results which are useful to managers and society in general.

Curriculum vitae

Hans Los

Curriculum vitae

Frederik Johannes (Hans) Los was born 13 April 1951 in Leiden (The Netherlands). He spend the first part of his childhood in Canada, returned to the Netherlands and went to the Christelijk Lyceum W.A. Visser 't Hooft High school in Leiden. At first he wanted to study civil engineering, but inspired by a number of books on threats to the natural environment, he later chose to study biology at Leiden University from 1969 through 1975. Following a bachelor in biomathematics, his main study areas during the master phase were theoretical ecology, behavioural biology and mathematical biology. Following an additional post academic year at the Leiden University developing population dynamical models, he joined Delft Hydraulics (now part of Deltares) in 1977. He was the first ecologist within an organisation, which by that time was still strongly dominated by civil engineers. Hence in stead of an engineer himself he could focus on the environmental impacts of engineering works. In close cooperation with the former Delta Service, a part of the Ministry of Public works and transportation, he started working on modelling phytoplankton using the 'Algae Bloom Model', which had been developed by the US Rand Corporation for the Policy Analysis of the Eastern Scheldt storm surge barrier. Eutrophication proved to be a persistent and common problem, but at the time there was a strong belief that (modelling) research could provide solutions to these problems. Hence several models, including BLOOM, were developed within the framework of the so called WABASIM project (1977-1991). Meanwhile these models were already applied within the framework of many advisory projects during this period. By the end of the 1980s development of the marine version of BLOOM started within the framework of what up to this day was the largest integrated North Sea policy analysis study called MANS.

During all these years he has been involved in a large number of highly diverse projects as a developer, modeller and code designer, as a project leader, scientific coordinator, advisor and reviewer. Most projects involved aspects of hydrodynamics, suspended matter, water quality and ecology. Among the water systems, which were regularly and thoroughly studies are the Lake IJssel and boder lakes, in particular Lake Veluwe, Lakes Loosdrecht and Breukeleveen, the Delta lakes in the south western part of the Netherlands, the channel and lakes system of Rijnland in the western part of the Netherlands, Venice Lagoon and the North Sea. In addition he has been involved in hundreds of other projects.

From 1998 through 2007 he was principal investigator of Ecology at Delft Hydraulics, which meant he advised the central management on issues regarding knowledge development with special attention on ecology. At the moment he is ecological specialist at Deltares.

List of Publications

Hans Los

List of publications

Los, F.J., 1980. Application of an Algal Bloom Model (BLOOM II) to combat eutrophication, Hydro-biological Bulletin Vol. 14, 1/2: 116-124.

Los, F.J, 1982. An Algal Bloom model as a tool to simulate management measures. In: Barica, J. and Mur, L.R. (eds.), 'Hypertrophic Ecosystems', Junk, Dr. W. BV Publishers, The Hague-Boston-London, 1982.

Los, F.J., 1982. Mathematical Simulation of Algae Blooms by the Model BLOOM II, WL | Delft Hydraulics, R1310-7.

Los, F.J., Smits, J.G.C. and Rooij, N.M. de, 1984. Modelling eutrophication in shallow Dutch lakes, Verh. Internat. Verein. Limnol. 22: 917-923.

Los, F.J. and J.J. Brinkman., 1988. Phytoplankton modelling by means of optimization: A 10-year experience with BLOOM II, Verh. Internat. Verein. Limnol., 23: 790-795.

Brinkman, J.J. , S. Groot, F.J. Los and P. Griffioen 1988. An integrated water quantity and quality model as a tool for water management. application to the province of Friesland, the Nehterlands. Verh. Internat. Verein. Limnol. 23: 1488-1494.

Bakema, A.H., W.J., Rip, M.W. de Haan, F.J. Los, 1990. Quantifying the Food webs of Lake Bleiswijkse Zoom and Lake Zwemlust, Hydrobiologica 200/201: 487 - 495.

Los, F.J.,1991. Mathematical Simulation of algae blooms by the model BLOOM II, Version 2, T68, WL | Delft Hydraulics Report.

Los, F.J., Bakema, A.H. and Brinkman, J.J.,1991. An integrated freshwater management system for the Netherlands, Verh. Internat. Verein. Limnol. 24: 2107-2111.

DeGroodt, E.G., Los, F.J., Nauta, T.A., Markus A.A. and Vries, I. de, 1992. Modelling cause-effect relationships in eutrophication of marine systems. In Colombo et al. (eds), Olsen & Olsen.

Los, F.J., M.T. Villars and M.R.L. Ouboter, 1994. Model Validation Study DBS in networks. WL | Delft Hydraulics, Research Report, T1210.

Los, Hans, Rene Jansen and Sandra Cramer, 1994. MANS Eutrophication Modelling System. National Institute for Coastal and Marine Management (RIKZ).

Van der Molen, D.T., F.J. Los, L. van Ballegooijen, M.P. van der Vat, 1994. Mathematical modelling as a tool for management in eutrophication control of shallow lakes. Hydrobiologia, Vol. 275/276: 479-492.

Los, Hans and Herman Gerritsen, 1995. Validation of water quality and ecological models. IAHR Specialist Forum on Software Validation, London, September.

Peeters, J.C.H., F.J. Los, R. Jansen, H.A. Haas, L. Peperzak and I. de Vries, 1995. The oxygen dynamics of the Oyster Ground, North Sea. Impact of eutrophication and environmental conditions. OPHELIA, Vol 42: 257-288.

Los, F.J., and B.F. Michielsen, 1996. Application of DBS to Lake IJssel. WL | Delft Hydraulics, report T1515 (in Dutch).

Michielsen, B.F., F.J. Los and D.T. van der Molen, 1996. Modellering van eutrofiering: toepassing op het beheer van het Veluwemeer van DELWAQ-BLOOM-SWITCH, H2O, Vol. 29, No. 12: 361-364 (in Dutch).

Uittenbogaard, R.E., M. Bokhorst, F.J. Los & J.A.Th.M. van Kester, 1996. Similarities between estuaries and reservoirs or lakes in terms of numericalmodelling of mixing and water quality. Paper in Int. Water Research Symp., Aachen Univ., Jan. 1996.

Los, F.J., and M. Bokhorst, 1997. Trend analysis Dutch coastal zone. In: New Challenges for North Sea Research, Zentrum for Meeres- und Klimaforschung, University of Hamburg: 161-175.

Vries, I. de , R.M.N. Duin, J.C.H. Peeters, F.J. Los, M. Bokhorst and R.W.P.M Laane, 1998. Patterns and trends in nutrients and phytoplankton in Dutch coastal waters: comparison of time-series analysis, ecological model simulation, and mesocosm experiments. ICES Journal of Marine Science, Vol. 55: 620-634.

Los, F.J., 1999. Ecological Model for the Lagoon of Venice. Modelling Results. WL | Delft Hydraulics report, T2162, November 1999.

Flipsen, E., F. Los, L. Mur, J. Stroom, 2000. Development of a blue green algae bloom model, H2O, Vol. 33, No. 14/15: 17-20 (in Dutch).

Los, Hans and Sharon Tatman, 2001, Description and modelrepresentation T0 situation, Part 2: Transport, nutrients and primary production Perceel 3, Deelproduct 2, project Z3030.10, MARE consortium. Prepared for Flyland project.

Van Duin, Elisabeth H.S., Gerard Blom, F. Johannes Los, Robert Maffione, Richard Zimmerman, Carl F. Cerco, Mark Dorth, and Elly P.H. Best, 2001. Modeling underwater light climate in relation to sedimentation, resuspension, water quality and autotrophic growth, Hydrobiologia, 444: 25-42.

Ibelings, Bas W., Marijke Vonk, Hans F.J. Los and Diederik T. v.d. Molen and Wolf M. Mooij, 2003. Fuzzy modelling of Cyanobacterial waterblooms, validation with NOAA-AVHRR satellite images, Ecological Applications, 13(5): 1456-1472

Los, F. J., S. Tatman & A. W. Minns, 2004, FLYLAND - A Future Airport in the North Sea? An Integrated Modelling Approach for Marine Ecology. 6th International conference on hydroinformatics. World Scientific Publishing Company ISBN 981-238-787-0.

Los, F.J., 2005. An algal biomass prediction model. In: Louks, D.P. and Van Beek, E. (Eds.), Water Resources Systems Planning and Management - an introduction to methods, models and applications. UNESCO.

Duel, H, F.J. Los, H. Kaas, A. Lyche-Solheim, N. Friberg, D. Krause-Jensen, S.Rekolainen, A-S. Heiskanen, J. Carstensen, D. Boorman, M.J. Dunbar, 2005. Relationships between ecological status of surface waters and both chemical and hydromorphological pressures. In: Lawson, John (ed.). River Basin Management.Progress towards implementation of the European Water Framework Directive. Institution of Civil Engineers, London, ISBN 0 0415 392200 4.

Prandle, D., Los, F.J., Pohlmann, T., De Roeck, Y.H., Stipa, T., 2005. Modelling in Coastal and Shelf Seas - European Challenges. European Science Foundation, European Marine Board Position Paper 7. www.esf.org.

Los, F.J. and Wijsman, J.W.M., 2007. Application of a validated primary production model (BLOOM) as a screening tool for marine, coastal and transitional waters. Journal of Marine Systems, 64: 201-215.

Rip, Winnie J., Maarten Ouboter, Hans J. Los, 2007. Impact of climatic fluctuations on Characeae biomass in a shallow, restored lake in The Netherlands, Hydrobiologia, Vol. 584: 415-424.

Los, F. J., M. T. Villars, & M. W. M. Van der Tol, 2008. A 3- dimensional primary production model (BLOOM/GEM) and its applications to the (southern) North Sea (coupled physical–chemical–ecological model). Journal of Marine Systems, 74: 259-294.

Erftemeijer P.L.A, van Beek J.K.L., Ochieng C.A, Los H.J., Jager Z., 2008. Eelgrass seed dispersal in the Dutch Wadden Sea: A Model approach. Marine Ecology Progress Series, Vol. 358: 115–124.

Blauw, Anouk N. , Hans F. J. Los, Marinus Bokhorst, and Paul L. A. Erftemeijer., 2009. GEM: a generic ecological model for estuaries and coastal waters. Hydrobiologia DOI 10.1007/s10750-008-9575-x

Los, F. J., M. Blaas. Complexity, accuracy and practical applicability of different biogeochemical model versions. Journal of Marine Systems, AMEMR 2008 Special Issue (Submitted).

Lenhart, Hermann-J., David K. Mills, Hanneke Baretta-Bekker, Sonja M. van Leeuwen, Johan van der Molen, Job W. Baretta, Meinte Blaas, Xavier Desmit, Wilfried Kuehn, Genevieve Lacroix, Hans J. Los, Alain Menesguen, Ramiro Neves, Roger Proctor, Piet Ruardij, Morten D. Skogen, Alice Vanhoutte-Brunier, Monique.T. Villars and Sarah L. Wakelin. Predicting the consequences of nutrient reduction on the eutrophication status of the North Sea Journal of Marine Systems, Journal of Marine Systems, AMEMR 2008 Special Issue (Submitted)

Erftemeijer, Paul L.A., Jan K. L. van Beek, Loes J. Bolle, Mark Dickey-Collas and Hans J. Los, Modelling the dispersal of fish larvae in the southern North Sea to assess the effects of a coastal reclamation scheme. Marine Ecology Progress Series (Submitted).

Salacinskaa, K., G. El Serafy, H. Los, A. Blauw. Sensitivity analysis of the 2-dimensional application of the Generic Ecological Model (GEM) to the North Sea with respect to algal bloom prediction (Submitted)